EARTH AND INDUSTRY

EARTH AND INDUSTRY

STORIES FROM GIPPSLAND

EDITED BY ERIK EKLUND AND JULIE FENLEY

© Copyright 2015
© Copyright of the individual chapters belongs to the respective authors.
© Copyright of this collection in its entirety belongs to the editors.
All rights reserved. Apart from any uses permitted by Australia's Copyright Act 1968, no part of this book may be reproduced by any process without prior written permission from the copyright owners. Inquiries should be directed to the publisher.

Monash University Publishing
40 Exhibition Walk
Monash University
Clayton, Victoria 3800, Australia
www.publishing.monash.edu

Monash University Publishing brings to the world publications which advance the best traditions of humane and enlightened thought.

Monash University Publishing titles pass through a rigorous process of independent peer review.

www.publishing.monash.edu/books/ei-9781922235046.html

Series: Monash Studies in Australian Society

Design: Les Thomas

Cover image: Tim Jones, *Gippland (sic)*, 1984, wood engraving, 28.5 x 21.9 cm. Latrobe Regional Gallery Collection. © Tim Jones.

National Library of Australia Cataloguing-in-Publication entry:

Title:	Earth and industry : stories from Gippsland / edited by Erik Eklund and Julie Fenley
ISBN:	9781922235046 (paperback)
Notes:	Includes bibliographical references.
Subjects:	Subjects: Sociology, Urban--Victoria--Gippsland--Anecdotes.
	Social change--Victoria--Gippsland--Anecdotes.
	Social ecology--Victoria--Gippsland--Anecdotes.
	Human ecology--Victoria--Gippsland--Anecdotes.
Other Creators/Contributors:	Eklund, Erik Carl, editor. Fenley, Julie, editor.
Dewey Number:	307.76099456

Printed in Australia by Griffin Press an Accredited ISO AS/NZS 14001:2004 Environmental Management System printer.

The paper this book is printed on is certified against the Forest Stewardship Council ® Standards. Griffin Press holds FSC chain of custody certification SGS-COC-005088. FSC promotes environmentally responsible, socially beneficial and economically viable management of the world's forests.

Contents

Contributors vii

Acknowledgements xiii

List of acronyms xiv

Introduction: Towards a new environmental history of Gippsland xv
Erik Eklund and Julie Fenley

1. Uncovering Gippsland history through reading evidence in the landscape . . . 1
   Jane Lennon

2. Forest conservation and the Gippsland press: A brief editorial survey, 1855–1962 20
   Stephen Legg

3. 'Nature is around us in her loveliness': Settler women in the Gippsland bush in 1930s Australia 37
   Ruth Ford

4. 'Trudging on manfully': Dr Andrew and the evolution of bushwalking . . . 55
   Julie Fenley and Kathy Lothian

5. Port Albert and its strategic role, 1841–1860 81
   Cheryl Glowrey

6. Northern Gippsland Aboriginal associations with country: The frontier period, 1834–1859 100
   Ruth E. Lawrence

7. 'It's just fly fishing against net fishing…' Commercial fishing, angling and the Gippsland Lakes, 1870s to the 1880s 124
   David Harris

8. The Currie family approach to land settlement 146
   Kerry Nixon

9. 'If the worst comes to the worst, you can always milk cows': Creating the 'dairy industry problem' in Gippsland 163
   Charles Fahey

10. Lyrical writing, lyrical campaigning: Jean Galbraith and the wildflowers of the Latrobe Valley 179
    Meredith Fletcher

11. 'We have to live here': Early days of forest activism in East Gippsland . . . 195
    Deb Foskey

12. Reams and reams of paper: The Strzelecki Forest campaigns 219
    Julie Constable

13. 'Nothing we liked better': An original homestead in a changing landscape 242
    Jillian Durance

14. Rural land use change and the Ridley paddock, 1897–2012 257
    Deirdre Slattery

15. People and forests in East Gippsland: Change and continuity in forest industries 277
    Helen Martin

16. 'The campers' powerful headlights are shining on the water': Motor tourism and the environment in far East Gippsland 298
    Sarah Mirams

    Afterword 314
    Erik Eklund and Julie Fenley

    Index 319

Contributors

Julie Constable

Julie Constable graduated from Monash University with a major in History and English and has taught in secondary colleges in Gippsland. Her prose and poetry have appeared in various publications including *Verandah, Hecate, Muse, Poetrix, Silverfish* and *200 Years of Australian Writing*. Her involvement in the campaign to oppose the privatization of the Strzelecki State Forest in the late 1990s led to the co-authoring of *A Proposal for a 30,000 hectare National Park in the Strzelecki State Forest* (1998) and further research into the history of the forest and past campaigns. Julie currently works in the public library sector.

juliemc@dcsi.net.au

Jillian Durance

Jillian Durance has been writing community history since 2003. *Still Going Strong: the Story of the Moyarra Honor Roll* won the Victorian Community History Prize in 2007. She is currently researching the role of the battalion bands in the Great War and has recently completed *A Shrine on the Mountain: the Story of St. Peter's Memorial Church Kinglake*. Her article *Nothing We Liked Better* forms part of her research on her own home, an early settler's homestead in South Gippsland.

Erik Eklund

Erik Eklund is a Professor of History at the Gippsland campus of Federation University. He has a strong interest in regional, community and labour history. He has worked at the Gippsland campus since 2008. His most recent book, *Mining Towns: Making a Living, Making a Life*, was published in 2012 by UNSW Press. His previous work, *Steel Town: The Making and Breaking of Port Kembla*, won the NSW Premier's Prize for Regional and Community History in 2003. He is currently the Keith Cameron Professor of Australian History at University College Dublin.

Erik.Eklund@ucd.ie

Charles Fahey

Charles Fahey teaches in the History Program at La Trobe University and is based at the Bendigo Campus. Charles Fahey's fields of research are the history of the central Victorian goldfields, the history of rural Victoria and labour history. In 2010 he published (with Alan Mayne) *Gold Tailings: Forgotten Histories of Family and Community on the Central Victorian Goldfields*. Charles is currently a member of team writing a history of Australian Mallee Landscapes funded by the Australian Research Council. His article written with André Sammartino, 'Work and Wages at a Melbourne Factory, The Guest Biscuit Works 1870–1921', was awarded the Sir Timothy Coghlan Prize for the best article published in the *Australian Economic History Review* in 2013.

C.Fahey@latrobe.edu.au

Julie Fenley

Julie Fenley is the Director of the Centre for Gippsland Studies, a regional research facility and repository at Federation University's Gippsland campus. She also lectures in history and conducts research in Aboriginal history, public history and museum studies. She has completed a number of heritage consultancies, including a study of the Dr Andrew and Andrew family collection at Old Gippstown.

Julie.fenley@federation.edu.au

Meredith Fletcher

Meredith Fletcher was director of the Centre for Gippsland Studies at Monash University's Gippsland Campus (now Federation University) for twenty years. She has published in local and community history and environmental history. Her most recent book, *Jean Galbraith: Writer in a Valley*, was published by Monash University Publishing in 2014.

fletchermeredith@gmail.com

Ruth Ford

Dr Ruth Ford is Senior Lecturer in History at La Trobe University. Her publications on rural women's relationship to the environment include '"The wattles are in bloom… crops are looking wonderfully well': settler women in the Victorian Mallee, 1920s–30s', in *Outside Country: Histories of Inland Australia* (2011) and '"I shut my eyes and picture our place": Gardens, farm landscapes and working-class dreams in 1930s–1940s Australia', *Studies in the History of Gardens and Designed Landscapes* (2011). She is currently completing a monograph titled '"Still Smiling Cockey's Wife": Women's letter writing in rural Australia, 1920s–1945'.

Deb Foskey

Deb Foskey became involved in the campaign to protect native forests in the late 1970s when clearfelling of iconic forests primarily for woodchipping extended to the ancient forests of the Errinundra and other montane areas near Bonang where she lives. Moving to Canberra in the mid-1980s, Deb became involved in broader environmental campaigns and green politics. Her academic work focused on the planning of Canberra and her PhD looked at the global politics of reproduction and population. In 2008, Deb returned to East Gippsland where she is a farmer and gardener, works as coordinator of the Tubbut Neighbourhood House and is a member of the Centre for Rural Communities.

Cheryl Glowrey

Cheryl Glowrey completed a PhD, titled 'An Environmental History of Corner Inlet, Victoria, Australia' in 2013. She has a background in researching and writing local history for the South Gippsland area, including a masters dissertation on the twentieth century decline of the hill communities of the Strzelecki Ranges and the publication of *Snake Island and the Cattlemen of the Sea in 2001*. Cheryl is a member of the Foster and District Historical Society and was employed for many years in secondary education in Gippsland.

cherylg@dcsi.net.au

David Harris

David Harris recently completed a PhD at La Trobe University. His thesis is concerned with how European ideas of nature shaped the exploitation of the Gippsland Lakes between the 1860s and 1900. Although his main interest is now with environmental history – and with fisheries in particular – he has previously published on working class housing and social reform in Melbourne during the early twentieth century.

da9harris@gmail.com

Ruth Lawrence

Ruth Lawrence B.Sc.(Hons.), Ph.D. is a geographer with a strong interest in environmental history, amongst other things. After completing her PhD on the environmental history of the alpine area of the Bogong High Plains, she started examining the interactions between Aboriginal people and the alpine environment, which led her to focus on connections between Aborigines and the Omeo environment, being the major source area for ethnohistorical records in the region. She regularly visits the Victorian high country for recreational, research and teaching purposes, and currently works as a Senior Lecturer in Environmental Studies in the School of Education, Outdoor and Environmental Studies at LaTrobe University in Bendigo.

Stephen Legg

Stephen Legg is a Senior Adjunct Research Fellow in the School of Geography and Environmental Science at Monash University. Stephen researches the environmental history and historical geography of Australasian natural resource management. He uses longitudinal and comparative studies at different scales to investigate various environmental issues since the early nineteenth century. These include the environmental impact of farming, forestry and mining; perceptions of and responses to environmental periodicity, change and disturbance; attitudes to nature and the development of conservation; and the political processes that shaped environmental policy. He maintains a particular interest in regional history including Gippsland's.

Jane Lennon

Jane Lennon AM, PhD is an historical geographer. She worked in national park planning and historic site management and currently is a heritage consultant in Brisbane, an honorary professor at the University of Melbourne and an adjunct at Deakin University. Jane is a founding member of Australia ICOMOS and an Australian Heritage Commissioner and Councillor (1998–2008). Her most recent publications are *Pastoral Australia: fortunes, failures and hard yakka: a historical overview 1788–1967*, CSIRO Publishing (2010) and *Managing Cultural Landscapes*, Routledge, (2012).

jlennon@hotkey.net.au

Kathy Lothian

Kathy Lothian holds a Master's degree in History from Monash University titled '"A Blackwards Step is a Forward Step": Australian Aborigines and Black Power, 1969–1972'. She taught and conducted research at Monash University for a number of years, and has particular interests in Australian Indigenous history and histories of Australian immigration. She is currently Senior Historian with Native Title Services Victoria, where she undertakes research and provides advice on native title claims in Victoria. With Dr Julie Fenley, she recently carried out a Heritage Victoria study of the Dr Andrew collection at Old Gippstown heritage park.

kathy.lothian@gmail.com

Helen Martin

Helen Martin is a geographer / regional planner with a strong personal and professional interest in East Gippsland's environment and heritage. She is a member of the Heritage Council of Victoria and an associate member of Heritage Network East Gippsland Inc. She has managed heritage studies for East Gippsland Shire, as well as overseeing preparation of the forest heritage installation at East Gippsland TAFE's Forestech campus, including the associated oral history project.

helen@shearwater-associates.com.au

Sarah Mirams

Dr. Sarah Mirams is a professional historian whose business Past and Future Perspectives specializes in history education, heritage and historical research. Sarah is a research adjunct in the History program at Monash University. She was awarded her PhD in 2011. Her thesis explores the life of writer and journalist E.J. Brady and his relationship with the environment of East Gippsland.

pastandfuture@bigpond.com

Kerry Nixon

Kerry Nixon is a mature-aged student, who repaired to university for want of a decent conversation. A return to study after the birth of her children ignited a passion for history, and her first encounter with the Currie Diary in 2008 was pivotal in driving her to better understand this wonderful document. Her PhD dissertation will explore the Currie Diary in greater detail particularly the issues of subjective representation, the family, marital relations, and the farm as a family project.

Kj2nixon@students.latrobe.edu.au

Deirdre Slattery

Deirdre Slattery is Adjunct Senior Lecturer at La Trobe University, Bendigo. Her research interests include the environmental history of Australia's alpine areas: she recently received the Australian Academy of Science's Moran History of Science Research Award from which she intends to extend this work using the Basser Library collections. Deirdre is retired from full time work and enjoys involvement in Connecting Country, a landscape restoration program in Castlemaine, bushwalking and local and family history.

deirdre_slat@aapt.net.au

Acknowledgements

This collection emerged out of two conferences hosted by the Centre for Gippsland Studies at Monash University's Gippsland Campus in 2010 and 2011 which were supported by the Department of Sustainability and Environment and the West Gippsland Catchment Management Authority. We wish to thank the contributors for capturing the diversity of applied history and academic work that is occurring on the subject of the environmental histories of Gippsland, and gratefully acknowledge the assistance of Cheryl Glowrey, Robyn Heckenberg, Jacqui Howell, Rebecca Jones, Stephen Legg, Rebecca Strating, and Cathy Trembath in planning these events.

We would also like to thank the following colleagues who acted as referees for this collection. We greatly value their contribution and thank them for the generous donation of their time and expertise. Nicholas Brown, Rebekah Brown, Fred Cahir, Georgina Clarsen, Nancy Cushing, John Dargavel, Peter Davies, Graeme Davison, Coral Dow, Helen Doyle, Bradon Ellem, Andrea Gaynor, Don Garden, Heather Goodall, Melissa Harper, John Hirst, Kate Hunter, Julia Horne, Rebecca Jones, Ruth Lane, Susan Lawrence, Alan Mayne, Patrick Morgan, Jayne Persian, Marian Quartly, David Roberts, Judith Scurfield, Mark Staniforth, Cathy Trembath, Barbara Webster, Keith Wilson, and Wendy Wright. All chapters in this collection have been refereed except for the contributions by Meredith Fletcher and Helen Martin.

List of acronyms

ANZAAS	Australian New Zealand Association for the Advancement of Science
ACF	Australian Conservation Foundation
ALP	Australian Labor Party
APM	Australian Paper Manufacturers
AWM	Australian War Memorial
BADRA	Balook and Districts Residents' Association
CARC	Cutting Areas Review Committee
CCV	Conservation Council of Victoria
CRB	Country Roads Board
CROEG	Concerned Residents of East Gippsland
CSIR	Council for Scientific and Industrial Research, 1926–1949
CSIRO	Council for Scientific and Industrial Research Organisation, 1949–
EEG	Environment East Gippsland
EES	Environmental Effects Statement
FCV	Forests Commission Victoria
FNCV	Field Naturalists' Club of Victoria
HVP	Hancock Victorian Plantations
LCC	Land Conservation Council
MAWTC	Melbourne Amateur Walking and Touring Club
NAG	The Nightcap Action Group
NFAC	Native Forests Action Council
NLA	National Library of Australia
NSW	New South Wales
ODEG	Orbost District Environment Group
PROV	Public Records Office of Victoria
RFA	Regional Forest Agreement
SLV	State Library of Victoria
SECV	State Electricity Commission of Victoria
UNESCO	United Nations Educational, Scientific and Cultural Organization
VNPA	Victorian National Parks Association
VPC	Victorian Plantations Corporation
VPRS	Victorian Public Record Office Series

Introduction

Erik Eklund and Julie Fenley

There is a large part of south-east Australia which has a fascinating history that deserves a wider audience. This collection explores the theme of the region's environmental history, and in particular how societies related to it and perceived it over time. Looking through the lens of multiple authors as well as different industries, locations and perspectives, we build up a picture of a complicated relationship to the natural world, which changes over time. The collection starts from the premise that history can tell us much about our evolving relationships to the earth and its ecosystems. We consider the tensions which exist between the earth – by which we mean the natural environment in its broadest sense – and human industry, highlighting the complex and diverse historical relationships between society and the environment. The term 'industry' refers in this context to the ways of earning a living that were established in the region from mining through to pastoralism and fishing, but industry can also be an adjective, referring to the active way in which people engaged with their environments. The Gippsland region in south-east Victoria, with all its diversity and fascinating history, offers an ideal case study of this interaction between humans and their natural surroundings.

Both Indigenous and non-Indigenous people have sought ways to make sense of the diverse Gippsland environments and have benefitted from its bounty. In the case of the Aboriginal people of the region, the Bunurong, Gunai-Kurnai and Yaitmathang, they developed sophisticated patterns of land use and exploitation, modifying the landscape to suit their needs, while also adhering to a complex spiritual worldview that revered the land, the ancestors, and the spirit beings. With colonial occupation after 1838, these Indigenous practices were challenged and then compromised, replaced by more exploitative land uses that have seen significant environmental destruction. From clearing the thick temperate rainforests of South Gippsland through to the fisheries of Lakes Entrance and the open-cut mining and electricity generation of the Latrobe Valley, new regional economies developed, underpinned by different views of the environment. The land was 'tamed', cleared and mined, forests were felled and rivers diverted, creating a patchwork of farms, villages and towns, though not all attempted to battle with nature and some questioned the exploitative practices which were carried out in the name of 'progress'. Many of these patterns of land use are still extant and remain controversial and contested. This then is the space in which this collection of essays seeks to make a contribution. The Gippsland region is the focus for a deeper and more considered historical overview of the sometimes-fraught relationships between society and the environment, and between different patterns of land use and exploitation.

Gippsland is a geographically and economically diverse region in south-eastern Australia. The current western boundary is the Bunyip River, just to the east of the urban fringe of Melbourne, near Pakenham, Cranbourne and Koo-wee-rup. The eastern boundary is at the New South Wales border, some 450 kilometres away, past Mallacoota and Lakes Entrance. From the south, small port towns interspersed by rainforest, farming country and mountain ranges grace the coastal areas. In the north the Great Dividing Range separates Gippsland from north-eastern Victoria. At the edge of this range Mount Baw Baw and the Baw Baw Plateau frames the Latrobe Valley (the apparent geographic 'centre' of the region), and the Strzelecki Ranges lie to the south. The boundaries of this large area in eastern Victoria were never definite but this can serve as a general working definition of the region. This land is Gunai-Kurnai country, cared for by the five clans that make up the Gunai-Kurnai nation, except in the west, past Warragul and Drouin, where we move into Bunurong country, and north Gippsland, near Omeo, which is ascribed to the Yaitmathang.

Gippsland is on the cusp of rapid economic and social change. The energy-production systems based on brown coal and its associated communities are the subject of considerable political and academic interest. In many ways the region is an exemplar of the technological and social challenges presented by the need for a low-carbon future as communities of all types and governments at all levels grapple with the question of sustainability. Under the present local government boundaries (largely formalised in 1994), the region includes the Bass Coast, South Gippsland, Baw Baw, Wellington and East Gippsland shires, as well as the City of Latrobe, with a total population in 2011 of 247,710, although Cardinia is often considered part of the region as well.

This collection addresses these contemporary issues by considering how individuals and communities have responded to change and have interacted with their environments in the past. In so doing the collection offers a more-encompassing portrait of the region, exploring its historical, social

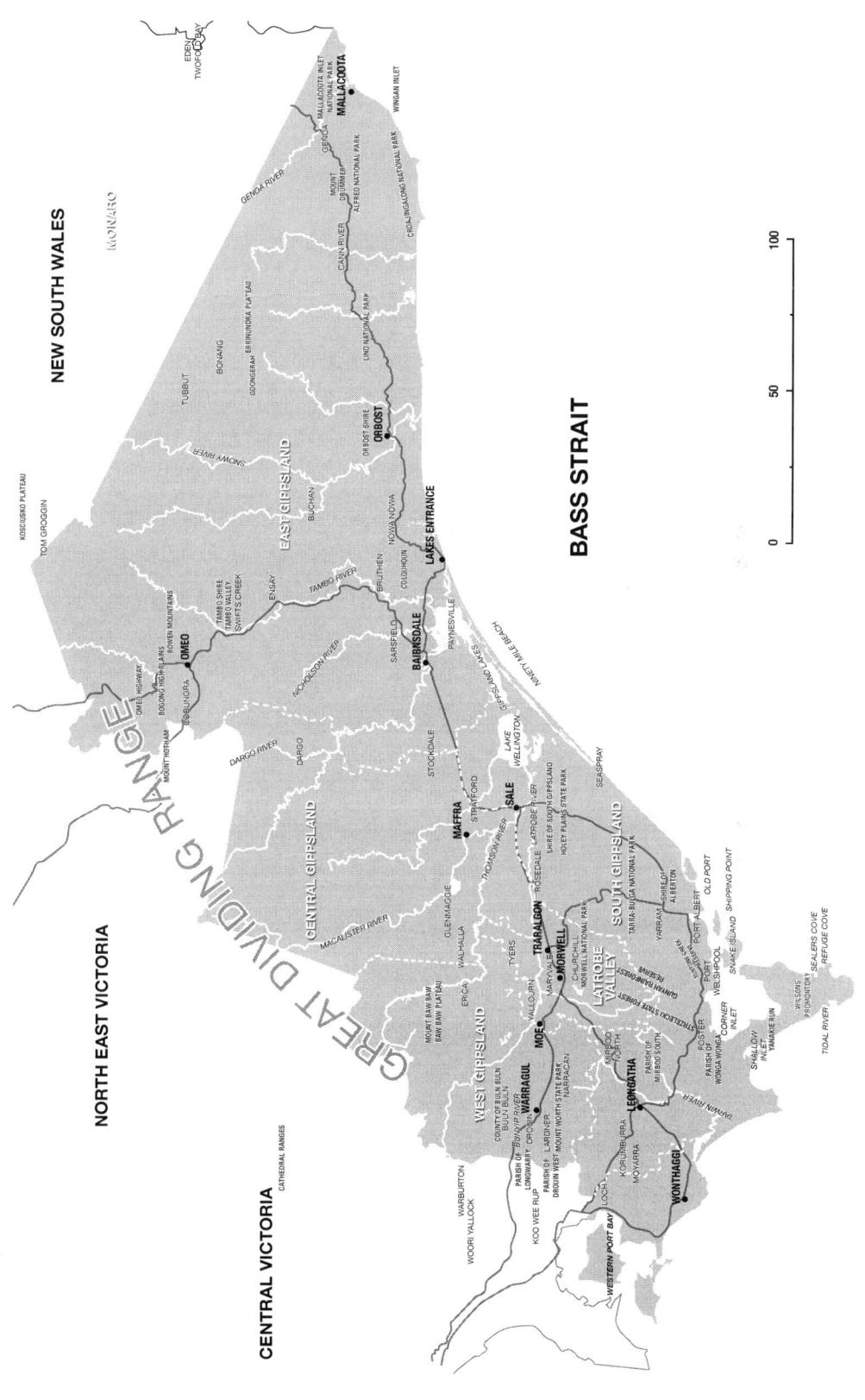

and geographical diversity. It takes us to places in the region that belie the dominant media representations of Gippsland in the early twenty-first century, which usually focus on the smoke stacks of the Yallourn, Hazelwood and Loy Yang power stations in the Latrobe Valley.

Throughout the collection the authors question how humans have impacted on the environment and how the environment has impacted on them. How did past generations make use of local resources and how has this altered the environment? What changes can be observed in how humans view the environment? How can we deal with habitat and species loss, and can we adopt more sustainable practices? In reflecting on early encounters with the Gippsland bush, water and land management, climate change, the exploitation of natural resources, and differing perceptions of the environment, the collection considers how we have engaged with our surroundings and our efforts to preserve the environment for future generations.

Having offered a general description of what constitutes 'Gippsland' at the outset, there still remains the challenge of providing a working definition of 'environmental history'. The field of environmental history is concerned with past biological and physical systems, and the history of human interactions with these environments. It primarily focuses on natural ecosystems, while also encompassing manufactured or technological environments in so far as these have been affected by the natural environment and have an impact on non-human surrounds. Environmental history does not regard humankind as separate from the rest of nature. It recognises that human experiences have occurred within the context of the surrounding environment and that our choices have environmental consequences. It seeks to broaden our understanding of the relationship between human and non-human settings by studying the human engagement with and response to nature.

Influential historians – such as William Cronon, Alfred W. Crosby, Donald Worster and Stephen Pyne – have questioned whether the non-human world can be ignored or presented as a background setting against which human activity takes place.[1] In seeking to write these natural systems into the scholarly literature, these American authors have demonstrated that nature is an active and dynamic force. They have emphasised the role of the natural world in shaping human organisations and traditions and being

---

[1] William Cronon, 1983. *Changes in the Land: Indians, Colonists, and the Ecology of New England*. New York: Farrar, Straus and Giroux; Stephen J. Pyne, 2012. *Fire: Nature and Culture*. London: Reaktion Books; Donald Worster, 1979. *Dust Bowl: The Southern Plains in the 1930s*. New York: Oxford University Press; Alfred W. Crosby. 2004. *Ecological Imperialism: the Biological Expansion of Europe, 900–1900*, 2nd ed. Cambridge: Cambridge University Press.

shaped by them – at times purposefully and at others as an unconscious by-product of human activity. Nature has also been portrayed as an active partner by scholars in this country.[2] In *How a Continent became a Nation*, Libby Robin considers the connection between nature and nation, and suggests that Australians might come to terms with this land by studying birds such as the Banded Stilt and other elements of the Australian natural world. For Robin:

> ... non-human actors like mountains and deserts have agency in history... If we consider the land from the perspective of a bird of the arid inland, we might find ourselves examining our cultural expectations, and perhaps thinking again about ways to manage our lands more like creatures that are already adapted to the country's variability.[3]

Similarly, Eric Rolls's *A Million Wild Acres* animates nature. In this remarkable history of the Pilliga scrub the trees 'move in passion', the forest 'takes over' from the cleared land and 'coarse foreigners' – cats, pigs, rats, rabbits, bees, horses, foxes – work to reproduce the Old Country environment. Rolls is interested in the transformative power of humans – he recognises that the open woodlands had been created by Aborigines and that the dense forests 'do not display the past as it was' but are modern forests produced by Europeans – as well as being concerned with the impact of these introduced, opportunistic species and their ability to unsettle the vulnerable Australian landscape.[4]

In acknowledging the presence of the natural world, some have been criticised for turning nature into its own agent, and not fully accounting for the role of humans within the environment. As J.R. McNeill writes:

> ... environmental history often arouses the indignation of readers who think that it leaves out people, or reduces them to abstractions. Human agency disappears into the shadows, while climate, or viruses, or technology hog the spotlight.[5]

---

[2] See, for example, Geoffrey Bolton, 1981. *Spoils and Spoilers: Australians Make their Environment 1788-1980*. Sydney: Allen and Unwin, 174 which maintains that Australians should view Australian earth as a mother and William J. Lines, 1991. *Taming the Great South Land: A History of the Conquest of Nature in Australia*. Sydney: Allen and Unwin, xviii, which celebrates the 'secondariness' of humanity and questions established views about the mastering of nature.

[3] Libby Robin, 2007. *How a Continent Created a Nation*. Sydney: UNSW Press, 2.

[4] Eric Rolls, 1981. *A Million Wild Acres* Melbourne: Penguin Books, 185, 269, 348, 399.

[5] J.R. McNeill, 2003. 'Observations on the Nature and Culture of Environmental History', *History and Theory*, 42(4), December, 36.

This criticism was directed in particular at Crosby's influential transnational study *Ecological Imperialism: The Biological Expansion of Europe, 900–1900*, which referred to the unwitting changes which humans wrought on the environment. Crosby questioned whether there was a biological or ecological explanation for the success of the European colonial venture. He detailed the triumph of European plants, livestock, pathogens and parasites in North America, southern South America, Australia and New Zealand, explaining that these organisms worked in concert with humans to disrupt indigenous species and spread across these regions.[6] In response, authors such as Tom Griffiths warned that environmental change should not be reduced to mere accident. Griffiths' concern was that this failed to recognise human responsibility and 'may present "ecological imperialism" as a latter-day "Social Darwinism", a way of denying human agency – for good or ill – on the frontier'.[7]

Mindful of these concerns, this collection focuses on the role of humankind in effecting environmental change. We take what might be regarded as a species-centric approach in reflecting on the ways in which human social, economic and political activity has modified one region within this continent. But this is not a one-way exchange. Just as this species has influenced nature, so too has humankind been influenced by it. We consider the interplay between humans and the rest of nature, and the willingness of past generations to acknowledge the extent to which their affairs have been informed by their surroundings. In directing our attention to the Gippsland region we are able to highlight the stories of past individuals and their interaction with these non-human actors.

In referring to the *human* relationship with nature, we deliberately focus on the Indigenous and non-Indigenous understanding and use of the environment. We acknowledge the work of Tim Flannery and Bill Gammage in directing attention towards the Aboriginal management of resources and creation of unnatural pre-European landscapes.[8] Indeed Flannery's *The Future Eaters* uncovers consistent patterns of human manipulation throughout Australasia, such as the Indigenous 'invasion' on the Australian continent – which he controversially insists brought about the extinction of

---

[6] Crosby. *Ecological Imperialism*, 7.

[7] Tom Griffiths, 1997. 'Ecology and Empire: Towards an Australian History of the World' in Tom Griffiths and Libby Robin eds, 1997. *Ecology and Empire: Environmental History of Settler Societies*, Seattle: University of Washington Press, 2.

[8] Timothy Flannery, 1994. *The Future Eaters: An Ecological History of the Australasian Lands and People*, Sydney: Reed New Holland; Bill Gammage, 2011. *The Biggest Estate on Earth: How Aborigines Made Australia*. Crows Nest: Allen and Unwin.

the megafauna – and later European introduction of new forms of resource use. However, Flannery argues that, unlike Aboriginal people, the non-Indigenous inhabitants have yet to learn to adapt to Australia's nutrient-poor environment, while Gammage turns the *terra nullius*–Aboriginal pairing on its head, insisting that the European attitude to country should correctly be termed *terra nullius*.[9] While these texts have uncovered a broad history of Indigenous land use, we share Stephen Dovers' view in relation to Flannery's work that other, more detailed accounts are needed in order to test these claims.[10] We argue that future studies of the Gippsland region and elsewhere might uncover geographical differences and changes across time.

The linking of Gippsland and environmental history has a strong tradition and many of Australia's pioneer scholars have contributed towards an understanding of the environmental history of the region. Gippsland features heavily in Tom Griffith's important work on the forests, while Jane Lennon, Coral Dow and Warwick Frost have all considered different aspects of Gippsland's environmental history.[11] Despite the efforts of these influential scholars, there is as yet no inclusive synthesis of the region's environmental history. Works remain focused on particular places or industries. More work also needs to be done on Gippsland's natural systems through the lens of science, biology and ecology, with this collection firmly on the historical and human side of the ledger.

Environmental history has often engaged with regional and local topics. In Australia, key practitioners have suggested that this focused spatial scale is one of its defining characteristics.[12] Despite this regional focus, the geographic extent of Gippsland is still very broad, covering some 41,524 square kilometres, and this can obscure sub-regional or local variations. Our contributors have covered many different environmental contexts from this diverse region. The chapters in and of themselves reinforce the point about regional diversity. The history of local government and administration originated in the distinct small towns and communities of the region, with

---

[9] Flannery, *The Future Eaters*, 231; Gammage, *The Biggest Estate*, 323.

[10] Stephen Dovers, ed. 2000. *Environmental History and Policy: Still Settling Australia*, South Melbourne: Oxford University Press, 9.

[11] Tom Griffiths, 2001 *Forests of Ash: An Environmental History*, Port Melbourne: Cambridge University Press; Jane Lennon, 1993. 'Cornerstone of the Continent: A History of Wilson's Promontory' *Park Watch*, (172); Coral Dow, 2005. 'Tatungalung Country: An Environmental History of the Gippsland Lakes', PhD thesis, Clayton: Monash University; Warwick Frost, 1997. 'Farmers, Government, and the Environment: The Settlement of Australia's "Wet Frontier", 1870–1920', *Australian Economic History Review*, 37(1), 19–38.

[12] Griffiths, 'Ecology and Empire', 12.

small local shires representing sometimes less than 1,000 ratepayers in what were very town-based societies. While these small shires, which appeared after the passage of the Victorian *Local Government Act* in 1874, have been amalgamated into larger local government units, town-based differences, and occasionally fierce intra-regional rivalry, remain. But there are some common themes that make assembling these studies into one edited collection a viable proposition. Appealing to or fighting off distant forces in the form of state governments, Melbourne-based interests, or national or multinational companies is a recurring theme. Another theme is that Gippsland has entered into the national imagination as a certain kind of place with particular characteristics.[13] This image has been dominated by the Latrobe Valley and its more recent industrialised past, with the areas to the west, east and south of the Valley being somewhat neglected.

Underpinning the economic, social and political frameworks that have shaped how the region has developed and how it is perceived are its environmental and geological features. For all the various definitions of environmental history, they all share one single characteristic, which expressed simply but succinctly, focuses on the interaction of the human and natural elements of any given system. The mountain ranges, the coastal flats and estuarine areas as well as the thick beech rainforests have already been alluded to. The region's heavy rainfall in the south (up to 1500 millimetres per annum in mountainous areas) and west (approx. 1000 millimetres) contrasts with the much lower levels in the east (approx. 650 millimetres). The soils are also diverse but on the whole are not very rich in nutrients, lacking iodine in particular. These environmental characteristics themselves are amenable to change. Climate change will alter the region's rainfall patterns and raise the sea levels. Offshore oil and gas drilling as well as coal seam gas extraction will also lower land levels in some areas by 50 to 90 centimetres by 2031.[14] Erosion is already steadily changing and reshaping the natural environment.

While there is a long tradition of writing on Gippsland's environmental history, there are nonetheless many gaps. Some of these gaps (but by no means all) are addressed in this collection. The chapters contained herein

---

[13] John Tomaney and Margaret Somerville, 2010. 'Climate Change and Regional Identity in the Latrobe Valley', *Australian Humanities Review*, (49), November. Accessed 1 May 2013 at http://www.australianhumanitiesreview.org/archive/Issue-November-2010/tomaney&somerville.html.

[14] 'Climate Change, Sea Level Rise and Coast Subsidence along the Gippsland Coast', Gippsland Coastal Board, Bairnsdale, 2008. Accessed 10 June 2013 at http://www.gcb.vic.gov.au/sealevelrise/sealevelriseSubsidence.pdf

first saw light of day through two conferences held at Monash University's Gippsland campus in Churchill, convened by Julie Fenley. Contributors were given a broad set of themes to address and there was no prescriptive sense of 'environmental history' imposed upon them. The 2010 and 2011 conference descriptions, for example, indicated that the conferences were about 'how the region has shaped the Gippsland people and how they in turn have shaped their surroundings', suggesting an interactive and connected sense of people and their worlds.[15] Both the 2010 and 2011 conferences also suggested five themes: Aborigines and early Gippsland; the use of natural resources; management of the environment; the landscape, flora and fauna of Gippsland; and conservation and representations of the environment. We received a number of contributions on the use of natural resources and management of the landscape themes. Unfortunately, we were not able to source contributions on the built or urban environment, and suggest that this area will require further work in the future. There are numerous specialist studies and local histories which cover the creation and development of Gippsland's regional centres and towns, but more needs to be done to assess their impact from an ecological standpoint.[16]

Much more work also needs to be done on the twentieth- and twenty-first-century industrial era of Gippsland's environmental history, dating from the beginnings of large-scale coal and electricity production in the 1920s, the opening of the Maryvale pulp mill near Traralgon in the 1930s and, more recently, the oil and gas production from nearby Bass Strait in the 1960s.[17] The bulk of the chapters in the collection have focused on nineteenth-century themes and experiences. Forestry certainly reached the level of a large industrial production system and that is covered in the chapters that follow, but Gippsland's industrial, mining and electricity generation era probably deserve separate treatment. At the same time, focusing on these aforementioned themes would have narrowed the geographical field of view

---

[15] See 'Gippsland Environments and Human Interaction: Past, Present and Future'. Accesssed 10 June 2013 at http://ceh.environmentalhistory-au-nz.org/2011/09/gippsland-environments.

[16] See, for example, Stephen Legg, 1992. *Heart of the Valley: A History of the Morwell Municipality*, Morwell: City of Morwell; Joseph White, 1989. *A History of the Shire of Korumburra*, Korumburra: Shire of Korumburra; Sally Wilde, n.d. *Forests Old, Pastures New: A History of Korumburra*, Warragul: Shire of Warragul; David Langmore, 2013, *Planning Power: The Uses and Abuses of Power in the Planning of the Latrobe Valley*. North Melbourne: Australian Scholarly Publishing.

[17] For recent work on the impact of the power industry in the Latrobe Valley see David Langmore, *Planning Power*.

to the Latrobe Valley at the expense of East, South and West Gippsland. Meredith Fletcher's chapter on Jean Galbraith and her responses to the industrialisation of the Latrobe Valley is a tantalising indication of the potentially rewarding research still to be done in this area.

Gippsland's history, in common with many other regions, has often been presented as a battle between settlers and a harsh unforgiving nature. The region has none of the challenges of dry land or desert climates, but the temperate rainforests and the wetlands near Koo-wee-rup or near what later became Moe were nonetheless formidable. For this reason, European settlement of what later became Gippsland occurred on the fringes of this region. The initial incursions took place from the north through the grazing plains of the Monaro and the Australian Alps and the relatively dryer plains of East Gippsland as pastoralists sought new grazing land. Squatters also took up runs in West and South Gippsland from the late 1830s and early 1840s, particularly as the best pastoral land had been acquired in the Port Phillip District from the late 1830s. However, the natural barriers of swampland and wet temperate rainforests slowed the settlement of South Gippsland, ensuring that the coastal ports, with readily accessible maritime trade routes, remained important gateways. Cheryl Glowrey's chapter provides more detail on this early colonial period and the importance of Corner Inlet, Port Albert and the maritime economy that provided an impetus for successful occupation. Following Glowrey's focus on Corner Inlet, and freed from our modern assumptions around the centrality of land and rail transport from Melbourne, we see a different kind of regional geography and different possibilities for Gippsland's history – a north–south coastal orientation for trade and commerce – which never quite came about.

Maritime footholds and pastoral incursions on Bunurong, Gunai-Kurnai and Yaitmathang country did not go unnoticed by the traditional owners. In common with other pastoral frontiers, Gippsland was a brutal site of frontier conflict. Conflict was inevitable as Europeans asserted control and sought to access and significantly modify large amounts of land. It is no surprise then that from the early 1840s the Gunai-Kurnai began to respond to European encroachment. The region experienced a series of hostile encounters and massacres, including those led by pioneering explorer Angus McMillan. Violence, or the threat of violence, conditioned many early encounters on the frontier.[18] Eventually, a policy of dispersal and separation

---

[18] Phillip Pepper in collaboration with Tess de Araugo, 1985. *What Did Happen to the Aborigines of Victoria Volume I: The Kurnai of Gippsland*, Hyland House, Melbourne,

of Gunai-Kurnai from their land onto designated reserves and missions, such as Lake Tyers Church of England mission in East Gippsland (founded in 1861), meant that the land was freely available for settler dominion from the 1860s.[19] Ruth Lawrence's contribution also adds to our understanding of the early colonial period. She focuses on a part of Gippsland where settler incursion occurred earliest. Lawrence charts the complex effects of European occupation in the Omeo region and the settlers' failure to learn from and adapt the Yaitmathang relationship with the land. She considers how rivalries with the Gunai-Kurnai, which may have traditionally existed as intercultural conflict or resulted from settler incursions, were exploited by the Europeans.

Passable roads were constructed from the 1860s, with enterprising settlers enduring great hardships as they forged new paths through sometimes thickly forested areas. Gold had already been discovered at Omeo (1851), but the newly cut roads encouraged rushes to the Jordan fields (from 1861), the Dargo-Crooked River fields (1860 and 1864) and the Walhalla area (1862). These attracted new migrants and established new forms of resource exploitation. Gold left significant scars on the landscape with sites for new towns cleared, watercourses altered and benches dug into hillsides. Former battery sites and mullock heaps can still be seen in these areas, most of which experienced periodic rushes and renewals of interest for the remainder of the nineteenth century. Jane Lennon observes that at Walhalla and elsewhere, evidence of human occupation and use in the landscape can be the source of conflict and debate. Lennon reflects on over 40 years of research in Gippsland, finding that the attempt to 'regreen' the landscape, create aesthetically pleasing tourist facilities or reconstruct particular histories works against conservation values, or conceals particular stories about disturbance to the whole landscape.

Gold seekers naturally attracted prospectors who found other payable minerals, such as copper, while lead-bearing ore was also discovered. Mills and timber cutters were not far behind the establishment of these new fields. Timber mills were often the basis of new towns, but it was in East Gippsland

---

1985; P.D. Gardner, 2001. *Gippsland Massacres: The Destruction of the Kurnai Tribes 1800–1860*, 3rd edition, Self published, Ensay.

[19] P.D. Gardner, 1991. 'Aboriginal History in the Victorian Alpine Region' in *Cultural Heritage of the Australian Alps*, ed. B. Scougall, Proceedings of a symposium held at Jindabyne, New South Wales, 16-18, Australian Alps Liaison Committee, 89-99; Sylvia Kleinert, 2012. 'Keeping up the Culture: Gunai Engagements with Tourism' In *Towards an Australia-wide anthropology*. G. Cowlishaw and L. Gibson, eds. *Oceania*, 82(1), 86-103.

where forestry and timber milling took hold most strongly. Helen Martin's chapter charts the operation of the logging industry in East Gippsland, which dates back to the second half of the nineteenth century. She utilises oral history records in referring to the resourcefulness and adaptability of forest industry operators and their attachment to the forest, but also explores the dangers, hardship and lack of stability inherent to this industry.

The rail line from Melbourne to Sale was completed in February 1878, effectively consolidating growth and prosperity through the central east–west axis of the region. Slowly but surely trade and commerce gathered to the spine of the Sale line with the coastal regions struggling to maintain or enhance their growth. This development was reinforced by the growing importance of the port at Lakes Entrance and Sale.[20] The South Gippsland railway line modified this picture somewhat but the line could never compete with the Sale line.[21] While affected by the expansion of the rail line, David Harris reveals that the opening up of commercial fishing on the Gippsland Lakes and increased availability of the Lakes as a holiday and recreational fishing destination was also reliant on the support of local administrators. He maintains that commercial fishers and angling clubs employed differing views of nature and attempted to influence government opinion and regulation in favour of their specific social and economic concerns.

From about 1870 through to the Great War and its immediate aftermath, extensive tracts of thick forests in West and South Gippsland were cleared to make way for dairying and closer settlement. This period and this region was the source of some of the most remarkable images from Gippsland's environmental history as ring-barked trees in cleared paddocks stood as silent sentinels to the destruction of the forest and the creation of new environments to support the dairying economy. Progress was often measured as a triumph over nature by both settlers and subsequent local histories. However, Stephen Legg challenges these accounts in his study of an often-forgotten set of concerns about forestry and forest management. His assessment of the role of regional newspapers nonetheless concludes that the regional press were relatively silent on these matters, apart from a few editors who were very much lone voices on forestry management and conservation. Similarly,

---

[20] Jane Lennon, 1971. 'The Development of Port Towns, Gippsland Victoria', Paper presented to the 43rd ANZAAS Congress, Brisbane 24-28 May.

[21] The furthest reaches of the South Gippsland lines were progressively closed in the twentieth century and the line was eventually closed to passenger and freight traffic at Cranbourne in the late 1990s. South Gippsland, however, certainly did not fade from the picture as a focus for settlement and development.

'Two men on springboards are cutting down a blue gum tree near Scanlon's property at Budgeree in 1913'.
Centre for Gippsland Studies Collection (http://arrow.monash.edu.au/hdl/1959.1/3834).

Ruth Ford's analysis of rural women's letters to the *Weekly Times* during the 1930s highlights that while many sought to transform the landscape – and were as invested as their husbands in cultivating the land and establishing a living – they also celebrated the beauty of the bush and demonstrated a sense of loss over the destruction of the environment. Beyond the simple description of particular farming activities, there was a complex and vibrant engagement by farming women in their farm economies, the environment around them and their hopes and dreams for life on the land.

By the late nineteenth century, government policy and state support encouraged new economic forms of engagement with Gippsland environments but the precise effects on individual settlers were often harsh and unpredictable. This is evident in Kerry Nixon's interrogation of the yeoman ideal – the myth of the independent farmer – which informed the Victorian Land Acts and impacted on landholders such as the Currie family, who selected land south of Drouin in 1874. Through Nixon's chapter we gain insights into the Currie family's attempts to satisfy the conditions of the colonial government's legislation, which prevented farmers from selecting larger and more economically viable plots and required that they 'improve' and cultivate the land, and their acceptance of the yeoman tradition's emphasis on industriousness and self-reliance. The extent to which government legislation underpinned settlement and the development of industry is also the subject of Charles Fahey's investigation of the origins of the 'dairy industry problem'. This problem was typified by small farm holdings and uncertain prices that prevented a sizeable proportion of mid–twentieth-century farmers from generating sufficient returns. Fahey traces the 'dairy problem' to government policies of selection and closer settlement and, drawing on the Currie family diaries and other records from surrounding parishes, maintains that these initiatives encouraged the establishment of small and unviable properties whose impact continued to be felt into the 1950s. Jillian Durance surveys the history of a dairy farm and homestead at Moyarra in South Gippsland, and finds that the success or otherwise of closer settlement was impacted by a combination of political, economic and social factors. Once part of the 'Great Forest of South Gippsland' the property, selected by John Gannon then purchased by William Rainbow in 1895, changed 'almost beyond recognition'. Durance demonstrates that landholders' engagement with the environment was influenced by government policy, economic conditions and personal fortunes, while decisions about land use and management not only had social implications, but also affected the region's environmental wellbeing.

East Gippsland presented a somewhat different prospect. There was a distinctive local environment with extensive freshwater lake systems from Marlo in the north to Sale and Seaspray in the south. This fertile land encouraged commercial activity, with cattle dominating the high country while sheep and cattle and mixed farming, were also important on the fertile low-lying areas.[22] In the late 1870s the railway line was completed and tourist traffic arrived in East Gippsland in greater numbers. In addition, since the lakes area was subject to flooding and regular inundation, it was soon altered by extensive reshaping of the land and water boundaries to suit settler needs. Nonetheless, there were some similarities across Gippsland and throughout Victoria, as Deirdre Slattery points out in her chapter on growing wool in East Gippsland. Slattery's chapter, 'Rural land use change and the Ridley Paddock 1897–2012', is a personal account of her family's association with and use of a single paddock in the region. But the Ridley

---

[22] Jane Lennon, 2009. 'Heritage on High, Reflections on Designating Alpine Heritage Places' In Heritage of the High Country Forum, Heritage Council et al, Omeo; *Census of the Commonwealth of Australia, 30th June 1933, Part II – Victoria Population Detailed Tables for Local Government Areas*, 220-231, Australian Bureau of Statistics, at http://www.abs.gov.au/AUSSTATS/abs@.nsf/DetailsPage/2110.01933?OpenDocument.

paddock also serves as an example of European settlement in Victoria and of the failure to identify a balanced approach to productive output, community resilience and conservation values. Likewise, while Sarah Mirams focuses on a small tourist community in East Gippsland, her assessment of alarm over modern technology and tourist consumerism in the 1920s finds some resonance with concerns over tourist exploitation elsewhere. However, the fears that motor tourism would lead to the 'ruination' of coastal East Gippsland did not come to pass, partly because the region is protected by its isolation, whereas other tourist areas closer to large metropolitan centres received greater exposure. Leisure and recreational pursuits evinced diverse reactions across Gippsland and in other areas, and not only for tourists. Julie Fenley and Kathy Lothian show how Gippsland medical practitioner James Moore Andrew interacted with and sought to make sense of his engagement with the Australian bush. Their examination of an extended walk he undertook in 1920, before the widespread acceptance of the term 'bushwalking' (and creation of the bushwalking industry), and his later leisure activities, encourage a reassessment of our definition of 'independent' bushwalking.

The use of Gippsland's resources continued apace in the twentieth century with the creation of a new timber mill and brown coal power-generation facilities. The Maryvale paper mill near Traralgon was in production by 1939, and as Julie Constable maintains, its proximity to the Strzelecki Ranges partly explains why widespread logging, the reclassification of regenerated and native forest as plantations, and the planting of non-indigenous species took place in this area. Constable evaluates the management of the Strzelecki State Forest and campaigns for increased forest protection. In two key periods in the 1970s and 1990s campaigners expressed concerns over these activities and the influence of the timber and pulpwood industry, and called for the enlargement of existing reserves and creation of new reserves in the Strzelecki Ranges. Constable contends that, despite these efforts, successive governments continued to act in the interests of logging concerns. Other campaigns also took place over the logging of significant forest areas. Deb Foskey's discussion of differing campaigns over the clear felling of the Errinundra forest between 1983 and 1984 reminds us that forest activism should not be reduced to a conflict between economic and environmental interests. While local protestors, radical activists and national and state conservation bodies might have been seen as working in unison, they had competing agendas, possessed varying degrees of influence over state and federal policy, and encountered different costs and outcomes.

Coal mining also became a central economic activity in South Gippsland from the 1890s and the Latrobe Valley more profitably after the Great War, and a state-owned power-generation industry was also begun at Yallourn in 1921. Coal and electricity helped people the Latrobe Valley in the twentieth century, especially during the postwar boom after 1945. Steady streams of migrants arrived, initially from Britain and Ireland, and later from Italy, the Netherlands, Greece, Yugoslavia and other European countries. Towns such as Moe, Morwell and Traralgon grew rapidly. By 1954 Morwell had a population of 13,033, Traralgon 10,033, and Moe 8770, dwarfing many of the rural towns in the region. New power stations, or expansion at existing plants, occurred at Yallourn, Morwell, Hazelwood and Loy Yang, all owned and operated by the State Electricity Commission of Victoria. Bass Strait oil and gas were significant in the 1970s and 1980s for Sale and South Gippsland but this was not as long-lasting or as wide ranging as the Valley's brown-coal industrialisation. But some disquiet was also expressed over these developments. Meredith Fletcher's study of Jean Galbraith indicates that the well-known botanist supported balanced land use through the creation of reserves next to cleared zones. Opposed to the unthinking and rampant destruction of remnant vegetation, Galbraith had some success in campaigning for areas to allow for the preservation of native flora and fauna.

Cruelled by the recession of the early 1980s and then again in the early 1990s, the period under Premier Jeff Kennett's administration (1992–98) was a difficult and tumultuous one for Gippsland as local government was overhauled, schools and services closed, and, most significantly for the Latrobe Valley, the electricity industry was sold off to private interests. A downsizing of industry prior to privatisation saw unemployment peaking at significantly higher levels in the Latrobe Valley than the rest of the state, while rationalisation also took place in Gippsland's timber industry, as Helen Martin explains with specific reference to East Gippsland. While Hazelwood Power Station has long been slated for closure, alongside Energy-Brix and Yallourn, an extension was granted to allow the facility, which produced around 600 million tonnes of carbon dioxide between 1971 and 2010, to continue operating until 2031.[23]

More recently, debates over climate change, a carbon-pricing scheme and the nation's mining resources have again highlighted the tension between earth and industry. In Gippsland, attention has focused once more on resource use and management. In 2009, following a period of prolonged drought,

---

[23] Langmore, *Planning Power*, 363.

the state's Labor government announced the construction of a desalinisation plant in Wonthaggi to secure Victoria's future water resources. Major emergency responses and recovery efforts were also required after the 2003 Alpine bushfire (which destroyed more than one million hectares of land in North-east and East Gippsland), the 2006–07 Great Divide Fires, and the 2009 Strzelecki and Black Saturday fires (in which 173 people lost their lives across Victoria – 11 of them in Gippsland). In 2014 the lack of adequate monitoring by Hazelwood mine owner GDF-Suez resulted in a mine fire at the open-cut mine in Morwell that blanketed the township in smoke and ash for 45 days, causing severe social, economic and environmental concerns. These problems suggest the need for more sustainable practices, as well as a rethink of our species-focused approach to our surroundings.

The collection takes a thematic approach and highlights the possibilities of future research, with chapters by Jane Lennon, Stephen Legg, Ruth Ford, and Julie Fenley and Kathy Lothian focusing on the way that we learn about, and represent, human–environment interactions. A second key theme explored by Cheryl Glowrey, Ruth Lawrence, David Harris, Kerry Nixon, Charles Fahey, Meredith Fletcher, Deb Foskey and Julie Constable is concerned with the activities of states or international organisations, and the participation of citizens or interest groups in political and legal struggle. Government policy was both an enabler and a hurdle for many Gippslanders, and the role of government in shaping the region and its environment would be another area ripe for sustained attention. A third theme, examined by Jillian Durance, Deirdre Slattery, Helen Martin and Sarah Mirams relates to production and technology. As Donald Worster explains, this field of inquiry focuses on 'understanding how technology has restructured human ecological relations, that is, with analysing the various ways people have tried to make nature over into a system that produces resources for their consumption'.[24] This section considers the complex and changing relationship between technology, socio-economics and the environment. The industrial era of the twentieth century, which so dominated the Latrobe Valley, would also repay further research effort. Individual chapters reveal a focused attention on single landholdings or single industries and suggest how this mosaic of experiences and industries interacted in a wider regional history.

---

[24] Donald Worster, 1990. 'Transformations of the Earth: Towards an Agroecological Perspective in History'. *The Journal of American History*, 76 (4) March, 1090.

Acknowledgements

We are indebted to Beth Edmondson, Fleur Gabriel, Elizabeth Hart, and Rebecca Strating for their feedback and encouragement on an early draft, to Keith Wilson for his invaluable assistance in facilitating the refereeing process, and to the two reviewers for their insightful comments.

References

Australian Bureau of Statistics. 2014. *Census of the Commonwealth of Australia, 30th June 1933, Part II – Victoria Population Detailed Tables for Local Government Areas*, 220–231, Australian Bureau of Statistics. Accessed August 2014 at http://www.abs.gov.au/AUSSTATS/abs@.nsf/DetailsPage/2110.01933?OpenDocument.

Bolton, Geoffrey. 1981. *Spoils and Spoilers: Australians Make their Environment 1788–1980*. Sydney: Allen & Unwin.

Crosby, Alfred W. 2004. *Ecological Imperialism: the Biological Expansion of Europe, 900–1900*, 2nd ed. Cambridge: Cambridge University Press.

Cronon, Willliam. 1983. *Changes in the Land: Indians, Colonists, and the Ecology of New England*. New York: Farrar, Straus & Giroux.

Dovers, Stephen ed. 2000. *Environmental History and Policy: Still Settling Australia*. South Melbourne: Oxford University Press.

Dow, Coral. 2005. 'Tatungalung Country: An Environmental History of the Gippsland Lakes', PhD thesis, Clayton: Monash University.

Flannery, Timothy. 1994. *The Future Eaters: An Ecological History of the Australasian Lands and People*. Sydney: Reed New Holland.

Frost, Warwick. 1997. 'Farmers, Government, and the Environment: The Settlement of Australia's "Wet Frontier", 1870–1920', *Australian Economic History Review*, 37 (1), 72–89.

Gammage, Bill. 2011. *The Biggest Estate on Earth: How Aborigines Made Australia*. Crows Nest: Allen and Unwin.

Gardner, P.D. 1991. 'Aboriginal History in the Victorian Alpine Region' in *Cultural Heritage of the Australian Alps*. B. Scougall ed. Proceedings of a symposium held at Jindabyne, New South Wales, 16-18, Australian Alps Liaison Committee.

Gardner, P.D. 2001. *Gippsland Massacres: The Destruction of the Kurnai Tribes 1800–1860*, 3rd edition, Self-published, Ensay.

Gippsland Coastal Board. 2008. 'Climate Change, Sea Level Rise and Coast Subsidence along the Gippsland Coast', Gippsland Coastal Board, Bairnsdale. Accessed 10 June 2013 at http://www.gcb.vic.gov.au/sealevelrise/sealevelriseSubsidence.pdf.

Gippsland Environments and Human Interaction: 'Past, Present and Future'. Accessed at 10 June 2013 at http://ceh.environmentalhistory-au-nz.org/2011/09/gippsland-environments

Griffiths, Tom. 1997. 'Ecology and Empire: Towards an Australian History of the World' in Tom Griffiths and Libby Robin eds. *Ecology and Empire: Environmental History of Settler Societies*, Seattle: University of Washington Press.

Griffiths, Tom. 2001. *Forests of Ash: An Environmental History*, Port Melbourne: Cambridge University Press.

Kleinert, Sylvia. 2012. 'Keeping up the Culture: Gunai Engagements with Tourism' in *Towards an Australia-wide anthropology*. G. Cowlishaw, and L. Gibson eds. *Oceania*,

82(1), 86–103.

Langmore, David. 2013, *Planning Power: The Uses and Abuses of Power in the Planning of the Latrobe Valley*. North Melbourne: Australian Scholarly Publishing.

Legg, Stephen. 1992. *Heart of the Valley: A History of the Morwell Municipality*, Morwell: City of Morwell.

Lennon, Jane. 1971. 'The Development of Port Towns, Gippsland Victoria'. Paper presented to the 43rd ANZAAS Congress, Brisbane 24–28 May.

Lennon, Jane. 1993. 'Cornerstone of the Continent: A History of Wilson's Promontory' *Park Watch*, (172), 5–8.

Lennon, Jane. 2009. 'Heritage on High, Reflections on Designating Alpine Heritage Places'. In Heritage of the High Country Forum, Heritage Council et al, Omeo.

Lines, William J. 1991. *Taming the Great South Land: A History of the Conquest of Nature in Australia*. Sydney: Allen & Unwin.

McNeill, J.R. 2003. 'Observations on the Nature and Culture of Environmental History', *History and Theory*, 42 (4), December, 5–43.

Pepper, Phillip in collaboration with de Araugo, Tess. 1985. *What Did Happen to the Aborigines of Victoria Volume I: The Kurnai of Gippsland,* Hyland House, Melbourne.

Pyne, Stephen J. 2012. *Fire: Nature and Culture*. London: Reaktion Books.

Robin, Libby. 2007. *How a Continent Created a Nation*. Sydney: UNSW Press.

Rolls, Eric. 1981. *A Million Wild Acres* Melbourne: Penguin Books.

Tomaney, John and Somerville, Margaret. 2010. 'Climate Change and Regional Identity in the Latrobe Valley', *Australian Humanities Review*, (49), November, 29–47. Accessed 1 May 2013 at http://www.australianhumanitiesreview.org/archive/Issue-November-2010/tomaney&somerville.html,.

White, Joseph. 1989. *A History of the Shire of Korumburra*, Korumburra: Shire of Korumburra.

Wilde, Sally. n.d. *Forests Old, Pastures New: A History of Korumburra*, Warragul: Shire of Warragul.

Worster, Donald. 1979. *Dust Bowl: The Southern Plains in the 1930s*. New York: Oxford University Press.

Worster, Donald. 1990. 'Transformations of the Earth: Towards an Agroecological Perspective in History'. *The Journal of American History*, 76 (4) March, 1087–1106.

Chapter 1

Uncovering Gippsland history through reading evidence in the landscape

Jane Lennon

Since the 1960s there have been marked changes in the ways in which researchers read the landscape. These shifts in understanding and the subsequent need for new approaches to cultural heritage have not been without controversy, with differences over how to manage the evidence of human use and occupation. A personal journey is presented here, to show how the Gippsland landscape taught me about the ways in which people responded to the environment and how they in turn have shaped their surroundings.

Three case studies examine how the Gippsland environment fashioned the settling and use of the land. The first assesses Wilsons Promontory, which was a cattle run at the time of its initial reservation as a national park. This study evaluates its contested use, involving conflict between officials and communities, whether local or metropolitan, farmers, miners, naturalists and conservationists or recreationists.

The second case study reflects on Walhalla, which was almost stripped bare of the vegetation binding the steep slopes of its valley during the mining era. Once regarded as the 'most picturesque' town in Victoria, it decayed into obscurity until Latrobe Valley workers – seeking relaxation in the mountain environment – 'rediscovered' the town and began building weekenders. Others came from the city to walk the tramways and considered rebuilding the ruins from the footings up. This presented a challenge for government agencies – as regulators of heritage and planning policies – and residents, whose interests could differ from these objectives.

The third case study considers our appreciation of the human impact on the Gippsland coast. Charting and understanding the Gippsland coast was essential for nineteenth century mercantile interests desiring a shipping outlet for their stock and timber and port settlement. The imprint of these activities is a legacy for today's planners who grapple with these landscape changes. However, little is known about this disturbance beyond government agencies and academic circles. Future environmental studies are required which utilise multidisciplinary techniques and various forms of evidence such as official documents from archival collections, photographs, aerial images and newspaper accounts, oral stories, and archaeological surveys. Through this evidence we can uncover the impact of human activities on the environment.

Beginnings

My late childhood years were spent on our family sheep farm on the steep southern slopes of Andersons Hill above Loch. I recall gazing through the fog to Western Port Bay and imagining the travails of Bass and Flinders in their whaleboat. Tarra National Park, with its magnificent fern gullies, was a favoured picnic destination of my parents, while I yearned for adventures at remote Wilsons Promontory. Years later, as senior planner in the National Parks Service, my staff were exhorted to 'know your patch'; that is, they should know its underlying geology, how that affected subsequent land form processes, its geomorphology, and the vegetation it was clothed in at particular times. Park rangers were also urged to become familiar with the dates when humans first occupied this landscape, the dates of first European exploration and areas of settlement. It was also necessary for them to understand the impact of the gold rushes, closer settlement, the Great Depression, World War One and later closer settlement policies or soldier settlement and swamp drainage, as well as more recent re-greening policies and rural residential subdivision.

As a geographer, I have used my eyes first to decipher the landscape: the 'seeing eye' taught to recognise the occupation layers in the landscape. Then I used parish plans for clues to these layers: settlement patterns via drainage lines, rising ground, accessible spurs, rugged hilltops and seemingly unscaleable ridges of mountain ranges. These layers influenced the pattern of settlements at creek or river crossings, at break-of-slope points before the long haul up a range, and place names which often reflect the settlers' cultural links. This provided a framework for a deeper investigation of place.

Training at the University of Melbourne (1965–68) in physical geography and the history of exploration and settlement led to a series of questions about the landscape characteristics of Victoria, particularly along the Gippsland coast. Examining the role of vegetation as a dune stabiliser led to increased knowledge of coastal vegetation and introduced the study of ecology, but what role did humans have in altering these coasts? Were the lines of shells we were seeing in cliffed dune faces stranded shorelines from an early period of higher sea level or Aboriginal 'kitchen' middens? Who had first surveyed this coastline? Had the surveys been updated? Surely these would show changes? Had pastoral or agricultural settlement affected the shoreline of the lakes?

There were no immediate answers to these questions. This surprised the university's academic staff, trained in English schools of geography where detailed studies of historical evidence in the English landscape abound. Our challenge was to find answers in both archival and field evidence. In 1940, Ernest Skeats had outlined this challenge in his foreword to Sherbon Hill's *Physiography of Victoria*:

> The eyes see only what they are trained to observe... 'The everlasting hills' are, in fact, but transient features in the landscape and slowly evolve into land forms of a different character... The geographer... should be able to interpret the face of nature and to recognize its slowly changing character.[1]

Our textbooks were English or American. Themes such as draining the swamplands, clearing the woodlands, reclaiming the heathlands, irrigating the deserts and building towns were discussed and referred to as vertical themes in contrast with the horizontal themes of chronology in historical geography. Closer to home, W.K. Hancock's brilliant 1972 study of the impact of humans on the Monaro environment of high country and tableland with its themes of squatting, spoiling and improving was still being researched.[2]

The challenge was to understand the detailed settlement patterns in the Gippsland landscape in south-eastern Victoria, my home territory, and the processes which formed and influenced them. Existing texts, such as Charles Daley's *Story of Gippsland* or ECF Bird's *Coastal Landforms*, were broad in coverage and a more detailed analysis of particular landscapes was required before authoritative statements about processes and their impacts could be presented.[3]

My research into Gippsland settlement (beginning in 1970) involved primary research as historians had not yet begun to use either the official public correspondence from first settlement – which in this case had not even been

---

[1] E.S. Hill, 1940. *Physiography of Victoria*. Melbourne: Whitcomb and Tombs.

[2] W.G. Hancock, 1972. *Discovering Monaro, A Study of Man's Impact on his Environment*. London: Cambridge University Press.

[3] Charles Daley, 1960. *Story of Gippsland*. Melbourne: Whitcomb and Tombs; E.C.F. Bird, 1964. *Coastal Landforms*. Canberra: ANU Press.

indexed in the State Archives, then located in the basement of the La Trobe Library – or family and other private papers deposited in the State Library. I was able to provenance the series of 16 Robert Russell pencil sketches in the Mitchell Library which were unsigned, but titled and sometimes dated, and catalogued as *Sketches in South-east Victoria, 1843*. These sketches, the first detailed ones for Corner Inlet and Wilsons Promontory, illustrated the primitive nature of the original European settlement and also showed the commercial links of entrepreneurs like the Imlay brothers, who were whalers and graziers of Twofold Bay, NSW, and Lagoon Bay on the south-east coast of Tasmania.[4]

Wilsons Promontory

Clues in the landscape are evident throughout Wilsons Promontory, which was reserved in 1898 and is the oldest and best-loved national park in Victoria. Designated a UNESCO Biosphere Reserve in 1977 due to its outstanding nature conservation values, 'The Prom' is an evolving cultural landscape, a park landscape created over the last 116 years with evidence of many occupations and rich in associations for many Victorians. The cultural values derived from these are not well understood and conserved. Matthew Flinders dubbed Wilsons Promontory the 'cornerstone of the continent', naming it for his London merchant friend, Thomas Wilson.

The Promontory is composed of a rugged granitic upland of Devonian age extending under Bass Strait to north-eastern Tasmania and the Yanakie isthmus composed of Pleistocene deposits connecting the Promontory to the mainland. Erosion of the granite has produced today's spectacular landscapes of high-rounded mountains, swampy valley floors, cliffs and offshore rocky islands. Archaeological evidence of Aboriginal occupation of the Pleistocene dune systems along the Yanakie isthmus 6500 years ago is found in middens. Occupation had probably been much earlier, given the land bridge to Tasmania and human occupation there dating from 38,000 years ago. Ethnographical accounts referred to the rugged mountains of the central Promontory harbouring the spirit of Loo-ern, a protecting deity for the Brataualang clan of the Kurnai.[5]

---

[4] Jane Lennon, 1975a. 'Squatters, Merchants and Mariners, an Historical Geography of Gipps' Land, 1841–55'. Melbourne: MA thesis, University of Melbourne.

[5] Jane Lennon, 1993. 'Cornerstone of the Continent, A History of Wilsons Promontory', *Parkwatch*, (March), 5–6.

European attempts at the commercial utilisation of the supposed and actual resources of Wilsons Promontory in the nineteenth century have been documented in various studies by Lennon.[6] Robert Russell's sketch, entitled 'Whalers Huts built by Dr Imlay's men' and dated 2 May 1843, was first published in my paper for the Royal Historical Society of Victoria in 1974. This sketch, together with the report by HB Morris in the *Hobart Town Courier* of 23 June 1843, was the first archival evidence of whaling activity at Wilsons Promontory.[7] Its publication prompted further research into the routes and ownership of whaling vessels (for example, by Cumpston, and Syme) and archaeological surveys of whaling sites along the Victorian coast by both the Victorian Archaeological Survey and academic researchers.[8] Previous uses of the land had an impact on local landscapes within the Prom and include: navigational refuge, with the establishment of a lighthouse on the southern point by 1859; timber milling in two phases at Sealers Cove (1849–59; 1898–1906); quarrying at Refuge Cove; sheep and stock grazing from 1850; gold prospecting in the 1860s; tin mining between 1920 and 1936; commercial fishing (especially by Chinese in the 1860s); and the campaign from 1887 to reserve the Promontory as 'a summer haunt of lovers

---

[6] Jane Lennon, 1974. 'Wilsons Promontory in Victoria, its Commercial Utilization in the Nineteenth Century'. *Victorian Historical Magazine*, 45, 179–200; Jane Lennon, 1988. 'Timeless Wilderness? The Use of Source Material in Understanding Environmental Change in Gippsland, Victoria'. In *Australia's Ever Changing Forests*, K.J. Frawley and N. Semple (eds), Canberra, 419–440; Lennon, 1993. 'Cornerstone of the Continent'; Jane Lennon, 1998b. 'Wilson's Promontory National Park – its Management as a National Park'. In *Celebrating the Parks: Proceedings of the First Australian Symposium on Parks History*, Mt Buffalo, April, 133–142; Jane Lennon, 1998. 'Whaling at Wilsons Promontory, Victoria in the 1840s'. In Susan Lawrence and Mark Staniforth (eds), *The Archaeology of Whaling in Southern Australia and New Zealand*, The Australasian Society for Historical Archaeology and The Australian Institute for Maritime Archaeology, Special Publication No. 10, NSW, 64–67.

[7] Lennon, 'Wilsons Promontory in Victoria', 184.

[8] J.S. Cumpston, 1973. *First Visitors to Bass Strait*. Roebuck Society Publication, No. 7, Canberra; Marten A. Syme, 1984. *Shipping Arrivals and Departures Victorian Ports*, vol 1, 1798–1845. Roebuck Society Publication No.32, Burwood, Victoria; Marten A. Syme, 1987. *Shipping Arrivals and Departures Victorian Ports*, vol 2, 1846–1855. Roebuck Society Publication No.39, Burwood, Victoria; Iain Stuart, 1989. 'An Historical Archaeological Survey of Wilsons Promontory'. In N. van Waarden and D. Ransom (eds), 'Wilsons Promontory Archaeological Project'. Unpublished report to Victoria Archaeological Survey, Melbourne; Karen Townrow, 1997. *An Archaeological Survey of Sealing & Whaling Sites in Victoria*. Heritage Victoria, Melbourne; Susan Lawrence and Mark Staniforth (eds), 1998. *The Archaeology of Whaling in Southern Australia and New Zealand*. The Australasian Society for Historical Archaeology and The Australian Institute for Maritime Archaeology, Special Publication No. 10, NSW, 64–67.

of nature, lovers of scenery'. This latter claim was in opposition to settlement by starving refugees –1000 crofters from the Isle of Skye, who were to be resettled and establish a fish-curing business. Place names on the east coast of the Prom can be seen as evidence of prior activities. Some were misspelt, so my newspaper research into shipping consignees led to the correct name: Johnnie Souey, a Port Albert fisherman. Subsequent La Trobe University archaeology department excavations of the Chinese fish-curing site at Port Albert confirmed the network of fishing sites.[9]

Despite its reservation as a national park, a second sawmill settlement operated at Sealers Cove until 1906 when a bushfire burnt out the mill and 16 buildings that were home to 40 people.[10] The jetty piles still remain in the cove. From 1913 the Park Committee of Management conducted a building program centred on Darby River and the Tourist Resorts Committee-funded road through Yanakie Run was officially opened in February 1939.[11] During World War Two the park was closed and Australia's first commando group was stationed at Darby River, where No. 7 Infantry Training Centre was located. Eight independent companies were based at various locations, including Tidal River, and their tracks and barbed wire were still evident 50 years later. Concrete footings remain of Darby Chalet, partially destroyed in an explosion in 1942 and finally demolished in 1949.[12]

Due to insufficient government funds to manage the park the Committee decided to grant grazing licences: 2000 acres on the Darby River and Oberon Bay were leased to the Lester Brothers of Leongatha from 1937 until 1959.[13] In May 1964 the National Parks Authority became the committee of management for the Yanakie Run and stock numbers were progressively reduced over 20 years until grazing finally ceased in 1992.

The landscape itself – forest structure, gaunt spars – tells a story. The impact of nineteenth century graziers' fires was mostly irregular despite hot burns occurring on the forest floor, compared with the devastating impact of sheep denuding vegetation on Yanakie isthmus in the nineteenth century causing massive sand erosion. The 1951 wild fires which raged over about 75 per cent of the Prom spared Tidal River, although nearby Lilly Pilly Gully,

---

[9] Alister Bowen, 2012. 'Archaeology of the Chinese Fishing Industry in Colonial Victoria'. *Studies in Australasian Historical Archaeology*, vol. 3, Sydney University Press.

[10] Lennon, 'Timeless Wilderness?', 432.

[11] John Noonan and Ann Fraser, 1970. *Of Wamman and Yanakie*. Foster, 20–22.

[12] Barry Collett, 1994. *Wednesdays Closest to the Moon, A History of South Gippsland*. Carlton: Melbourne University Press, 249, 255.

[13] National Parks Authority file 7/6/9.3

'a fern filled cathedral', was burnt open to the sky.[14] In the 1980s fires caused by lightning strikes effected the north end; in 2005 the bottom half of the Prom was engulfed by wild fire, which also created a new landscape.

Landscape conservation implications

Despite all this evidence from the published history – collections of historical photographs such as those from the Field Naturalists Club of Victoria depicting their annual activities at the Prom from the early 1900s, and archaeological surveys and records such as those from Victoria Archaeology Survey's Operation Raleigh in the early 1990s – the current management plan promotes the re-greening of the landscape. The environmental nationalism which demands blanket protection of the bush in its 'pristine condition' does not promote informed debate and discussion about what constitutes natural condition. Currently 43 per cent of the park is zoned wilderness 'to protect the essentially unmodified condition of the area… and minimal interference to natural processes'.[15] There is a case for maintaining a small mosaic of modified cultural landscapes within the larger 'wild' park landscape.

In 1988, the bicentennial of European impact on the east coast of this continent, I had argued that historical research at the local level would help in understanding ecological change – the Hoskins approach of landscape reading as 'a new history', one acknowledging the far-reaching impact of human activity while giving pride of place to rural scenery:

> Since the catastrophic state-wide bush fires of 1939… much of the Victorian landscape has greatly mellowed under its regenerated clothing of native vegetation… This has led viewers to perceive a 'virgin bush', untouched and unsullied, and demand legislation and keepers to protect its virtue. But this virtue of 'pristine condition' may be an illusion born of ignorance of the detailed history of the place. A case study of Wilsons Promontory National Park illustrates this.[16]

The Wilsons Promontory landscape is a result of management over the last century, which allowed natural processes to continue. Regrowth, fire and recreation activities have contributed to the obliteration of surface evidence

---

[14] Lennon, 'Timeless Wilderness?' 435.

[15] Parks Victoria, 2002. *Wilsons Promontory National Park Management Plan*. East Melbourne: 10.

[16] Lennon, 'Timeless Wilderness?', 421. For details of Hoskins approach to reading the landscape see W.G. Hoskins, 1955. *The Making of the English Landscape*. London: Penguin.

of nineteenth century sites of occupation, now heritage sites. Historical research is an essential component of the natural resource management planning process. To this planning process long-term climate change as an agent affecting all our landscapes must be examined. This is particularly noticeable in periods of drought and high temperatures when fire weather may result in lightning-ignited wild fires, a natural process following its course despite our culturally conditioned urge to suppress it. However, history repeated itself 54 years on when the 'wilderness' of the southern 15 per cent of the Prom was burnt in a very hot wild fire in April 2005. Natural factors dominated in this latest phase of cultural landscape making: the park management culture of controlled burning had failed.

Wilsons Promontory remains as the cornerstone of the continent, a place of refuge for weary urban dwellers in this century just as it was for Aboriginal tribes and storm-tossed seafarers in the nineteenth century. It provides an anchor for Gippsland. Nonetheless, we must rethink our approaches to the management of the park to allow for a more nuanced understanding of changes to this supposed wilderness area.

Walhalla

The differing claims over heritage management are illustrated by a study of the Walhalla region. The Walhalla Historic Area, covering 2500 hectares, is north of the Latrobe Valley in steeply dissected mountain country. Walhalla Township, covering 72 hectares, is located within this larger area. In 1982 the historic landscape, one of the richest nineteenth century mining areas in Victoria, attracted 20 permanent residents, many part-time residents with holiday houses, and 80,000 visitors. Today there are 15 permanent residents and 324 in the 2011 Census Walhalla district, and while there are no reliable figures on tourist numbers, it is estimated by Walhalla Board of Management Inc that 100,000 visit the region each year.

The miners at Walhalla were also its first European explorers. The isolated terrain of steep mountain gullies and plentiful water shaped a distinctive technology using waterwheels and incline tramways. Tramways ferried ore from the mine to riverside batteries and extensive networks delivered wood for boilers and timber for the mines. Italian woodcutters congregated at Poverty Point. The planned relationship of timber and gold has left interesting clues in the forest today. Man-made features such as tramways, mullock heaps, creek diversion, stone retaining walls and terraced house sites are clear evidence of mining-period activity, even though associated structures may have gone. In the late 1970s I spent many days exploring along the old tramways and staying at the former hospital high above the valley floor.

Gold discoveries in 1862 led to a rush of miners staking small claims at creek level. Gold at deeper levels presented problems which were solved by companies introducing batteries and treatment plants from 1864. Between 1885 and 1895 the population accelerated to a peak of 4500, nearly half of whom lived in settlements above the narrow valley. In November 1888 fire destroyed 30 buildings and within a year a new central Walhalla had risen from the ruins. However, devastating floods in August 1891 killed four people, washed away bridges and houses, buried the mine winding gear under mudslides, and carried away wood stacks. This caused havoc in the lower township and beyond, where waterwheels and fluming were wrecked and coarse sands stacked for treatment were washed down to the Thomson River valley. Recovery was expensive and slow. Poor road access was a source of constant complaint and the population demanded a railway to the Latrobe Valley.

Cohen's Reef, with 15 mines producing one and a half million ounces of gold, was the richest gold-bearing reef in the world at the time and was worked to depths of over 3000 feet. The peak gold yield was 25,198 ounces in 1886.[17] The Long Tunnel Company built 40 miles of tramway. Timber cutting denuded the Walhalla Ranges of its dense forest cover. Horse tramways extended in every direction, even across the Thomson River. Lack of new gold deposits combined with falling gold prices, increased costs and shortage of labour at the outbreak of World War One led to closure of the mines by the end of 1914. The population declined from over 2000 in 1901 to 235 in 1921. Remaining buildings included the railway station, the Star Hotel, two banks, 17 shops and the hospital, which operated until 1945. Because of the demise of the town, Walhalla Shire merged with the Shire of Narracan in 1918 and, in late 1994, the area was incorporated into Baw Baw Shire.

Small-scale gold mining continued until World War Two. Sawmilling became the most significant activity in the district to 1970. The railway line to Walhalla closed in 1941 and the population fell to 54 in 1961. Many historic relics were destroyed in the numerous fires and floods which continued to plague the town. In the early 1970s the National Trust opposed road construction through the town as it entailed widening the existing

---

[17] Raymond Paull, 1967. *Old Walhalla*. Carlton: Melbourne University Press, 102.

alignment, replacing the timber bridges and demolition of buildings. At the same time, the Walhalla Improvement League was established and purchased and restored buildings, while volunteers built a locomotive and terminus and operated a tourist steam train from Walhalla to the Thompson River. As well as this private activity, the local Shire of Narracan and a Committee of Management were appointed in 1976 to progressively restore the Long Tunnel Extended Gold Mine.

The planning process for conserving Walhalla's heritage

In 1977 the Land Conservation Council recommended that 2500 hectares be set aside as the Walhalla Historic Area 'to preserve the various sites (and the surrounding environment) that are associated with Victoria's early mining history'. The Council also adopted a controversial conservation approach to management:

> … co-ordinated planning of both public and private land be undertaken to ensure the preservation of this part of Victoria's history, and to create an atmosphere that will allow visitors to appreciate the historical significance of the town.[18]

Photographs of Walhalla in 1977 show dense forest-clad hill slopes.[19] Did the government and community really want to re-create an historic environment of denuded hills and fouled streams? Historically, the hills surrounding Walhalla had been stripped of trees for a distance of about 12 miles to feed the mine boilers and for household fuel. Regrowth over the past 60 years produced an open forest with an overstorey of messmate-brown stringybark, manna gum and silver top, an understorey of wattles, blackwood and tree ferns, and sassafras in the wet gullies. However, re-creation of the historic landscape was not desirable as serious erosion would result from clearing the steep valley sides. Instead, natural regeneration of open forest was encouraged and exotic trees confined to Walhalla and Maidentown.[20] All identifiable historic places remaining from the mining period (1863–1915) were regarded as significant to the history of Walhalla. Any development works would be designed to retain the original fabric. Reconstruction of

---

[18] Land Conservation Council, 1977. *Final Recommendations, Melbourne Study Area*. Melbourne: Land Conservation Council, 87.

[19] G.F. James, and C.G. Lee, 1970. *Walhalla Heyday*. Melbourne: Robertson and Mullens Pty. Ltd.

[20] Department of Conservation, Forests and Lands (CFL) and Shire of Narracan, 1988. *Walhalla Historic Area Management Plan*. Melbourne: Government Printer, 32.

some buildings had ensured their conservation, notably Eliots Bakery, which was in ruins in 1975. There was considerable debate about design controls. Because of the large number of freehold allotments available, new building could proceed in three basic forms: re-creation ('faithful' reconstruction to the pre-existing form); sympathetic design (consistent with the historical mining character); or no-control design.

The local population wanted replicas so that private property owners could present tourists with an image of the physical shape and density of Walhalla at its peak of development.[21] This was in conflict with the Australia ICOMOS Burra Charter principles for conservation of places of cultural significance and immediately pitted professional against local interests. The ICOMOS principle that 'reconstruction is limited to the completion of a depleted entity and should not constitute the majority of the fabric of the place' was not acceptable to the majority of the Walhalla Historic Area Advisory Committee.[22] The Committee wanted one of the hotels re-created to provide commercial facilities, as occurred with the Star Hotel in the 1990s, and limited re-creation allowing visitors to appreciate the 'once famous mining town with the few remaining in-situ original historic places in an isolated picturesque valley'.[23]

However, the main driving force for re-creation was tourism, not conservation. This was the majority view held passionately by the local and elected representatives. Tourists are attracted to Walhalla because of its landscape setting and its historical atmosphere. It offers not only the experience of walking around the narrow streets past the historic buildings but also the opportunity to enter a genuine historic gold mine, at the Long Tunnel Extended Mine. This combination is Walhalla's major asset.

Walhalla residents had a keen sense of the history of mining in this difficult area, but were not convinced of the need to leave archaeological layers of historic evidence and construct new structures that were sympathetic to but not confused with previous ones. Nor did they wish to reflect on the environmental history of the area and its phases of timber felling, fire and flood.

---

[21] Jane Lennon, 1993. 'Presenting History *in situ*: Historic Site management on Public Land in Victoria', in C. Michael Hall and Simon McArthur, *Heritage Management in New Zealand and Australia: visitor management, interpretation and marketing*. Auckland: Oxford University Press, 154.

[22] Australia ICOMOS (article 18, 1981 version).

[23] CFL and Shire of Narracan, 1988. *Walhalla Historic Area Management Plan*, 24.

The management plan sought to reduce threats to the visual physical resources. The Burra Charter guidelines assisted in identifying places, assessing their significance, and prescribing policies for their conservation. That a place has historic significance may be irrefutable, but the planning process cannot ensure continuity of use or function. In adapting to new uses, a sympathetic design in or for that historic place may be more honest and appropriate than a re-creation.[24] The good intentions but conflicting values implied in a legislative objective for professional protection of a static past, an archaeological layer, had very different results given residents' views of the same past.

Certainly at Walhalla the democratic view favouring replicas and reproductions overruled the more purist architectural conservation view. An exception was the careful restoration of the Walhalla Post Office, including its historic wallpaper. The social values expressed by the majority resulted in structural reconstruction as the means of maintaining the heritage value in the Walhalla landscape, even to the extent of reconstructing the railway as a tourist attraction. However, contemporary appreciation of the bush overruled any attempted re-creation of the historic landscape, except for cottage gardens in the township and the resolution of the dilemma of conserving one historic resource (cemetery headstones) at the expense of another (senescent conifers falling on graves, a public safety issue).[25] The irony may well be that wild fire through the regenerated bush might again destroy the Historic Area's assets.

The Shire of Baw Baw is the responsible local government authority and in response to a decade of bushfires in the high country of Gippsland it recently commissioned a *Walhalla Bushfire, Heritage and Overlay Review* for its planning scheme.[26] This review examined the conflict between bushfire prevention and heritage concerns. It found that the planning policy framework relating to Walhalla is complex with numerous overlapping policies, guidelines, and provisions, mostly warranted due to the heritage, environmental and landscape characteristics. Significant development is

---

[24] Jane Lennon, 1991. 'Whose Heritage? Memory, Nostalgia and Good Intentions: The Politics of Conserving Mining Heritage at Walhalla, Victoria'. Heritage and Memory Conference, Humanities Research Centre, A.N.U. July 1991.

[25] Jane Lennon, 1995. 'Conservation Policy and Practices in Victoria: Review from an Administrative Viewpoint'. In Cultural Conservation: Towards a National Approach, Sharon Sullivan (ed.), Australian Heritage Commission Special Australian Heritage Publication Series Number 9, Canberra 1995, 442.

[26] Meinhardt Infrastructure & Environment Pty Ltd, 2013. *Walhalla Bushfire, Heritage and Overlay Review, Summary Report;* for Shire of Baw Baw, Warragul.

unlikely to occur in Walhalla in the foreseeable future, and given that any changes to planning controls would only apply to new development, the existing *Design and Development Guidelines* should be clarified and used to conserve and enhance the heritage character of the town.[27]

While Walhalla is located in a high fire-risk landscape its buildings have not been destroyed by bushfire. Bushfire analysis and mapping has identified that development may be possible under the provisions of the Bushfire Management Overlay in limited locations, primarily already developed sites in the central areas of the township.[28] The review further recommended making the town a Cultural Landscape of State Significance where Heritage Victoria would then be the Responsible Authority for heritage matters.[29]

Walhalla provides a classic case study of an historic area with keen interest groups wanting a living landscape which caters to tourists. Two excellent forums guided its initial conservation: the Land Conservation Council and the Advisory Committee. However, the outcome was that social values overrode the strict conservation of historic values in archaeological sites and allowed new development of replica structures. The intangible values of meanings about and associations with the landscape triumphed.

Coastal changes

In contrast to the previous case studies, changes to the Gippsland coast around Corner Inlet have been little studied beyond academic or regional planning circles. Cook, Bass and Flinders sailed along the coast of Gippsland charting its outlines but the first detailed description of the shores of Corner Inlet and Port Albert harbour came from Captain Lewis in 1841. He took a relief crew to the steam ship *Clonmel* that was grounded on sand banks near Corner Inlet, and was enthusiastic about the 'safe and commodious harbour'.[30] At the same time an official general survey by Commander Stokes in HMS *Beagle* charted Corner Inlet waters. However, Stokes' charts were

---

[27] Tract Landscape Architects, 1999. *Walhalla Township Design and Development Guidelines*, for Baw Baw Shire and Department of Natural Resources and Environment.

[28] Meinhardt Infrastructure, *Walhalla Bushfire*, 73.

[29] Conflicting values continue within conservation frameworks. In 2010 a fish ladder was planned at the Horseshoe Bend tunnel in the Historic Area in order to preserve an endangered species of fish, but local stakeholders argue that the tunnel generates tourist dollars and is also a historically significant site which should be preserved [http://www.horseshoebendtunnel.com/home.html].

[30] Jane Lennon, 1975. 'Our Changing Coastline'. *Victorian Historical Journal*, 46, 478.

not available until 1846, so firsthand navigational knowledge by masters of coasting vessels was essential. The first official survey of the shores of Corner Inlet and adjacent waters was completed by Surveyor Townsend in December 1841.

Despite this detailed surveying and naming of places and sites, there was much confusion at both popular and official levels over place names. Newspapers advertised vessels as 'laid on for Corner Inlet', whereas they were sailing to the settlement in Port Albert, the name Townsend had given to the estuary of the Albert River. Established in 1841, the Old Port was one of the very earliest legal settlements in Victoria outside of the Port Phillip region. In 1841 about 100 individuals were already living in tents and cottages near the beach, and an 1843 watercolour by Robert Russell shows at least two boarded houses and a number of huts. Business transferred to Shipping Point and the Old Port was officially abandoned in 1844.[31] This important archaeological site is currently endangered by waterfront erosion.

By 1847 there were problems with navigation at Corner Inlet. Between Shipping Point and Snake Island it was indeed 'a floating harbour, the sands shift according to the winds'.[32] In April 1847 stockholders wrote to La Trobe about the dangerous approach to the harbour, as shown by the loss within 12 months of three schooners that were totally wrecked, while others were damaged by grounding.[33] In June 1848 Captain Haig, surveying Port Albert for three leading Hobart Town insurance companies, described the inlet as 'a terror to everybody'. Captain Bentley recommended the most suitable location for navigation buoys, which were finally installed in October 1850.[34] Lighthouses at Wilsons Promontory and at the entrance to Port Albert harbour were operating by 1859 and greatly assisted shipping in avoiding the shifting shoals extending two to three miles off land from the entrance.

Nevertheless, between 1860 and 1900 a further 23 vessels were wrecked in the vicinity of the Port and Corner Inlet, many due to changes in the

---

[31] Jane Lennon, 1979. 'Port Albert', *Historic Places of Australia*, 2, 208–216.

[32] Parliamentary Papers (Victoria), 1856–57, vol. 4. *Report from the Select Committee Upon Shipping Point*, p. 21.

[33] Lennon, 'Our Changing Coastline', 481; Cheryl Glowrey, 2013. 'An Environmental History of Corner Inlet, Victoria, Australia'. Melbourne: PhD thesis, Monash University, 111.

[34] Lennon, 'Our Changing Coastline' 482–3.

channels and sand banks.[35] Jenkin described the physical processes involved in the evolution of tidal features, channel systems and outer barriers along the Corner Inlet – Port Albert – Shallow Inlet coastlines.[36] As students in the 1960s we also surveyed some of this coastline and were involved in drift card research which measured currents flowing along the Ninety Mile Beach. In June 1961 the outer barrier was breached by a storm, giving residents of McLoughlin's Beach their own entrance to the sea.[37] These sands are ever-moving and define the environment of this part of Gippsland, a reminder that natural forces dominate.

Local newspapers provided much comment on attempts to navigate the Gippsland Lakes for commercial use and on the campaign to create a permanent artificial entrance from the sea. These formed the basis for the 1973 book I wrote in conjunction with Dr Eric Bird, *The Entrance to the Gippsland Lakes*, which was updated in 1989 and renamed *Making an Entrance: The Story of the Artificial Entrance to the Gippsland Lakes*.[38] Provision of an artificial entrance led to a century of modifications to the lake shoreline, dieback of reed swamps and the gradual build up of salinity in the eastern lakes.[39] These major environmental changes illustrate the indirect and unforeseen consequences of human interference with a natural system. Shoreline surveys undertaken in 1959 and subsequently re-measured were also essential illustrations for this study, and the combined evidence from historical archives and field measurements helped to answer questions of landscape change in this coastal lake system.

Local historians have not favoured researching change in physical features of their landscapes. Dramatic events such as shipwrecks and cutting canals have been recorded while events of regular occurrence, such as bushfires and floods, were often overlooked, possibly because the early settlers accepted these hazards as part of nature and beyond their control. Even less notice was taken of events which led over several decades to incremental landscape changes, such as the invasion of coastal scrub by sand, shallowing of streams, clearing the forests and the retreat of native

---

[35] J.K. Loney, 1968. *Wrecks of the Gippsland Coast*. Geelong.

[36] Jeffrey J. Jenkin, 1968. 'The Geomorphology and Upper Cainozoic Geology of South-east Gippsland, Victoria'. *Geological Survey of Victoria*, Memoir 27, 32–35.

[37] *Yarram News*, 21 June 1961.

[38] E.C.F. Bird and Jane Lennon, 1973. *The Entrance to the Gippsland Lakes*, Gippsland Studies No. 2, Bairnsdale, Republished 1989.

[39] Lennon, 'Our Changing Coastline', 486.

fauna.[40] This has been addressed more recently by Coral Dow and Cheryl Glowrey.[41]

Today, Australian scholars and heritage practitioners understand that historical activities leave both physical evidence and meanings, and attachment to places which may change over time. Charting the coastline, selecting a port, establishing a stock export trade, forming a settlement, testing the utility of natural resources (such as the durability of timbers easily felled close to deep water), and using water power to drive stamper batteries were all pioneering activities which left imprints on the landscape. Some of these historical activities have been researched. But what impact did they have on our contemporary landscape? And what is the long-term effect of this impact?

Australian scientists in the 1970s were beginning to grapple with the behaviour of complex environmental systems, and knowledge of historical processes should have been a main consideration. The Victorian Ministry for Conservation, which was undertaking major studies in both Port Phillip and Western Port Bay, hosted the third international environment conference, *Environment '76*. My paper at the opening session covered necessary historical processes – 'Fire, Fell and Drain: Does History have any Value for Environmentalists?' – and advocated use of historical records to enable us to recreate the scene from which we could assess the extent and rate of change.[42] This includes the history and pattern of clearing the forests, ploughing the native grasslands, draining the swamps, altering the coasts and introducing alien species.

Thirty years on it is extremely pleasing that most local government conservation surveys now commence with an environmental history and a timeline detailing key dates and events impacting on the local landscape. However, much of this research remains hidden in the 'grey literature' of unpublished reports, and the public continues to remain uninformed.

---

[40] Ibid., 476.

[41] Coral Dow, 2004. 'Tatungalung Country: An Environmental History of the Gippsland Lakes'. Melbourne: PhD thesis, Monash University; Glowrey, 'An Environmental History of Corner Inlet'.

[42] Jane Lennon, 1976. 'Fire, Fell and Drain: Does History Have Any Value for Environmentalists?' *Environment '76 – Proceedings of Third International Environment Conference*, Melbourne, 31–37.

Conclusion

The case studies all have a recurring theme: that cultural heritage evidence was seen as historic islands in a sea of nature, rather than as key reference points from which to measure disturbance to the whole landscape, with human activity whose extent, duration and impact can be measured as ecological agents of change. It has been a long struggle to have historic places – sites and their surroundings – recognised as a separate and legitimate area for management in the public land and parks system, which would provide an integrated approach to these areas. In 2003, Parks Victoria's *Heritage Management Strategy* directed integration into park management of the 40 major historic buildings and complexes and 2500 recorded historic places and working cooperatively with the community to protect, present and promote heritage places.[43]

The new reading of the Gippsland landscape that I had been advocating was made difficult by the professional disciplines of those involved: the botanists examining individual species requirements while unsure of the ecosystem dynamics; the archaeologists discovering Aboriginal 'antiquity' and wanting to dominate site identification and management; the historians researching a new social history but unable to read the physical evidence remaining in the landscape; and architects who understood the evolution of structures but regarded wrecks and ruins in the bush as having little value.

There is no one correct approach to deciphering the landscape but a multi-disciplinary one ensures that a variety of sources are examined from the ground itself, its forms and vegetation, its place names, explorers, painters, official recorders, private recollections, maps and charts, photographs and oral testimonies. Excavated artefacts can also tell more and confirm missing pieces of the settlement story.[44] But, first and foremost, one must put on solid boots and inhabit the place: examine its lineaments, smell its air and breathe in its vegetation while looking for clues about its occupation and past use.

---

[43] Parks Victoria, 2003. *Heritage Management Strategy*, Melbourne.
[44] Susan Lawrence, Alasdair Brooks and Jane Lennon, 2009. 'Ceramics and Status in Regional Australia'. *Australian Historical Archaeology*, 27, 67–78.

References

Australia ICOMOS, 1981. *The Australia ICOMOS Charter for the Conservation of Places of Cultural Significance (The Burra Charter)*. Sydney: Australia ICOMOS.

Bird, E.C.F., 1964. *Coastal Landforms*. Canberra: ANU Press.

Bird, E.C.F. and Lennon, Jane. 1973. *The Entrance to the Gippsland Lakes*, Gippsland Studies No. 2, Bairnsdale. Republished 1989.

Bowen, Alister, 2012. 'Archaeology of the Chinese Fishing Industry in Colonial Victoria'. *Studies in Australasian Historical Archaeology*, Vol. 3, Sydney University Press.

Collett, Barry, 1994. *Wednesdays Closest to the Moon, A History of South Gippsland*. Carlton: Melbourne University Press.

Cumpston, J.S., 1973. *First Visitors to Bass Strait*. Canberra: Roebuck Society Publication No. 7.

Daley, Charles, 1960. *Story of Gippsland*. Melbourne: Whitcomb and Tombs.

Department of Conservation, Forests and Lands (CFL) and Shire of Narracan, 1988. *Walhalla Historic Area Management Plan*. Melbourne: Government Printer.

Dow, Coral, 2004. 'Tatungalung Country: An Environmental History of the Gippsland Lakes'. Melbourne: PhD thesis, Monash University.

Glowrey, Cheryl, 2013. 'An Environmental History of Corner Inlet, Victoria, Australia'. Melbourne: PhD thesis, Monash University.

Hancock, W.G., 1972. *Discovering Monaro, A Study of Man's Impact on his Environment*. London: Cambridge University Press.

Hill, E.S., 1940. *Physiography of Victoria*. Melbourne: Whitcomb and Tombs.

Hoskins, W.G., 1955. *The Making of the English Landscape*. London: Penguin.

James, G.F. and C.G. Lee, 1970. *Walhalla Heyday*. Melbourne: Robertson and Mullens Pty. Ltd.

Jenkin, Jeffrey J., 1968. 'The Geomorphology and Upper Cainozoic Geology of Southeast Gippsland, Victoria', *Geological Survey of Victoria*, Memoir 27, 32–35.

Land Conservation Council, 1977. *Final Recommendations, Melbourne Study Area*. Melbourne: Land Conservation Council.

Lawrence, Susan, Brooks, Alasdair and Lennon, Jane, 2009. 'Ceramics and Status in Regional Australia', *Australian Historical Archaeology*, 27, 67–78.

Lennon, Jane, 1974. 'Wilsons Promontory in Victoria, its Commercial Utilization in the Nineteenth Century', *Victorian Historical Magazine*, 45, 179–200.

Lennon, Jane. 1975. 'Squatters, Merchants and Mariners, an Historical Geography of Gipps' Land, 1841–55'. Melbourne: MA thesis, University of Melbourne.

Lennon, Jane, 1975. 'Our Changing Coastline', *Victorian Historical Journal*, 46, 476–488.

Lennon, Jane, 1976. 'Fire, Fell and Drain: Does History Have Any Value for Environmentalists?' *Environment '76 – Proceedings of Third International Environment Conference*, Melbourne, 31–37.

Lennon, Jane, 1979. 'Port Albert'. *Historic Places of Australia*, 2, 208–217.

Lennon, Jane, 1988. 'Timeless Wilderness? The Use of Source Material in Understanding Environmental Change in Gippsland, Victoria', in *Australia's Ever Changing Forests*, K.J. Frawley and N. Semple (eds), Canberra, 1988, 419–440.

Lennon, Jane, 1991. 'Whose Heritage? Memory, Nostalgia and Good Intentions: The Politics of Conserving Mining Heritage at Walhalla, Victoria', Heritage and Memory Conference, Humanities Research Centre, A.N.U. July 1991, 19 pp.

Lennon, Jane, 1993. 'Cornerstone of the Continent, A History of Wilsons Promontory'. *Parkwatch*, (March), 5–8.

Lennon, Jane, 1993. 'Presenting History *in situ*: Historic Site Management on Public Land in Victoria', in *Heritage Management in New Zealand and Australia*, ed. Michael Hall and Simon McArthur, Auckland: Oxford University Press, 147–156.
Lennon, Jane, 1995. 'Conservation Policy and Practices in Victoria: Review from an Administrative Viewpoint', in *Cultural Conservation: Towards a National Approach*, Sharon Sullivan (ed.), Australian Heritage Commission Special Australian Heritage Publication Series Number 9, Canberra 1995, 439–449.
Lennon, Jane, 1998. 'Whaling at Wilsons Promontory, Victoria in the 1840s'. In Susan Lawrence and Mark Staniforth (eds), *The Archaeology of Whaling in Southern Australia and New Zealand*, The Australasian Society for Historical Archaeology and The Australian Institute for Maritime Archaeology, Special Publication No. 10, NSW, 64–67.
Lennon, Jane, 1998. Wilson's Promontory National Park: its Management as a National Park, in *Celebrating the Parks: Proceedings of the First Australian Symposium on Parks History*, Mt Buffalo, April, 133–142.
Loney, J.K., 1968. *Wrecks of the Gippsland Coast*. Geelong.
Meinhardt Infrastructure & Environment Pty Ltd 2013. *Walhalla Bushfire, Heritage and Overlay Review, Summary Report*, for Shire of Baw Baw, Warragul.
National Parks Authority, 1959, file 7/6/9.3, Yanakie grazing leases.
Noonan, John and Fraser, Ann, 1970. *Of Wamman and Yanakie*. Foster.
Parks Victoria, 2002. *Wilsons Promontory National Park Management Plan*. East Melbourne.
Parks Victoria. 2003. *Heritage Management Strategy*. Melbourne.
Parliamentary Papers (Victoria), 1856–57, vol. 4. *Report from the Select Committee Upon Shipping Point*.
Paull, Raymond, 1967. *Old Walhalla*. Carlton: Melbourne University Press.
Stuart, Iain, 1989. 'An Historical Archaeological Survey of Wilsons Promontory'. In N. van Waarden and D. Ransom (eds), 'Wilsons Promontory Archaeological Project'. Unpublished report to Victoria Archaeological Survey, Melbourne.
Syme, Marten, A., 1984. *Shipping Arrivals and Departures Victorian Ports*, vol 1, 1798–1845. Burwood: Roebuck Society Publication No.32.
Syme, Marten, A.,1987. *Shipping Arrivals and Departures Victorian Ports*, vol 2, 1846–1855. Burwood: Roebuck Society Publication No.39.
Townrow, Karen, 1997. *An Archaeological Survey of Sealing & Whaling Sites in Victoria*. Melbourne: Heritage Victoria.
Tract Landscape Architects, 1999. *Walhalla Township Design and Development Guidelines*, for Baw Baw Shire and Department of Natural Resources and Environment.
Yarram News, 21 June 1961.

Chapter 2

Forest conservation and the Gippsland press
A brief editorial survey, 1855–1962

Stephen Legg

This chapter investigates the contribution of the Gippsland press to the enduring debate on forest conservation during a period of a little over a century. To explore spatial and temporal variation in the press coverage, a range of settlements was chosen to study the potential influence of environment, land use, geography and settlement age, with the towns having a similar Anglo-Celtic and later Australian-born heritage. Seventeen newspapers were chosen from a dozen settlements across the region: from Port Albert and Sale in the south to Korumburra and Warragul in the west; Woodspoint, Walhalla and Omeo in the north; Morwell, Mirboo North, Traralgon and Maffra in the centre; and Orbost in the far east. To see if, and why, positions on the debate varied within each town pairs of newspapers were chosen in Warragul, Traralgon and Mirboo North and three in Sale. Nine of the newspapers were relatively long-lived (10–93 years), the remainder ranging from 1–10 years. A biographical study was also undertaken to determine who was waging the press campaigns and to explore what factors may have shaped their conservation attitudes. The press items were identified by systematic digital searches using the State Library of Victoria's Trove newspaper archive supplemented by manual searches of selected papers not digitally recorded.

The study period

On the eve of European land settlement in the mid 1830s, when the pastoral frontier moved through Aboriginal territory southward from the Monaro down the narrow alpine valleys and on to the extensive coastal plains, Gippsland was Victoria's most densely forested region.[1] It remains

---

[1] Stephen M. Legg, 1984. 'Arcadia or Abandonment? The Evolution of the Rural Landscape in South Gippsland, 1870–1947'. Melbourne: MA thesis. Dept of

so, despite 175 years of forest clearance and the exploitation of various forest products, thanks mainly to limited accessibility and the poverty for agriculture of most of its ancient upland soils. The study period begins with the establishment of the press in Gippsland in 1855, approximately two decades after European occupation in a region where newspapers developed relatively late and slowly compared to other parts of Victoria.[2] The mid 1850s was only a few years before a major international awakening to the need for forest conservation, beginning in the early 1860s when forest clearance in the New World alarmed a growing number of observers, including some Australians, who began calling for 'scientific forestry'. This involved the regulation of indiscriminate deforestation and the establishment of a reserve-based multiple-use, sustained yield, public forestry system managed by professional foresters. Amidst widespread public apathy and antagonism, elements of the press articulated the growing concerns, presented examples from Australia and overseas, debated the evidence, pleaded with the public to manage forests more wisely, and lobbied politicians to introduce enabling legislation.

In Gippsland, the local deforestation that stimulated the debate grew rapidly during successive waves of agricultural settlement in the 'Selection before Survey' era of the 1870s and 1880s. This was followed soon after by the dairy revolution from 1888 when newly constructed butter factories and their surrounding dairy farms in freshly cleared forest lands became connected by refrigerated railcars to the metropolis. From here, the butter was freighted in the refrigerated holds of ships to the London butter market during the European off-season after the cows there had 'dried off' and prices were high due to reduced supply. This period of rapid technological change was followed, soon after the turn of the century in Australia, by the closer, soldier and group settlement eras, which focused on subdivided estates as well as clearing the few remaining accessible forested Crown lands. The same rail expansion that sped the dairy revolution also greatly stimulated inland development of what had earlier been a predominantly coastal trade in timber. In some parts of Gippsland as early as the 1860s forest devastation was caused by the local gold mining industry pursuing its enormous demands for wood as mining fuel and timber, and Gippsland timber was also sent as far afield as Victoria's central goldfields. In other

---

Geography, Monash University: Victoria; Patrick Morgan, 1997. *The Settling of Gippsland*. Leongatha: Gippsland Municipalities Association.

[2] E. Morrison, 1989. 'The Newspapers of Gippsland'. *Gippsland Heritage Journal*, (6) June, 3–10.

areas, the problem was bushfires used for clearing by settlers or by graziers encouraging grass growth, while elsewhere timber-getters depleted forests.

The end of the study period encompassed enormous changes to forestry and forest utilisation due principally to four consecutive developments generally supported by the Gippsland press. The first was the establishment of a large paper-pulp mill in the Latrobe Valley in central Gippsland in 1936. The second was the rationalisation of timber milling in Victoria after the devastating 1939 bushfires that forced a geographical shift of Victorian timber-getting into central and east Gippsland. The third was the use of these forests in a concerted effort to focus the timber industry on export orientation.[3] The fourth occurred from the late 1940s, when parts of south and central Gippsland began to be re-afforested by the Victorian Government (and later by Australian Paper Manufacturers Pty Ltd), largely for pulpwood production. These included extensive areas of once densely forested steep hill country farmland settled from the 1870s but later abandoned because of the encroachment of vermin and noxious weeds after the Great War and particularly after the collapse of the London butter market in 1921.[4]

Results

Forest conservation was the focus of 329 newspaper items published in the sampled newspapers from 1865 to 1960.[5] Of the 17 sampled newspapers all except three (the *Great Southern Advocate* and the relatively short-lived *Gippsland Chronicle* and *Woodspoint Times*) published items on forest conservation. The historical distribution of the 329 items is shown in Figure 2.1. It reveals the episodic nature of the overall press coverage on all aspects of the forest conservation debate, its longevity, and the dominance of a few periods including the late 1880s, late 1890s and 1904. These major peaks coincided with threats to forests from the local 'land rushes', agitation for a Royal Commission on Forests in Victoria, and debate leading to the 1904

---

[3] Stephen M. Legg, 1977. 'The Location of the Log-Sawmilling Industry in Victoria, 1939–77'. Melbourne: MA prelim. thesis. Dept. of Geography, Monash University.

[4] Legg. 'Arcadia or Abandonment?'

[5] The sampled newspapers include; *The Age*, *The Great Southern Advocate*, *Gippsland Chronicle*, *Woodspoint Times*, *Gippsland Mercury*, *Omeo Standard*, *Morwell Advertiser*, *Gippsland Guardian*, *Gippsland Times*, *The Maffra Spectator*, *The Mount Alexander Mail*, *Traralgon Record*, *Gippsland Farmer's Journal*, *Warragul Guardian*, *West Gippsland Gazette*, *Walhalla Chronicle* and the *Snowy River Mail*. All titles had multiple name changes and some were the result of mergers between smaller titles.

Victorian Forests Bill respectively. Six much smaller peaks occurred later. These comprised: advocacy of forest conservation during the Great War to secure Imperial timber supplies and to support the 1918 Forests Bill that, when passed, established the independent Forests Commission; conflicts in the 1920s over renewed excisions of forest reserves for closer settlement; support for forest improvement and planting as unemployment relief during the Great Depression; support from 1944 for Victoria's Save the Forests Campaign (which often supplied carefully-crafted press releases); and widespread support for the economic development promised by industrial forestry and afforestation in the 1950s. The *Times*, *Mercury* and *Omeo Standard* occasionally pleaded for action to preserve threatened native mammals from the fur trade, and after the Second World War the *Times* and *Morwell Advertiser* published a few brief articles on forests as sanctuaries for wildlife and plant preservation. Nevertheless, wildlife preservation was rarely mentioned in relation to forest conservation in the survey before the early 1950s.[6]

Very few items dealing with forest conservation were written in the early years of the three oldest newspapers in the survey: the Port Albert *Gippsland Guardian* 1855 to 1867, or the Sale newspapers the *Gippsland Chronicle* in 1866 and *Gippsland Times* from 1861 to 1872. Items noting the damaging environmental effects of deforestation began to proliferate in the surveyed newspapers from 1869, although one editor noted retrospectively that local concern with timber scarcity had been expressed from the early 1850s but that no action followed what was then merely 'campfire talk'.[7]

Two hundred and ninety-seven items directly advocated forest conservation, the remaining 32 being split fairly evenly between criticism and neutrality. Most criticism came in letters to the editor or in local correspondents' reports complaining that nearby timber reserves should be 'thrown open' for settlement. The location and frequency distribution of the pro-forest conservation items is shown in Figure 2.2, which also locates the newspapers against the backdrop of the 1869 distribution of forest lands.[8]

---

[6] Wildlife preservation was not a major element of forest conservation in the Victorian press until the 1920s; see Stephen M. Legg, 2004, 'Bunyips, Battues and Bears: Wildlife Portrayed in the Popular Press, Victoria 1839–1948'. In *Conservation of Australia's Forest Fauna*, 2nd edn, edited by D. Lunney. Royal Zoological Society of New South Wales. Mosman: 150–174.

[7] *Gippsland Times*, 2 November 1872.

[8] Derived from A. Everett. 1869. 'Victoria: Distribution of Forest Trees (revision of colour map engraved by William Sligh under the direction of R. Brough Smyth FGS in 1866, for the Hon. John McGregor MLA, Minister for Mines)'.

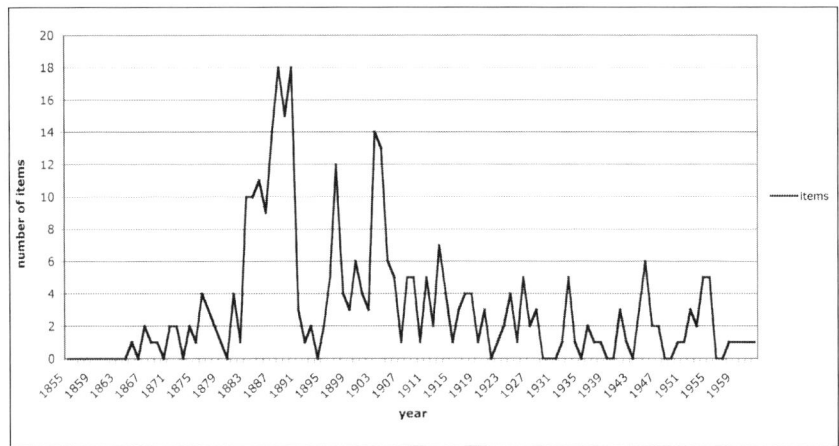

Figure 2.1: Historical distribution of 329 items canvassing forest conservation: Gippsland newspaper survey, 1855–1962.

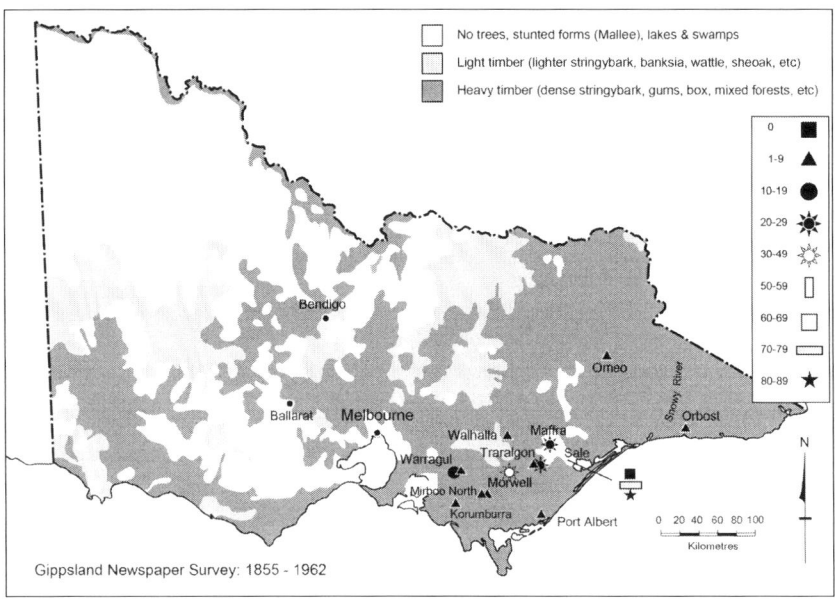

Figure 2.2: Distribution of 297 newspaper items advocating forest conservation.

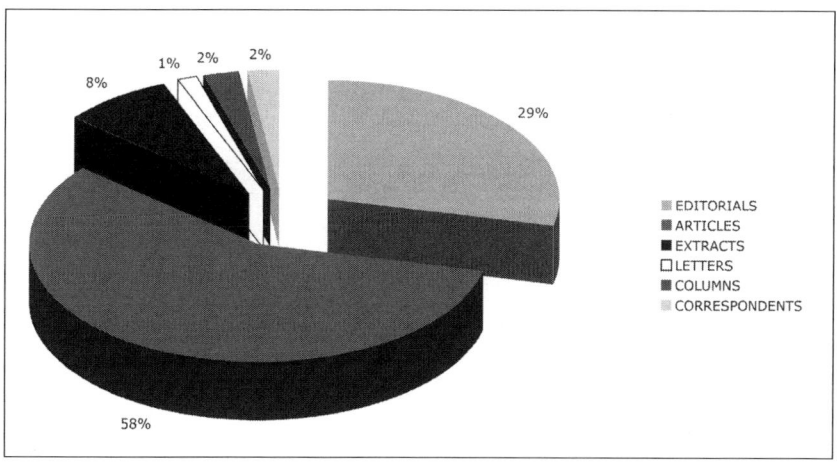

Figure 2.3: Type of item advocating forest conservation, Gippsland newspaper survey, 1855–1962.

Figure 2.3 shows the distribution by type of these 297 items. Fifty-eight per cent were articles often written by the editor, chief reporter, or ghostwriter synthesising various extracts into a short essay form and commonly based on stock articles purchased from press agencies, affiliated newspapers or syndicates.[9] A further 8 per cent were direct extracts from the press or public reports with no local context or commentary, 2 per cent regular columns (typically on agricultural or scientific matters), 2 per cent from correspondents' reports (mainly the regular metropolitan coverage of Victorian Parliamentary bills and debates, and local correspondents' reports from Gippsland or beyond), and only 1 per cent in the form of letters to editors. Finally, 29 per cent of the items comprised 85 editorials, the contents and authors of which are discussed below.

Editors and editorials

Only eight of the 17 sampled newspapers published editorials on forest conservation, and these were limited to a period ranging from 1872 to 1920 (17 years after the beginning, and 42 years before the end, of the survey

---

[9] Ghostwriting refers to the practice known as 'the system' whereby anonymous expert articles and editorials could be purchased at a fee through the Victorian press in the late nineteenth century; see E. Morrison, 1991. 'The Contribution of the Country Press to the Making of Victoria, 1840–1890'. Melbourne: PhD thesis, Dept. of History, Monash University, 134.

(see Figure 2.4)). Remarkably, all 85 of these editorials promoted forest conservation.[10] Three peaks in editorial activity are evident: the mid 1880s at the height of the free selection land rush; the late 1890s in the lead up to and early coverage of the 1897–1904 Royal Commission on Forests; and in the early 1900s, with bitter disputes over the release of forested Crown Lands, including timber reserves, for closer settlement purposes.

Remarkably, there were no editorials on the issue published in the surveyed newspapers during the period 1921–62. The three newspapers that had dominated the debate during the nineteenth century (the Sale *Mercury* and *Times* and *The Maffra Spectator*) continued to publish articles on forest conservation but virtually stopped editorialising on the issue. These were family businesses whose earlier firebrand conservationist editors (the *Times*' Robert Stanton Overend, *Mercury*'s Henry Alfred Luke, and *Spectator*'s James Ryan) had journalist sons who did not continue their fathers' editorialising on the forests debate into the twentieth century (the sons being Stanton James Overend, Henry Alfred Luke Jnr., and Percival Ryan, respectively).[11]

A number of factors were implicated in this generational shift in editorialising past the turn of the century: widespread alarm that timber reserves were harbouring pests and weeds in the early 1900s; rapid growth in demand for forest reserves to be alienated for closer settlement from 1904; the notion that the forests had been 'saved' from 1918 by the establishment of the powerful Forests Commission; and after 1920 a rejection of their fathers' earlier faith that forest clearance caused desertification, drought and long-term climate change. Figure 2.5 shows a little opposition (anti) and equivocation (mixed), but overwhelming press support (pro) for the contention that forests influenced climate. The end of the editorial campaigns is all the more stark when it is noted that the *Mercury*'s Henry Alfred Luke had significantly expanded his own father's, Henry Luke's, forest conservation editorialising from 1885. The

---

[10] In a general survey of Victorian editorials, (B. Trevena, 1981. 'The Country Press Editorial in 1888 in the Colony of Victoria'. Melbourne: BA Hons thesis, Dept. of History, University of Melbourne, 27), Trevena contrasts activist 'critical' editors from those 'printers' who rarely wrote any editorials. All six of the key editors promoting forest conservation in this Gippsland sample were 'critical' editors, overtly political, and active in a range of campaigns including forest conservation.

[11] Percival Ryan did contribute two editorials and Henry Alfred Luke Jnr. one on the forest conservation issue and it is likely both published a little more than a dozen articles each on the subject, but this was comparatively small compared to their fathers.

Forest conservation and the Gippsland press | 27

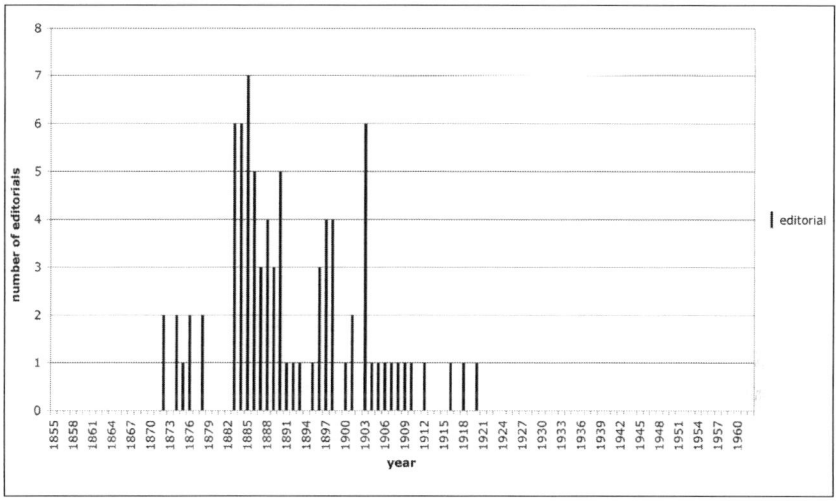

Figure 2.4: Historical distribution of editorials advocating forest conservation, Gippsland newspaper survey, 1855–1962.

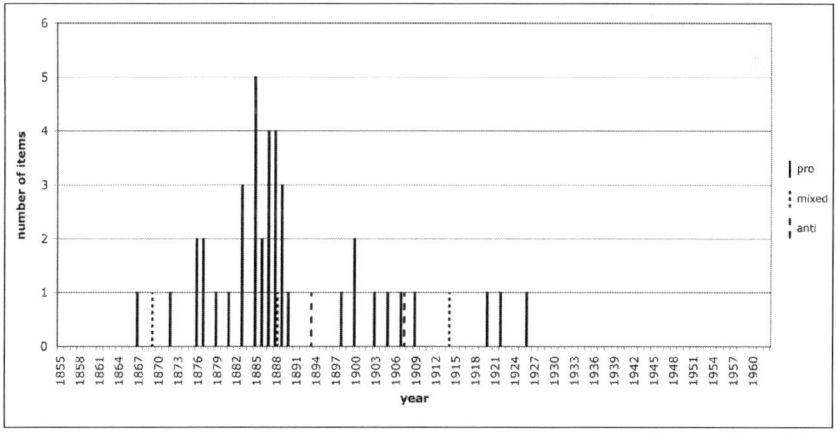

Figure 2.5: Debate on the climatological influences of forests, Gippsland newspaper survey, 1855–1962.

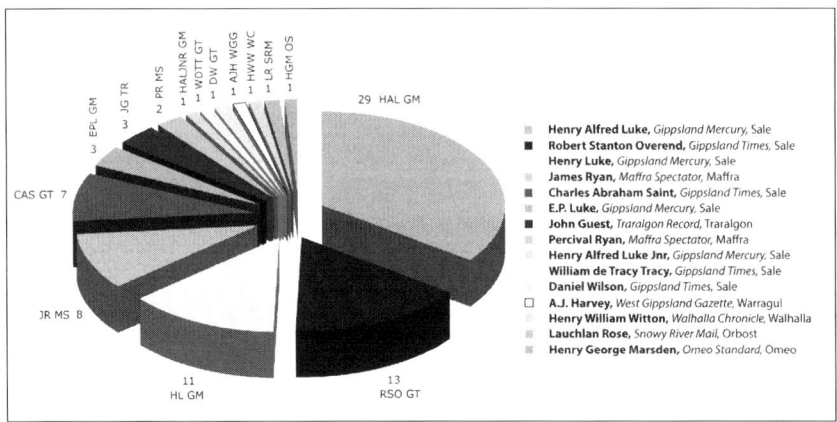

Figure 2.6: Distribution by author of editorials advocating forestry conservation, Gippsland newspaper survey, 1855–1962.

editorial silence also reflected a shift from what Travena described as 'critical' activist editors to mere 'printers' happy to publish articles but not editorials, as well as the general decline in political activism in the press noted by Morrison.[12]

The survey revealed that the 85 editorials on forest conservation were written by ostensibly 15, but most likely 14, editors across eight different newspapers (see Figure 2.6).[13] This represented only about 10 per cent of all editors who worked on all of the sampled Gippsland newspapers during the study period, so the editors who editorialised on the debate were only a tiny minority of the press, despite the fact that the non-editorial coverage was geographically widespread. Moreover, 87 per cent of the 85 editorials were authored by only six individuals representing only three of the 17 sampled newspapers: the *Gippsland Times*, the *Gippsland Mercury*, and the *The Maffra Spectator*. This also meant considerable spatial concentration, with 90 per cent of the editorials emanating from only two towns: Sale, home of the *Times* and *Mercury*; and Maffra, home of the *Spectator*.

---

[12] Morrison. 'The Newspapers of Gippsland'; Trevena. 'The Country Press Editorial.'

[13] The editorials assigned to E.P. Luke of the *Gippsland Mercury* were most likely penned, or at least influenced, by his older brother, managing director and long-time editor Henry Alfred Luke. The young compositor E.P. Luke was ensconced as 'editor' for 18 months by his brother while the latter evaded the payment of fines from a libel case by legally disassociating himself from the firm's assets (*Gippsland Times*, 26 August 1891).

Editorial positions

Charles Abraham Saint of the Sale *Gippsland Times* editorialised on the debate between 1874–78. He advocated improved rail transport to help native forests appreciate in value and compete with imports. He used foreign examples to warn of deforestation, supported political agitation by the Mining Boards to conserve mining timber, and complained that conservation was being prevented by vested interests in parliament. Generally supportive of a well-managed timber industry, and local control of forests, he criticised government timber regulations when they became increasingly harsh in 1878.

Gippsland Mercury's Henry Luke monopolised the years 1883–84. He regularly warned of climatic disturbance from forest clearance, supported political agitation to secure mining timber, reiterated the Railway Timber Board's and other critiques of the parlous state of forest management in Victoria, and called for new forests bills. He advocated European scientific forestry, especially that of state forestry in Germany, but warned that only local forest management boards would bring immediate, practical results. He often promoted South Australian and New Zealand forestry, and supported calls for the preservation of forests on hillsides 'to show our grandchildren the importance of protecting mountain forests'.[14]

The *Mercury*'s Henry Alfred Luke (Henry Luke's son, and father of later editor Henry Alfred Luke Jnr) dominated the years 1885–86, 1888–97 and 1901–04. He was by far the most prolific editor in the survey. His major themes were promoting the climatological benefits of forests and attacking the critics of the theory, and supporting the Mining Boards' campaigns to conserve mining timber. He was critical of parliamentary moves to alienate forest reserves and called for the use of native timbers in government works rather than continued reliance on imports. He reiterated the importance of forest protection on watersheds. He often quoted examples of the environmental costs of deforestation from the American press, such as to warn of clearing central Victoria's Mount Macedon forest in 1885. In later years he supported the Bendigo-based Northern District Forest Conservation League's brief political campaign, called for tree planting in 'every little village' in Gippsland, promoted Arbor Day, lamented the lack of trained professional foresters in Victoria, and was alarmed at the threat to the forests from the 'Land Hunger' during 1893. He supported the scathing attacks on

---

[14] *Gippsland Mercury*, 16 February 1884.

forest management by Howitt, Vincent and Ribbentrop in the mid 1890s, regularly publicised evidence from the Forests Royal Commission, and praised the Melbourne *Age* campaign for a new forest bill in 1897.[15] In the early 1900s Luke lamented the lack of government action on the Royal Commission's recommendations, supported the political campaign from 1903 by the Maryborough-based National Forests Protection League, and called for employment in scientific forestry as a means of slowing the drift to the cities.

The *Gippsland Times*' Robert Stanton Overend editorialised on the issue persistently between 1883 and 1918. He warned that lax government regulations over timber reserves would allow a rush for forests lands with dire climatological consequences and supported the Forests Royal Commission's late 1890s calls for a new forests bill. In the early 1900s he promoted arboriculture, warned that vested interests continued to prevent progress in forest conservation legislation, and railed at the excesses of closer settlement raids on the forests. He argued that Forest Conservator Perrin's replacement would need sufficient power to overcome these interests. He noted the long history of parliamentary abuse of forests and acknowledged but rejected criticism of climatic theories. He supported the National Forests Protection League. During the Great War he warned of the international timber famine and called for greater federal intervention.

The *Maffra Spectator*'s James Ryan wrote his editorials between 1887 and 1898. He pleaded for a rational forest management system and stated that 'the most important topic to Gippslanders is the forestry'.[16] Ryan focused on the need for public education and harsher regulations for forest conservation and called for new forests bills as recommended by the 1885–94 Vegetable Products Royal Commission. His focus was often highly localised given the imminent threats to Maffra's forests from settlement, but he often used international evidence to justify his fierce support for saving forests for climatic benefits. He roundly criticised local critics of the theory, such as the South Gippsland Shire.

John Guest of the *Traralgon Record* penned his editorials between 1908 and 1920. He discussed European forestry models that might be appropriate for Australian conditions, and supported the 1910 British Empire survey that emphasised Australia's shortcoming in forest reservation and management.

---

[15] The Melbourne *Age* and *Argus* ran prominent forest conservation campaigns from the 1860s. Both were widely quoted in the Gippsland press but the Gippsland *Guardian* later sided with the *Age* after losing a bitter legal battle against the *Argus*, which sued Luke for publishing its articles without permission or payment.

[16] *The Maffra Spectator*, 14 November 1887.

Guest was appalled that forest destruction would continue as long as Victoria remained import dependent and didn't value the characteristics of our native timbers. Like all of his contemporaries, Guest was supportive of the timber industry and expanded settlement but wanted them tempered to sustain the forests and forestry.

A biographical investigation

All six editors were sole proprietors or co-owners during the period in which these editorials were written. Patchy information on education is available for some, but no significant common features were found. Similarly, their religious affiliation and observance varied, as did their politics – the latter including a mix of protectionists, free traders, liberals, conservatives, and an avowed 'anti-Socialist'. Two were members of the Order of Foresters and one a prominent Freemason. A couple were keen hunters and fishermen who led campaigns to secure tougher game laws to sustain their sport, but this was not unusual among their contemporaries. All were Victorian-age 'Improvers' and active in a number of campaigns, but this was common in the progressive press. Their personalities varied from the shy and pious to the ebullient, and their editorial style differed from the gently insistent to the provocative. What other factors may have shaped their attitudes?

Age

Elizabeth Morrison has noted that the political influence of Victoria's rural press peaked in the 1860s and was still powerful in the 1870s thanks largely to a generation of British-born pressmen who came to Victoria during the initial gold rushes of the 1850s.[17] In this Gippsland survey, only Charles Abraham Saint and Henry Luke came to the colony during that golden decade, and neither had press experience when they arrived. Four of the Gippsland six were born before the 1850s, ranging from 1823 to 1844 (Saint, Henry Luke, Ryan and Guest in chronological order) with the younger Sale contemporaries Henry Alfred Luke and Robert Stanton Overend being born in 1859 and 1861 respectively. Four of the six wrote their first editorial advocating forest conservation in the brief period between 1883 and 1885, but the other two were very different – much earlier in the case of Saint, in 1874, and much later with Guest, in 1908. There are very few similarities in

---

[17] Morrison.'The Contribution of the Country Press', 283.

their age when they were involved in their editorial campaigns. Two started young – Overend when 22 in 1883, and Henry Alfred Luke, when 25 in the same year – but the former maintained his editorials until aged 33 in 1918, while the latter much longer until aged 45 in 1904. Three others commenced in their middle age (Saint at 51, Henry Luke at 47 and Ryan at 44), but they maintained their advocacy for different periods: Saint for four years until aged 55 in 1878, and Henry Luke for three years until aged 50 in 1884, but Ryan for 21 years until aged 61 in 1898. John Guest was unusual among the group, starting when comparatively elderly at 64 in 1908 and writing for 12 years until age 76 in 1920. Thus there is little in common about their birthdates, age or the duration and timing of their forest conservation campaigns other than the stimulus to write during the early 1880s 'land rushes' shared by four of the six.

Birthplace

Birthplaces were similarly varied: the two youngest, Henry Alfred Luke and Robert Stanton Overend, were Australian-born (Sale and Melbourne respectively); there were the two Irish-born, James Ryan and John Guest (Counties Tipperary and Armagh respectively); and two English-born, Charles Abraham Saint and Henry Luke (the latter from Highgate, London). Four of them supported the Australian Natives Association (ANA), which was itself instrumental in public agitation for forest conservation and the preservation of native flora and fauna around the turn of the century, but support for the ANA was also common among journalists who were silent on, or opposed to, saving the forests.

Previous occupation

Not surprisingly, their previous occupations reveal similarities. Guest, Ryan and Saint were all described as 'pioneer pressmen'. Guest worked in the press between 1862 and 1890 in two regions of Victoria: the central goldfields (*Ballarat Evening Post*, *Bendigo Independent* and managing director of the Castlemaine *Mount Alexander Mail*) before moving to Hamilton in the Victorian western district (printer and joint publisher of the *Hamilton Spectator* and then proprietor and printer of its affiliated *Western Agriculturalist*) before coming late in life to *The Record* in Traralgon.

James Ryan and his eldest brother worked on the *Nenagh Gazette* in Limerick before both migrated to Australia in 1862. Ryan became heavily involved, working in three regions of Gippsland from the early 1860s, notably in Sale (working for the *Gippsland Times* in 1863, and as printer, publisher and proprietor of the *Gippsland Chronicle* 1866) and the north-Gippsland goldfields at Dargo (where he established the short-lived *Crooked River Chronicle* 1865), and Walhalla (printer-publisher of the *Walhalla Chronicle* 1870–1915) before moving to full-time proprietorship of his earlier established *The Maffra Spectator* 1883–1902.

Charles Abraham Saint was a cavalryman in the Lancers before migrating to Victoria, where he then worked for seven years as printer and joint proprietor on the Castlemaine *Mount Alexander Mail* 1854–63. He then moved to Hong Kong as editor and part-owner of the *China Mail*, where he earned a reputation opposing slavery in the Portuguese colonies in the late 1860s, before taking up the *Gippsland Times* as printer, publisher, co-owner and later sole proprietor.

I have no details on Robert Stanton Overend's prior occupation, but as he commenced in the Gippsland press in his early twenties it is likely that it lay in the press.

Henry Alfred Luke worked as a boy in his father's *Gippsland Mercury* in Sale and so was the fourth pressman of the group. His father, Henry, became a gold miner at Forest Creek (Castlemaine) in 1856 and Emerald Hill in north-west Gippsland before moving to Rosedale on the lucrative transport route between Port Albert and the Walhalla goldfields. At Rosedale he became a hotelier, general store manager, postmaster, district coroner, justice of the peace, and Rosedale Roads Board member before moving to Sale and purchasing the *Mercury* in 1871 at the age of 37. Thus, two of the six men definitely had other careers before they came to Gippsland (Saint as a soldier and Henry Luke as a miner and businessman) but five (including Saint) had early employment in the press.

Goldfields experience

One possible link between the four foreign-born pressmen was their experience on the Victorian goldfields, where deforestation was most marked and where political agitation for forest conservation most prominent. But such a connection was not unusual with their Gippsland contemporaries, 90 per cent of whom wrote no editorials on forest conservation. Furthermore, even though each of the six key editors gave prominence to the shortages of mining timber and its link to political agitation as part of the typical forest conservation narrative in late nineteenth century Victoria, their involvement with the issue was not especially strong. For example, James Ryan was essentially silent on the matter until he came away from the Dargo and Walhalla goldfields (after losing heavily on his Walhalla mining shares) down to the relatively open woodlands at Maffra, and then only two years after he took up the *Spectator* there full time. The *Walhalla Chronicle* under James Ryan, then Henry William Witton and finally James Ryan's son Herbert H Ryan, put its faith in rail connection as a solution to the timber shortages that plagued the mines there.[18]

Similarly, Charles Abraham Saint's paper the *Mount Alexander Mail* had nothing to say on the matter during the final year of his editorship there in 1863 (probably two years too early to engage with the mining timber crises from 1865). Saint's last editorial at Castlemaine spoke glowingly of the wonderful transformation of nature that had been wrought from the landscape that providence had provided and that the 'Great Law of Progress' had engendered.[19]

So too with John Guest who, despite his experience at Ballarat, Bendigo and Castlemaine, spent many years on the *Traralgon Record* before commencing his agitation to save the forests. Consequently, even though most lamented mining timber shortages were in Gippsland, there is little direct evidence in their editorials that their early experiences on the goldfields profoundly shaped their later conservational attitudes.

Other business involvement

The men's obituaries reveal that at least four of them were heavily engaged in other business ventures while at the same time being newspaper proprietors and editors. Overend was involved in insurance and butter factories; Henry Luke seemed independently wealthy from property ownership and his earlier ventures; his son Henry Alfred Luke was involved with a steamboat company, Sale's first motor garage, and a butter factory; and Ryan had 'financial interests' in mining companies. Unlike some of their contemporaries, none of the six appear to have been engaged directly in timber companies, but all promoted them generally as part of their local development advocacy.

---

[18] The need to save forests for the mining industry was periodically noted by the Gippsland Mining Board, whose monthly minutes were published in the press. The Board's agitation was often a stimulus to editorials on forest conservation in Sale and Maffra. The *Walhalla Chronicle* regularly lamented the depletion of nearby forests and the shortages of mining timber as the forests there were depleted for a distance of 65 kilometres by 1900. These concerns were not translated into major forest conservation campaigns as the Walhalla community fought for rail connection, which it secured in 1910, ironically only three years before the gold petered out (see Stephen M. Legg, 2008. 'Mining and the Timber Question – Early Forest Conservation Movements in Victoria before 1918'. *Asia-Pacific Economic and Business History Conference*, Melbourne University).

[19] *Mount Alexander Mail*, 30 December 1862.

Henry Alfred Luke and Robert Stanton Overend were both heavily involved in municipal politics as council members and Orbost Shire President and Mayor of Sale respectively, both on two occasions. They were also leading citizens in their long involvement on many and various social, cultural and sporting organisations. John Guest's obituary noted that because of his retiring nature he respectfully declined engagement in civic duties, and Charles Abraham Saint was deeply religious and prominent in youth work, the Church of England and Voluntary Religious Instruction.

Conclusion

The survey showed little evidence that a town's location significantly affected its press coverage on the forests debate. Comparisons of the press in the four towns with selected pairs of newspapers were limited by differences in sample size and comparability across time. Nevertheless, the fact that Sale's *Times* and *Mercury* were both strongly conservationist whereas the papers in the densely forested timber milling town of Mirboo North were silent on the issue tells little about either environmental or economic determinism, especially when the contrasting results from other heavily timbered towns are considered. Thus, at least one of the papers in the timber, dairying and rail depot town of Traralgon (the *Record*) strongly advocated forest conservation while its compatriot (the *Gippsland Farmer's Journal*) was virtually silent. By comparison, the town of Warragul – which had similar economic, geographic, environmental and historic features to Traralgon – had very little difference between the moderate conservation coverage of its press pair (the *Warragul Guardian* and *West Gippsland Gazette*). Moreover, a few of the most prescient press items came from the younger or more remote towns where forests were more abundant and the struggle for 'development' more pressing. The latter is evident not only from John Guest's work in Traralgon between 1890 and 1920, but also in the rare but impassioned editorials of Henry Witton in Walhalla (1885), Lauchlan Rose in Orbost (1890), Henry George Marsden in Omeo (1896), and A.J. Harvey in Warragul (1903).[20] Harvey also published many expert articles promoting forest conservation between 1903 and 1922.

---

[20] Henry Whitton in *Walhalla Chroncile*, 24 April 1885; Lauchlan Rose in *Snowy River Mail*, 13 September 1890; Henry George Marsden in *Omeo Standard*, 24 April 1885, Lauchlan Rose in *Snowy River Mail*, 13 September 1890, Henry George Marsden in *Omeo Standard*, 11 December 1896, and AJ Harvey in *West Gippsland Gazette*, 21 April 1903.

Silence was the typical way for parts of the Gippsland press to register its acquiescence of the prevailing forest destruction. But the collective discussion of forest conservation became remarkably widespread, persistent and supportive across the region. Nevertheless, editorial coverage in the sample was highly concentrated in the hands of probably only 14 editors, with only six being prolific in the debate (and two of them from the same family).

The editorials were concentrated in three newspapers in two towns: Sale and Maffra. The *Mercury* and *Times* developed what might be loosely termed a pro-conservationist culture, but this became less confrontational after the Great War as their editorials on the matter disappeared. No definitive alignment was revealed in relation to geography, environment, land use, or age of settlement to suggest why some towns supported conservation and others did not. Finally, the biographical study indicated no major features shared by the six key conservation advocates that might distinguish them from the majority of editors who avoided editorialising on the issue, let alone help explain or predict their otherwise individualistic passion for forest conservation.

References

Legg, Stephen M., 1984. 'Arcadia or Abandonment? The Evolution of the Rural Landscape in South Gippsland, 1870–1947'. Melbourne: MA thesis, Dept of Geography, Monash University.

Legg, Stephen M., 2004. 'Bunyips, Battues and Bears: Wildlife Portrayed in the Popular Press'. Victoria 1839–1948. In *Conservation of Australia's Forest Fauna*, 2nd edn. Edited by D. Lunney. Mosman: Royal Zoological Society of New South Wales: 150–174.

Legg, Stephen M., 2008. 'Mining and the Timber Question – Early Forest Conservation Movements in Victoria before 1918', *Asia-Pacific Economic and Business History Conference*. Carlton: Melbourne University.

Legg, Stephen M., 1977. 'The Location of the Log-Sawmilling Industry in Victoria, 1939–77'. Melbourne: MA prelim. thesis, Dept. of Geography, Monash University.

Morgan, P., 1997. *The Settling of Gippsland*. Leongatha: Gippsland Municipalities Association.

Morrison, E., 1989. 'The Newspapers of Gippsland'. *Gippsland Heritage Journal* 6 (June): 3–10.

Morrison, E., 1991. 'The Contribution of the Country Press to the Making of Victoria, 1840–1890'. Melbourne: PhD thesis, Dept of History, Monash University.

Trevena, B., 1981. 'The Country Press Editorial in 1888 in the Colony of Victoria'. Melbourne: BA Hons thesis, Dept. of History, University of Melbourne.

Chapter 3

'Nature is around us in her loveliness'
Settler women in the Gippsland bush in 1930s Australia

Ruth Ford

In the midst of the Great Depression, in summer 1931, Adele Ayres wrote to 'Miranda' at the *Weekly Times* Women's Bureau under the pseudonym 'Columbine':

> I am a selector's wife and live in the Gippsland bush. The land seemed to be the only way out for us in these days of depression, and we were fortunate enough in securing a block. But oh, Miranda, the work! The scrub and the undergrowth is so dense in places that one can scarcely penetrate it, even crawling on hands and knees. However, my husband has cleared three acres and fenced it with pickets which all had to be cut. It took several thousand pickets to complete the fence. We have a nice garden although we have only been here nine months, we are eating our new potatoes and have 1000 tomatoes just coming into bloom. Nature is around us in her loveliness and the birds are singing beautifully, so we have a lot to be thankful for.[1]

Adele was one of many Gippsland women who wrote to the women's pages of the *Weekly Times* about their experiences settling, clearing the land and farming. Like other correspondents, Adele highlighted the hopes, hard work and pride of selectors' families in clearing the Gippsland forests to create a farm. And like other correspondents she described her great pleasure in nature and the bush. Her dream to transform the landscape was combined with an intense love of the bush and its bird life; at the same time as clearing the land, she wandered through it, spotting orchids and other wildflowers with delight, listening to bird calls, observing the changes throughout different seasons.

Letters like Adele's provide a rich source for examining rural women's lives in Gippsland during the 1930s. This was a significant period in which government closer settlement schemes on difficult terrain intersected with the Depression (and associated falling butter-fat prices) and drought, amidst ongoing clearing of the forests and timber and pulp mill expansion, and an

---

1   *Weekly Times*, 19 December 1931.

emerging conservation movement.[2] Through descriptions of intense pleasure, mundane utility and despair, Gippsland women's letters encapsulate tantalising fragments of their lives and suggest the complexity of relationships to landscapes undergoing significant change. This chapter analyses white rural women's relationships to Gippsland environments during the 1930s through examining their letters to Miranda at the *Weekly Times*.

The project of land settlement by small selectors in Gippsland was initially written about in either heroic or disparaging terms.[3] Within both these narratives, women were often absent. Recent work has increasingly explored the diversity of women's and men's connection to Gippsland environments.[4] I wish to examine farmwomen's complex and shifting relationship to the environments in which they were settling, clearing and living. Libby Robin has reminded us of the need to give land agency.[5] If we do this, we have a far more complex picture of women's relationships to Gippsland: relationships tied to the seasons and climate and affected by the land in different parts of Gippsland. I argue that their letters show a complex and shifting relationship to the landscapes in which they were settling, clearing and living. Specifically, I examine the tension between their involvement in the settler project of clearing the bush and their desire for, and enjoyment of, a cultivated and tamed landscape, and their developing love of the forest and its flora and fauna.[6] I consider letters written by women from diverse class backgrounds,

---

[2]   Stephen Legg, 1986. 'Challenge or Tragedy? Farm Abandonment in South Gippsland's Strzelecki Ranges'. *Gippsland Heritage Journal* 1 (1):14–22.

[3]   Pioneer and early local histories glorified the hard work of (male) selectors in clearing the forests (Committee of the South Gippsland Pioneers' Association, 1920. *The Land of the Lyre Bird: A Story of Early Settlement in the Great Forest of South Gippsland*. Melbourne: Gordon & Gotch). Initial environment histories tended to emphasise farmers' destruction – or even hatred – of the environment. 'The invaders hated trees', wrote W.K. Hancock (1930. *Australia*. London: E. Benn, 33). Geoffrey Bolton argued that 'to most settlers trees were simply a nuisance to be cleared to make room for building or farming', in his chapter 'They Hated Trees' (Geoffrey Bolton. 1992. *Spoils and Spoilers: A History of Australians Shaping Their Environment*. 2nd ed. Sydney: Allen & Unwin, 37).

[4]   Tom Griffiths, 2001. *Forests of Ash: An Environmental History*. Cambridge: Cambridge University Press; Warwick Frost, 2002. 'Did They Really Hate Trees?: Attitudes of Farmers, Tourists and Naturalists Towards Nature in the Rainforests of Eastern Australia.' *Environment and History* (8):3–19; Meredith Fletcher, 2009. 'Becoming "Correa": Jean Galbraith and "Australian Native Flowers".' *The La Trobe Journal* (84), (December):11–22; Patrick Morgan, 1997. *The Settling of Gippsland: A Regional History*. Leongatha: Gippsland Municipalities Association; Patrick Morgan, 2010. *Foothill Farmers: The Literature of Gippsland*. Ensay: Ngarak Press.

[5]   Libby Robin, 2007. *How a Continent Created a Nation*. Sydney: University of New South Wales Press, 2.

[6]   Little work has considered this tension within 1930s Australia, although Tim Bonyhady's groundbreaking work on the nineteenth century shows the ways that

including women from prosperous farms on the fertile plains, from foothill dairy farms, and from closer settlement blocks and unemployment farms on marginal land high in the hills or on the ranges.[7] In doing so, I am interested in the diverse ways women responded to and interacted with Gippsland environments and how this was affected by stage and length of settlement, economic position, and the geology of different localities.

'Dear Miranda'

The Melbourne based *Weekly Times* was Australia's largest rural newspaper, with its main readership based in Victoria.[8] In the depths of the Depression, on 5 September 1931, the *Weekly Times* launched a new section dedicated to women's issues entitled 'The Women's Bureau' edited by Miranda [Sonia Hardie]. Letters to Miranda poured in from readers seizing the opportunity to express their feelings and views, to share their lives, gain support from others and find pen friends.[9] From 1931 to the end of the decade, well over 4300 letters were published, including at least 250 letters by women who identified as being from Gippsland. Of these, 125 letters (50 per cent) mentioned Gippsland environments – farm and/or bush landscapes or gardens – and it is these letters I examine in this chapter.[10] For many women, reading other women's letters and writing was in part a response to the isolation they faced living in the Gippsland bush. 'I think the Woman's Section is splendid', wrote Adele in her first letter,

---

preserving endangered species and protecting forests were concerns of early settlers. He argues that 'the environmental aesthetic is as deeply embedded in the culture as is resistance to putting environmental ideals into practice' (Tim Bonyhady, 2000. *The Colonial Earth*. Carlton: Miegunyah Press, 11).

[7] On history of settlement in different areas of Gippsland, see Morgan, *The Settling of Gippsland*, 108–110, 117–119.

[8] The *Weekly Times* cost 4d plus 2d postage in the 1930s.

[9] The Women's Bureau started as one page and in addition to 'The Woman's Point of View – Letters from *Weekly Times* Readers', it included a reader's exchange column featuring recipes, hints and requests. Within 12 months, it had expanded to five pages, with over 600 letters being published each year in 1932 and 1933.

[10] Letters to the press are obviously different from private letters. They are written with the awareness – and desire – that they would be published in a public paper. Women wrote for an imagined community of women readers and writers, but also sometimes addressed fellow-letter-writers by name. The letters were also in part influenced by Miranda's suggestions. As the original letters have not survived it is unclear how much they were edited, and whether most or all letters were published, or whether the selection of letters included particular types of letters and excluded others. Prior published letters certainly influenced what women wrote about and their tone; 'cheery' letters were common, although letters relating hard times and loneliness were also published. They were clearly aimed at a readership of women readers and accounts of children, recipes, domestic failures, queries and hints and concern for other readers' health and welfare were prevalent.

in December 1931, several months after moving to a selection. 'I do enjoy the cheery letters and look forward greatly to the *Weekly Times* each week.'[11]

'We are busy clearing our land'

Clearing the land and creating the domesticated landscape of a cultivated homestead with lawn, flowerbeds, vegetable garden and orchard is central to the narrative of successful settling. While some writers, such as Adele Ayres, attributed the clearing of the bush and the cultivation of crops to their husbands, many women wrote of clearing in terms of 'we', although most did not describe their own role in physically clearing the bush. 'We have only been here eight months and came to the bush from the city', wrote 'Poly-Ann' in 1932. 'We have had a lot of work to do clearing and cutting the everlasting bracken ferns, but we are getting on. We have one cow and a few fowls and ducks. In a little while we hope to have more cows.'[12]

While closer settlement government officials and historians have predominantly attributed the clearing of the land to men, women's letters emphasise the family farm nature of the 1930s settlement project.[13] Women who had only recently moved onto the land recounted their progress, their hopes and the work ahead. 'We are busy clearing our land', wrote 'Kia-Ora Third'. Her dream of a productive farm and homestead sustained her through the arduous and slow clearing:

> I shut my eyes and picture our place at some future date. Around the house will be our flowering gardens, out front will be fruit trees, at the back will be our oats, potatoes, onions… and still further back will be a grass paddock for goats and cows to graze. There will be rows of fowl sheds where numerous fowls will lay numerous eggs.[14]

Other letter-writers reflected similar dreams of transforming 'the scrub' into a farm. 'We live in the South Gippsland hills on a block of land, and are hoping to make a nice farm of it', wrote 'Mum Outback' in 1932.[15]

---

[11] *Weekly Times*, 19 December 1931, 19.

[12] *Weekly Times*, 7 May 1932, 23.

[13] There are exceptions to this focus on men's work. See for example Marilyn Lake, 1985. 'Helpmeet, Slave, Housewife: Women in Rural Families 1870–1930.' In *Families in Colonial Australia*, Patricia Grimshaw, Chris McConville and Ellen McEwen (eds). Sydney: Allen & Unwin.

[14] *Weekly Times*, 8 October 1938, 41.

[15] *Weekly Times*, 10 December 1932, 25.

Some writers who referred to 'we' in settling and establishing their farm also described their own role in clearing the land. 'We came here seven years ago', wrote 'Winsome Wayback' from near Orbost, in 1938, 'and in that time, with little money or implements, have cleared 20 acres of land, rebuilt all out-houses and sheds, made waterholes for stock, fenced and cultivated even'. 'I think the pioneer spirit still lived in me', she explained, before writing, 'I have helped to clear, fence, and build… mend harness, and any job that goes, even branding cattle… Now we are after more land adjoining'.[16]

The transformation of diverse Gippsland landscapes to farm landscapes with a particular aesthetic is also evident in these 1930s letters. Many writers described clearing the forest of indigenous trees and planting pine trees instead. 'Kia-Ora Third' wrote of planting pines after clearing the 'scrub'. 'Just Minne-Ha-Ha' described their home as 'surrounded by pine trees'.[17] 'Bracken-Thwaite', who lived on 'a farm up in the Gippsland hills' 20 miles from Warragul, had been 'in a city office, with a good salary, prior to marriage', and became 'a struggling settler's wife'. They attempted to cope with the winds by their homestead plantings. 'We live in a very windy place, so have big pines planted all around, but not too close to the house… they have been out five years and have grown wonderfully… Our small orchard will soon be bearing'.[18] Her description emphasises her personal transformation from a modern single working woman on good city wages to a poor farmer's wife, but also the transformation of the landscape through the clearing of the forest in the hills, the planting of pines around the home paddock and the establishment of an orchard.

'The struggling settler of today'

Women's letters reflected the very different Gippsland environments that they inhabited, their length of settlement and their different economic situations. Some recent settler women in the hills narrated with pride their family's success in surviving through producing enough food to eat. In winter 1932, 'Bracken-Thwaite' wrote:

> When I read of the hard times so many others are having, I feel we are, indeed, fortunate in having so much of our living grown in the place, but it is only the outcome of hard and continual work that give it… 'As we sow so shall we reap'.[19]

---

[16] *Weekly Times*, 11 June 1938, 47.
[17] *Weekly Times*, 28 May 1938, 52.
[18] *Weekly Times*, 20 August 1932, 19.
[19] *Weekly Times*, 20 August 1932, 19.

Their survival meant she could draw on protestant notions of reward from hard work. Other women wrote less triumphantly about their progress in clearing the land, describing their failure to generate income and/or adequate food from their farm. 'We are struggling along with a family of eight children, trying to get our block cleared', wrote 'Brightside' in January 1932, but 'it is really a problem trying to exist on 16' [shillings] sustenance'. Her letter made a plea for food and clothing:

> We must keep going somehow, and the children must have food. We are most anxious and willing to keep on with our work, but the shortage of food and clothing is a serious matter and I would be grateful if any of your readers could assist me in our need.[20]

Five months later, in mid-winter, 'Brightside' reported, 'since my last letter to you we have done quite a lot of clearing', but emphasised the slow progress due to her husband's ill-health from war wounds.[21] 'We must keep going, trying always to do our best to get our land cleared, if only small sections at a time', she wrote, perhaps trying to reassure herself, as well as attract material assistance from readers. 'We are most anxious to stay on our block, and continue working… rather than go back to Melbourne and walk about doing nothing' she emphasised, noting that her war-wounded husband would 'rather stay in the country'.[22]

Her letter also stressed that they were not afraid of hard work, implicitly answering possible judgements that people asking for assistance were to blame for their situation.

> It is not the work that is worrying us, but as we are on a sustenance farm and on our probation period, we find that the problem of sufficient food and clothing is a very serious matter. We now have seven children to keep and the amount of sustenance is not sufficient to allow us to purchase footwear.[23]

She provided further details to potentially sympathetic readers that she and her daughters, aged 16 and 18, who were 'helping clear' were 'compelled to wear bags on their feet, sewn to the shape of their feet'. It appears that her family was on an 'unemployment farm', for which they received a weekly

---

[20] *Weekly Times*, 23 January 1932, 18.
[21] *Weekly Times*, 11 June 1932, 19.
[22] *Weekly Times*, 11 June 1932, 19.
[23] As were her younger children, aged 2, 4, 6 and 9½ years old. *Weekly Times*, 11 June 1932, 19.

sustenance payment of 16 shillings.[24] Their drive to continue on their farm, despite insufficient food and clothing, emphasised the lack of options during the Depression in cities such as Melbourne, and perhaps also the desire to own land and live on a farm, escaping high urban unemployment and the uncertainty of a wounded returned soldier being able to work. Her pen-name, 'Brightside', speaks of her attempt to be positive amidst the harsh conditions and poverty she was struggling with on a small block on marginal land in the Gippsland ranges.

'Our beautiful Gippsland'

Many Gippsland women's letters indicate a strong connection to place. This is evident in both their descriptions of Gippsland environments and in their pen-names. Taking a name and creating an identity is a powerful act.[25] Many of their pen-names relate to the physical environment in which they lived: Another Bushwoman, Boggy Creek, Bracken-Thwaite, Ferny Nook, Forest Queen, Hill-Top Second, Green Timber, Green Woodlands, Janie Possum, Orchids. While some names draw on the indigenous vegetation of ferns, forests and wildflowers or the fauna, others invoked the timber resources and the dreaded bracken and bogs. Other pen-names drew on place names: Waratah Dell, Yallmay, The Gippsland Wanderer, Anxious Gippslander.

Some women wrote very affectionately of Gippsland. 'Barberry Bush' referred to 'our beautiful Gippsland', indicating both her love of it and her connection to it – and her sense of white settler possession. 'I can quite understand your intense longing for the sight of the lovely hills and gullies, tree ferns and green grasses of South Gippsland', wrote 'Queenie 2', who used to live near Wonthaggi.[26] While some women loved the forests, other

---

[24] Sustenance payments, known as 'susso', comprised 8s 6d per week for man and wife and 1s 6d per week for each additional child (not in employment) up to a maximum of 20s 6d. (This suggests that only five of her eight children were considered dependent or unable to find work.) The sustenance scheme was established during the Depression for the relief of individuals able and willing to work but unable to find employment. 'Susso' was not paid in cash but via an identification card, which enabled recipients to gain groceries, meat, bread and milk from local shopkeepers to the value of their approved sustenance. *Victorian Year Book 1930–31*, 1932. Melbourne: Government Printer, 220–221. See the Unemployment Relief Act 1930; Stamps (Unemployment Relief) Act 1930, and the Unemployment Relief Amendment Act 1930.

[25] The almost exclusive use of pen-names also provided women some opportunity to say things they might not have otherwise said. A total of 97 per cent of writers identifiable as being from Gippsland used pen-names.

[26] *Weekly Times*, 25 March 1933, 19.

women loved the Gippsland plains – variously formed by geology, Indigenous fire-stick farming and white settler clearing.[27] 'Boggy Creek' who lived near Buchan admitted:

> Although I live in the hills at present and think the views are lovely, I can never understand someone saying they 'love the hills'. I maintain that all hills are alike, and I am sure I would be a bad bushwoman in them. Picture the lovely plains just after the grain is sown, and another patch where it is green, and so on. I think it a lovely sight. I could drive for miles and enjoy scenery like that.[28]

While recent settlers wrote of struggles in clearing the forest in the hills, women on longer established and more affluent farms on the plains wrote very differently of the landscape. 'This district is one of the best in central Gippsland', wrote 'Easter Daisy No. 2'. 'It is undulating, with good wide flats between the hills, and good roads out from the main township of Warragul.' She goes on to describe not the ferns or the forests but the crops and soil and its present and future productivity: 'The country about here is looking well now with the grass and oats so nice and green and the patches of brown-ploughed ground between, laying in fallow ready for potato and maize crops later on'.[29] While she favoured the plains, she also described her view of the surrounding ranges.

That many writers identified as living in the hills or on the plains suggests the ways that identities and communities were formed by different Gippsland environments, and points to the importance of bio-regional identities as well as regional ones. 'Forest Queen' not only loved the landscape of the Gippsland hills where she lived, she believed that hill life was superior. She contended of children at the local school, where her brother was the teacher, that:

> Being hill children with a horizon that includes the ocean, their imaginations are not dwarfed, as are the minds of some children of the plains, where the horizon is only a log fence. Wherever they look they see billowy blue forests, but it is a beautiful panorama, and the distant ocean with its white-capped rollers… speaks to them more plainly than the best picture book in the world.[30]

---

[27] On firestick farming see Bill Gammage, 2011. *The Biggest Estate on Earth: How Aborigines Made Australia*. Sydney: Allen & Unwin, 322–23.
[28] *Weekly Times*, 8 October 1932, 21.
[29] *Weekly Times*, 12 November 1932, 19.
[30] *Weekly Times*, 30 January 1937, 43.

Forest Queen believed that the environment was a powerful influence, not just on citizen's bodies, but on their minds and emotions. Perhaps her notion of dwarfed minds and stunted imaginations formed by living on the plains was also an implicit challenge to the views of established prosperous families on the plains towards hill settlers.

'Towering gums, graceful ferns, clear streams'

While 'fernmania' within the urban middle and upper classes and the promotion and popularity of fern gullies in sightseeing and tourism has been considered by historians, less is known of rural women's relationship to ferns.[31] These letters give insights into how Gippsland women – both wives in wealthy farming families and poverty stricken struggling settlers – regarded ferns in the 1930s. How did women who helped with the hard physical labour of clearing the 'dread ferns' regard 'the feathery fronds'?

Ferns featured in many Gippsland women's letters. 'There is nothing to equal Gippsland with its towering gums, graceful ferns, clear streams and bright birds', wrote 'Ferny-Nook' in winter 1938. Her pen-name and her letter reflected her love of, and connection to, the Gippsland forests and its ferns. She described 'one lovely gully some miles away far from the works of man':

> A lovely clear stream tinkles along between banks covered with the clothes of nature. Pretty maidenhair goes to the water's edge. Dozens of tall tree ferns stand sentinel to the smaller ferns, and the sun flittering through their fronds cause a lacy network pattern to be thrown on the earth. Gums and wattles line the creek, too, and in spring when the trees are a golden mass of glory the mountain lory parrots perched on the boughs a bright splash of red against the background of yellow.[32]

While not using botanical names, 'Ferny-Nook' identified both tree ferns and maidenhair ferns, describing the majesty of tree ferns and the delicacy

---

[31] Tim Bonyhady (*The Colonial Earth*, 101–124) examines conservation and tourism in relation to fern gullies close to Melbourne from 1860s to the 1890s. Julia Horne (2005. *The Pursuit of Wonder: How Australia's Landscape Was Explored, Nature Discovered and Tourism Unleashed.* Melbourne: Miegunyah Press, 254) highlights that by 1900 'the feathery fronds of the ferns were well-established as a symbol of middle-class taste'.

[32] *Weekly Times*, 11 June 1938, 50.

of smaller ferns in the undergrowth and close to the stream.[33] She took great delight in their beauty, and nature in the Gippsland forests provided solace to her 'when I am down in the dumps, that is where I love to go'. Her explanation that 'among the wonderful works of God our troubles seem to get lighter and life takes on a brighter hue' drew on literary romanticism and notions of the divine in nature. The fern gullies also provided respite from another aspect of 'nature': the heat. 'In the hottest day of summer this ferny nook offers a cool retreat.'[34]

Readers wrote enthusiastically – and often poetically – of the diverse ferns, which were prevalent in the wet sclerophyll forests, along creek beds and in gullies, and which also existed at high altitudes.[35] 'It is lovely with the fern gullies and cool stream', wrote 'Old Lace', an English immigrant – with a young boy – who had recently 'got settled' in her 'mountain home'. The beauty of the fern gullies perhaps helped her cope with the isolation: 'it is nice here, but rather lonely as we are the only ones'.[36]

It was not just the feathery fronds in fern gullies that women wrote about. They wrote of the hard physical labour in clearing bracken ferns. Ferns were also a commodity of exchange for Gippsland women. Adele Ayres (now writing under a new pen-name 'Crocus') exchanged ferns with *Weekly Times* readers for items she needed. She wrote asking if any readers had size 10 shoes: 'I would exchange ferns for it'.[37] Other women gave ferns away to women's bureau readers. 'I have some ferns. I would send any reader for freight', wrote 'Bettyite'.[38] Giving away or exchanging ferns indicates their prevalence, but also women's regard for ferns as aesthetic garden plants which would be valued by other *Weekly Times* readers.[39] By the 1930s, other indigenous plants

---

[33] Nature writers of the time discussed different species of ferns. Edward Pescott's *Native Flowers of Victoria* included a section on ferns, which outlined several different species found in Gippsland (Pescott, *Native Flowers*, 95–98). Charles Barrett who wrote for the *Sun*, as well as the *Weekly Times*, was the author of several nature books including one book on ferns (Charles Barrett and Richard Bond. 1934. *Victorian Ferns: Descriptions of All the Species Occurring in the State*. Melbourne: Field Naturalists' Club of Victoria).

[34] Bonyhardy (*The Colonial Earth*, 112) notes that Ferntree Gully in the 1870s was more popular than beach resorts during the hottest months and described as 'a refreshing retreat'.

[35] *Weekly Times*, 11 June 1938, 50.

[36] *Weekly Times*, 29 April 1933, 21.

[37] *Weekly Times*, 23 April 1932, 22.

[38] *Weekly Times*, 3 Sept 1932, 25.

[39] It is unclear from their descriptions which species of ferns they were exchanging – common tree ferns or bracken ferns or rarer species. For accounts of fern species, see Pescott (*Native Flowers*, 95–98).

were desirable and could be used for financial gain. 'Kia Ora Third' collected bunches of pink heath and gathered gum moss from the nearby bush to sell in order to help her family survive on their small subsistence holding.[40]

Wattle featured strongly in many women's letters. 'The silver-wattle grows very thick in the gullies around here, and there are some beautiful sights in spring when it is in flower', wrote 'Ames' who lived on a small farm, and whose income was supplemented by her father being 'employed at sleeper-cutting'.[41] 'South Gippsland is the garden of Victoria', wrote 'Happy Wife'. 'In a week or so the creek banks will be a picture with the wattles all in bloom.'[42] The repeated references to wattle are not surprising given it was a national emblem.[43] However, for Gippsland women in the hills, perhaps the coming of the wattle blossom was particularly significant as it heralded the end of bitter cold winters and isolation through impassable roads. Adele Ayres' spring 1933 letter rejoiced in the coming of spring:

> Spring is with us once again in the Gippsland mountains, and how lovely it is, bringing as it does, the wealth of golden bloom and new life to everything, even to us poor humans. I feel cheered, just to look upon the lovely silver wattle, that lines the river banks here, and the golden, which dots the hills. I love the silver wattle best; it seems to be a fairy tree and has a more beautiful perfume.[44]

Wildflowers were fondly mentioned, although by fewer writers than were ferns. 'Kia-Ora Third' wrote eloquently of orchids and wildflowers over several years.[45] The disappearance of wildflowers is also acknowledged implicitly. 'Another Bushwoman' reminisced of her childhood 50 years earlier in which she went 'rambling in the bush after the most glorious wildflowers'.[46]

---

[40] *Weekly Times*, 15 October 1938, 41. Ruth Ford, 2011. '"I Shut My Eyes and Picture Our Place": Gardens, Farms and Working-Class Dreams in 1930s–1940s South-Eastern Australia.' *Studies in the History of Gardens and Designed Landscapes* 31 (2):109–120.

[41] *Weekly Times*, 19 August 1933, 21.

[42] *Weekly Times*, 15 October 1932, 19.

[43] Libby Robin, 2002. 'Nationalising Nature: Wattle Days in Australia.' *Journal of Australian Studies* 26 (73), 15–18; Katie Holmes, 2011. 'Growing Australian Landscapes: The Use and Meanings of Native Plants in Gardens in Twentieth-Century Australia.' *Studies in the History of Gardens and Designed Landscapes* 31 (2):122; Katie Holmes, Susan Martin, and Kylie Mirmohamadi, 2008. *Reading the Garden: The Settlement of Australia*. Melbourne: Melbourne University Publishing, 99–100.

[44] *Weekly Times*, 07 October 1933, 19.

[45] Ford, 'I Shut my Eyes and Picture Our Place'.

[46] *Weekly Times*, 15 October 1932, 17.

'Alas they went with the scrub'

The letters narrate the clearing of the forests for farms with pride, but also reminisce about the ferns, forest and birds and express nostalgia at their loss. 'Chizzie', who identified as 'a Narracan pioneer's daughter', proudly described her father as 'the first pioneer to select' and recounted him clearing 'the patch by the creek for the house'. Yet she also described their 'beautiful block with the Narracan River and falls at the foot of the homestead… These were lovely ferneries with lyre birds in dozens'. 'Chizzie' recounted her father saying he would never forget 'the singing of the hundreds of different kinds of wild birds in the scrub around the house the month she was born', and reminisced nostalgically, 'alas they went with the scrub'.[47]

'Wyworri' had admired the beauty of Gippsland parks and noted the large trees 'in this part of Gippsland', but emphasised the transformation: 'there are several mills here and timber cutters are working for the paper pulp mill… in the adjoining town'. She, like others, was writing at a transitional moment, when vast forests had been destroyed but clearing was still happening and pulp and paper mills were proliferating. She acknowledged, but did not criticize, the loss of towering ancient trees and forests. These increasingly existed only in people's memories, aided by photographs. She noted, 'I have a photo of a giant tree which grew near here 37 feet in diameter and 11 feet in circumference, and it will hold 11 horses.[48]

Other writers more explicitly acknowledged the destruction resulting from clearing the forests. After making 'a new home in the vast Gippsland forest', 'Anthotis' wrote acknowledging early settlers' need to clear the forest in the name of progress, but expressed regret at the death of animals and birds:

> Fires were necessary in early days to clear the land, but only too often they left ruin in their trail, and what a toll they took of animal and bird life. How sad to think that these happy, inoffensive creatures had to pay such a price in the cause for progress.[49]

Both 'Anthotis' and 'Chizzie's' regret was at the death of fauna and birdlife, rather than the loss of species of flora and forest eco-systems.[50]

---

[47] *Weekly Times*, 6 October 1934, 24.
[48] *Weekly Times*, 4 June 1938, 50.
[49] *Weekly Times*, 30 July 1938, 42.
[50] Ecosystem thinking wasn't popularised until the 1960s.

'Lucky Reader', who lived 'right on the falls' at Turton's Track and made 'a little tiny living out of tourists by running a tearooms' had a different relationship to the bush landscape to that of farming women. She regretted the destruction of the old growth forest by axe and fire:

> The wild growth is coming back, but it will be years before it is as of old, if ever, as man so often comes along with axe and cutter, then the match, and all is a black picture of misery. The bracken soon puts a green mantle on, but the fern trees do not come back for twelve months and young trees never get a chance.[51]

However, despite her love of the bush at the falls – particularly the ferns – her cultivated landscape aesthetic did not allow her to simply leave the 'natural' landscape as it was. Instead she attempted to create her own landscape blending indigenous species with familiar British and exotic plants:

> I have planted lots of young blackwoods and wattles interspersed with weeping willows, pine trees, cherry plum trees, along the creek in front of my house and above the falls, and hope some day to view them at their best.[52]

Her choice of plants suggests that she saw wattles and blackwoods as the most aesthetic indigenous trees and as complementing her favorite exotic trees.

'I have not many friends, as we live in the hills'

Gippsland women writing letters to the paper was, in part, a response to the landscape they lived in and to their physical isolation. Getting the *Weekly Times* on mail day was for many women their contact with the outside world. 'I have not many friends, as we live in the hills and have few neighbours', wrote 'Hendy', who had migrated from 'the Old Country' three years earlier. 'Like others, I find life lonely in the country, but love it all the same'.[53] Loneliness was central to women's delight in reading the women's pages and part of their motivation for writing. 'Green Woodlands', who had one 19-month-old baby and was six months pregnant, begged for pen friends: 'I wonder are there other readers who would care to write to me? I would answer all letters.

---

[51] *Weekly Times*, 2 January 1937, 47.
[52] *Weekly Times*, 2 January 1937, 47.
[53] *Weekly Times*, 16 January 1932, 19.

I get very lonely here with no one to talk to, only the children. My nearest neighbours are three and four miles away, so I don't see much of them'.[54]

For 'Darbienne', an English war bride from Derby, her isolation was seasonal with impassable access for half the year to the nearest town:

> We are dairy farming in the hills of Gippsland, with only a bush track to use to cover the five miles from our homestead to the nearest township; and this bush track is impassable for at least six months of the year.[55]

Here nature is isolating and hostile. Women's perceptions of the uncleared forests as being threatening related to their physical isolation (due to the dense bush), their visual and emotional isolation (as the forests robbed them of views of cultivated, inhabited valleys) as well as to ideas of a tamed landscape.

'When I have loathed the bush'

The bias of the women's pages was certainly for cheerful letters, but not all women wrote of loving the bush and living in Gippsland.[56] In autumn 1938, 'Orchids' wrote of her intense loneliness:

> On arriving at our dairy farm in the bush (after living in Melbourne) I felt very lonely. I was listening one day to a concert over the air given by members of a friendly circle. They all sounded so carefree and happy that I found myself in tears. I felt so lonely. My husband came in, in the middle of it, to make things worse, and, after trying to comfort me and murmuring something about being a brute to bring me to this. I felt really ashamed and resolved to keep a stiff upper lip from then on, which I have done.[57]

Orchids' admission of resolving to keep a stiff upper lip – and other women's accounts of 'looking for the silver lining' – emphasise the silences in the letters. Writers were more likely to wax eloquent about the landscape than write of despair and loathing. Although some letters do acknowledge disliking the

---

[54] *Weekly Times*, 19 March 1938, 47.

[55] *Weekly Times*, 12 August 1933, 17.

[56] This was created by women's self-censorship of not writing of their despair and by a culture of cheery letters fostered by both Miranda and women letter-writers – and possibly by Miranda's selection of letters.

[57] *Weekly Times*, 16 April 1938, 48.

bush, they are rare. 'Orchids' admitted 'there have been times when I have loathed the bush' but immediately qualified it by writing 'fortunately they have not been frequent'. She then recounted a positive experience:

> Today I had to go through the bush with a message for a neighbour. It was a beautiful still morning, with just sufficient tang in the air to be pleasant. There was such a lovely fresh smell, and with the birds chirping in the trees I realised the bush has its compensations after all. Tonight as I sit writing this in front of a log fire, I realise it more than ever.[58]

Significantly, it is the utilitarian aspect of the bush – clearing its timber for firewood and warmth – which she concluded her account with.[59] It is the cleared land which gives pleasure, alongside the living bush.

'He had beaten back the jungle'

The sense of clearing and settling being a battle waged against nature is evident in some of the letters. 'Forest Queen' wrote lyrically of the nature scenes hill children could view, while at the same time glorifying in the clearing of the bush. She noted the school 'in the little clearing among the huge logs felled by their parents' and recounted that 'on the hillside just beyond the schoolhouse an elderly man guided the plough as it cut into the chocolate earth reclaimed from the forest':

> He had done his share towards winning 'the dominion over palm and pine'. He had beaten back the jungle. My brother remarked, 'You have your place nicely cleared now'. 'There's a good deal more to do yet,' he said, viewing the forest beyond the clearing, and the plough began another furrow.[60]

Both the language used – 'winning', 'dominion', 'beaten', 'jungle' – and the anecdote emphasise battling against, and conquering, nature. These metaphors were central to the settler colonial project of taming the land and making it productive.

---

[58] Ibid.

[59] In a similar way to Mallee women writing of their enjoyment in sitting in front of a roaring fire of mallee stumps (Ruth Ford, 2011. '"The Wattles Are in Bloom. Crops Are Looking Wonderfully Well": Settler Women in the Victorian Mallee, 1920s–1930s.' In *Outside Country: Histories of Inland Australia* edited by Alan Mayne and Stephen Atkinson, 64–94. Adelaide: Wakefield Press).

[60] *Weekly Times*, 30 January 1937, 43.

Alongside the descriptions of land being successfully cleared and controlled or tamed were letters that emphasised the power of nature and of nature as uncontrollable. In the same letter in which 'Hendy' had described the 'lovely gums, wattles and fern trees', she wrote of the closer settlement blocks: 'oh, how hard the men have to work to clear their blocks of the dread fern'. 'Hendy' emphasised the battle between settler and nature, with nature winning in the absence of vigilance: 'even when the land has been cleared, unless well looked after they will soon find ferns creeping again'.[61] Her letter highlighted how settler attitudes to indigenous plants varied according to where they were placed. Fern trees in the national park were beautiful and peaceful, but bracken ferns on blocks were dreaded both because of the hard labour in clearing them and because of their invasive nature.

While Victorian women in the Mallee wrote repeatedly of drought, heat and dust storms, Gippsland women wrote far more frequently of floods and bushfires.[62] There was a sense of nature as wild and uncontrollable in the rivers and forests beyond their selections, however successful they had been taming and defeating it on their farms. Farms could be destroyed in minutes by fire, or stock and crops lost in floods. Women's descriptions of Gippsland environments drew on their own experiences of being in nature, but they located these experiences within the stock of stories which surrounded them in 1930s Gippsland.[63]

* * *

The letters to Miranda provide glimpses of rural women's relationships both to cultivated farm landscapes and to the bush in Gippsland at a time of continued transformation of the environment through clearing, and of severe economic hardship. While the women were involved in the settler colonial project of clearing the forests – as the lease conditions dictated 'substantial and permanent improvements' – their letters also reveal a strong attachment

---

[61] *Weekly Times*, 26 November 1932, 24.

[62] On the Mallee women see Ford, 'The wattles are in bloom'.

[63] These included national stories of the brave pioneer woman and the woman alone in the bush, as well as local accounts of successful settlers. On the pioneer legend generally, see John Hirst, 1992. 'The Pioneer Legend.' In *Intruders in the Bush: The Australian Quest for Identity*, edited by John Carroll, 14–37. Melbourne: Oxford University Press; on the pioneer legend in Gippsland, see Committee of the South Gippsland Pioneers' Association; Stephen Legg, 1987. 'Portraits of South Gippsland Pioneers.' *Gippsland Heritage Journal* 2 (1). 33–38; Morgan, *The Settling of Gippsland*.

to and love of the flora and fauna of the Gippsland bush, particularly its forests, ferns and wildflowers.

Their letters exhibit a tension between an economic need and cultural desire to clear the land, and a delight in the bush. Settler women impacted upon the landscape through the family farm project of clearing and farming. But they were also changed by the landscape and in their growing attachment and connection to Gippsland environments; there were moved both by majestic tree ferns and small delicate ferns but also overwhelmed by the extent of forest, which created their isolation, particularly in the winter months.

Although women worked to create a cleared farm and homestead garden, another side existed to their relationship to the landscape. Women cherished a love of nature, awe at the imposing forests, delight in delicate tree ferns and wildflowers, and pleasure in indigenous birdcalls. This relationship was created by their physical experiences of the environment, as they cleared the land, created vegetable and flower gardens, milked cows, and looked to the uncleared bush beyond. It was framed in cultural assumptions of the time. It was also influenced by nature tourism promoted by photographs of forests and ferns; by nature study in schools, nature writing in newspapers; and by literary romanticism.

And as for Adele Ayres (whose moving account I opened with), her family lost their 640-acre selection at Hickey's Creek, near Glenmaggie, as they could not meet the annual payments. Despite their diversity, these letters written for the public realm only tell part of the story. Struggling settler women described hard work, loneliness and hardship, but personal pride coupled with local and national pioneer legends and rural yeoman mythology meant women rarely wrote publicly of failure and the loss of their farms.

Acknowledgements

I'd like to thank the anonymous reviewers and Gippsland Environments Conference participants for their perceptive comments and questions, Kyla Cassells for her research work, and the Australian Research Council for its funding.

I'd also like to thank the Ayres family, particularly Una Killeen and Valerie Northrop, for sharing their memories of their mother Adele and the Ayres family story; and ABC Gippsland for an interview which led me to contact the Ayres family, who revealed that 'Columbine' in the *Weekly Times* was Adele Ayres.

References

Barrett, Charles and Bond, Richard, 1934. *Victorian Ferns: Descriptions of All the Species Occurring in the State*. Melbourne: Field Naturalists' Club of Victoria.

Bolton, Geoffrey, 1992. *Spoils and Spoilers: A History of Australians Shaping Their Environment*. 2nd ed. Sydney: Allen & Unwin.

Bonyhady, Tim, 2000. *The Colonial Earth*. Carlton: Miegunyah Press.

Committee of the South Gippsland Pioneers' Association, 1920. *The Land of the Lyre Bird: A Story of Early Settlement in the Great Forest of South Gippsland*. Melbourne: Gordon & Gotch.

Fletcher, Meredith, 2009. 'Becoming "Correa": Jean Galbraith and "Australian Native Flowers".' *The La Trobe Journal* (84, December): 11–22.

Ford, Ruth, 2011. '"I Shut My Eyes and Picture Our Place": Gardens, Farms and Working-Class Dreams in 1930s–1940s South-Eastern Australia.' *Studies in the History of Gardens and Designed Landscapes* 31 (2):109–120.

Ford, Ruth, 2011. '"The Wattles Are in Bloom. Crops Are Looking Wonderfully Well": Settler Women in the Victorian Mallee, 1920s–1930s.' In *Outside Country: Histories of Inland Australia* edited by Alan Mayne and Stephen Atkinson, 64–94. Adelaide: Wakefield Press.

Frost, Warwick, 2002. 'Did They Really Hate Trees? Attitudes of Farmers, Tourists and Naturalists Towards Nature in the Rainforests of Eastern Australia.' *Environment and History* 8:3–19.

Gammage, Bill, 2011. *The Biggest Estate on Earth: How Aborigines Made Australia*. Sydney: Allen & Unwin.

Griffiths, Tom, 2001. *Forests of Ash: An Environmental History*. Cambridge: Cambridge University Press.

Hancock, W.K., 1930. *Australia*. London: E. Benn.

Hirst, John, 1992. 'The Pioneer Legend.' In *Intruders in the Bush: The Australian Quest for Identity*, edited by John Carroll, 14–37. Melbourne: Oxford University Press.

Holmes, Katie, 2011. 'Growing Australian Landscapes: The Use and Meanings of Native Plants in Gardens in Twentieth-Century Australia.' *Studies in the History of Gardens and Designed Landscapes* 31 (2):121–130.

Holmes, Katie, Martin, Susan and Mirmohamadi, Kylie, 2008. *Reading the Garden: The Settlement of Australia*. Melbourne: Melbourne University Publishing.

Horne, Julia, 2005. *The Pursuit of Wonder: How Australia's Landscape Was Explored, Nature Discovered and Tourism Unleashed*. Melbourne: Miegunyah Press.

Lake, Marilyn, 1985. 'Helpmeet, Slave, Housewife: Women in Rural Families 1870–1930.' In *Families in Colonial Australia*, edited by Patricia Grimshaw, Chris McConville and Ellen McEwen. Sydney: Allen & Unwin.

Legg, Stephen, 1986. 'Challenge or Tragedy? Farm Abandonment in South Gippsland's Strzelecki Ranges.' *Gippsland Heritage Journal* 1 (1):14–22.

Legg, Stephen, 1987. 'Portraits of South Gippsland Pioneers.' *Gippsland Heritage Journal* 2 (1): 33–38.

Morgan, Patrick, 1997. *The Settling of Gippsland: A Regional History*. Leongatha: Gippsland Municipalities Association.

Morgan, Patrick, 2010. *Foothill Farmers: The Literature of Gippsland*. Ensay: Ngarak Press.

Pescott, Edward Edgar, 1914. *The Native Flowers of Victoria*. Melbourne: George Robertson.

Robin, Libby, 2002. 'Nationalising Nature: Wattle Days in Australia.' *Journal of Australian Studies* 26 (73).

Robin, Libby, 2007. *How a Continent Created a Nation*. Sydney: University of New South Wales Press.

Victorian Year Book, 1930–31. 1932. Melbourne: Government Printer.

Weekly Times, 1930–40.

Chapter 4

'Trudging on manfully'
Dr Andrew and the evolution of bushwalking

Julie Fenley and Kathy Lothian

Introduction

It was raining very soon after we arrived in Warburton and we tramped along a tram track fringed with tall dripping scrub that rained merrily on us as we passed in single file one umbrella up and packs cutting into unaccustomed shoulders. The distance to the sawdust [ex-logging camp that had previously been located] seemed longer than by day… [but] eventually we found it. The night was as dark as ebony and a misty rain falling… By the effort of several newspapers and a good pair of lungs a fire was eventually induced to feed on green bracken and wet sticks as tho' it enjoyed such fuel. The amount of bad language used in lighting that fire (it took about an hour) was phenomenal… That sawdust got everywhere down your neck and into boots amongst blankets into the butter and it seemed that there are certain disadvantages about a sawdust heap.

(Frank Macfarlane Burnet, 6 December 1920.[1])

On 6 December 1920, James Moore Andrew and his friends Keith MacKenzie and Frank Macfarlane Burnet set off for a three-week, 205-mile (330 kilometre) walking trip from Warburton to Woori Yallock. Over the next several decades, Andrew was to undertake dozens of other walking and camping trips, many of them in Gippsland. Andrew's identity as one

---

[1] Frank Macfarlane Burnet, 1920. 'Diary'. Frank Macfarlane Burnet Collection, 1989/0034 Series 2/5 Melbourne University Archives.

of Gippsland's early medical practitioners, as well as his contribution to many aspects of the township of Yallourn, in Gippsland's Latrobe Valley, makes him an important figure in the region. This, combined with his active participation in bushwalking over a number of decades, also makes him a significant part of Gippsland's bushwalking history. Gippsland is a region with a number of important and remote natural sites. It has also been the site of significant contestation about the value and meaning of these areas. However, we know little about the history of individuals who engaged with these regions as independent bushwalkers, those whose walking in this region contributed to a growing awareness and appreciation of these places, and who perhaps unwittingly assisted in developing a recreational model that focused on an entitled access to 'unspoiled' tracts of wilderness.[2]

Andrew's personal effects are now lodged with Old Gippstown Heritage Park in Moe. Together, the often unique items in this collection – which include an original 1930s Paddy Pallin backpack, several old billies, old skis and stocks, numerous maps, detailed packing lists for many of his trips, and a diary of the trip made by Andrew in 1920 – help us to gain an understanding of how one individual negotiated the Australian bush.[3] It was the items in this collection, which we examined as part of a significance assessment for Heritage Victoria – particularly his bushwalking diaries and equipment lists – that inspired our interest in Andrew's walking experiences.

As Melissa Harper has written, during the 1920s and 1930s a discernable bushwalking 'movement' emerged, and the practice of bushwalking began to take on a more formalised identity.[4] We argue that since this period there has been a tension between formal ideas about bushwalking that developed via organised bushwalking societies, and the ways in which individuals have related to the bush. We can observe how formal ideas of bushwalking, partly encapsulated in concepts of self-sufficient, active bushwalking, were evident in some of the activities that Andrew and his friends undertook during

---

[2] Deirdre Slattery, 2009. 'Bushwalking and Access: The Kosciusko Primitive Area Debate 1943–6'. *Australian Journal of Outdoor Education*. 13 (2). 14–23.

[3] There are few remaining object-based collections which document early recreational activities in Australia. They include a large collection associated with bushwalker and conservationist Milo Dunphy at the National Museum of Australia, some of which was handed down by his father, bushwalking pioneer Myles Dunphy, and a significant collection of early skiing equipment at the Upper Murray Historical Society – Man from Snowy River Museum, which includes a range of items formerly owned by champion skiers Tom and Elyne Mitchell.

[4] Melissa Harper, 2007. *The Ways of the Bushwalker: On Foot in Australia*. Sydney: UNSW Press, 215–216.

these decades. However, his recreational walking trip in 1920 displayed significant differences to what we might now think of as bushwalking, and his walking activities in the decades after 1920 show that he continued to adopt a range of ways of engaging with the bush. Our purpose here is not to assess whether bushwalking clubs from the 1920s developed a more 'authentic' or 'evolved' view of bushwalking, in comparison to Andrew's undertakings. Yet walking clubs did develop particular views about the correct way to engage with the bush and Andrew, for his part, had no formal connection to bushwalking clubs in his lifetime and resisted categorisation as a walker. Indeed, as his daughter has recalled, he was both unconventional and highly independent.[5]

* * *

Andrew was born in 1899 in Kyneton, Victoria, but spent much of his childhood in a small farming community west of Rushworth. After winning a scholarship to Ormond College at the University of Melbourne, Andrew graduated with medicine and surgery degrees in 1922 (famously beating his Gippsland-born friend Frank Macfarlane Burnet, who later won the 1960 Nobel Prize in medicine), and as a Doctor of Medicine in 1924.[6]

Though his talents meant that he could have taken his pick from any number of rewarding medical careers, Andrew chose deliberately the role of general practitioner at the company township of Yallourn, which serviced the State Electricity Commission's open-cut coal mine and energy facility. A major part of Andrew's work involved treating workers from the mine and power station. Andrew held this position from 1926, when a hospital or even a properly equipped medical clinic were yet to be built, and practised in the region until his death in 1972.

Despite his busy working life, Andrew had an extensive range of interests and took an active role in community affairs. He was a founding member of the Alpine Rover crew, helping to build the Rover hut at Mount Erica.[7] Because Andrew participated in bushwalking in a variety of ways, ones that

---

[5] Margaret Andrew, 1976. 'The Life and Times of Dr and Mrs J. M. Andrew. A monograph produced for the opening of the "Andrew Memorial" at the Gippsland Folk Museum, 7 March 1976'. Unpublished manuscript.

[6] Bachelor of Medicine and Bachelor of Surgery (MBBS).

[7] Julie Fenley and Kathy Lothian, 2012. Dr J.M. Andrew and Andrew Family Collection. Old Gippstown Significance Assessment. Heritage Victoria, Department of Community Planning and Development, 14–18.

were sometimes at odds with the growing formalisation of bushwalking from the 1920s, a consideration of his activities provides us with greater insight into alternative, and richer, histories of walking.

Early bushwalking

The history of extended, leisure bushwalking in Australia is a relatively short one, although it can be linked to early exploration, travel and tourist walking.[8] The term 'bushwalking' made its first formal appearance only in 1927 with the creation of the Sydney Bush Walkers club; however, there was a growing appreciation of the bush in the nineteenth century.[9] This was manifested in a variety of ways. It was part of an emergent Australian nationalism, which led to greater forays into the country for picnics and photographic expeditions. Concerns about the dirt and disease of the cities led to a greater appreciation of the clean and pure air of the bush and, from at least the middle of the nineteenth century, the collection and categorisation of Australian plants and animals had been a popular activity for the leisured middle classes.[10]

The first walking clubs were formed in Australia in the closing decades of the nineteenth century. Both the Melbourne Amateur Walking and Touring Club (MAWTC) and Melbourne's Wallaby Club (whose members appreciated walking and stimulating conversation in equal measure) were formed in 1894, and the Warragamba Walking Club was established in Sydney in 1895.[11] However, at the time Andrew and his companions undertook their walk in 1920 these were the only major clubs in existence. Consequently, club members could not (though there were exceptions) hope to draw upon the long walking experience of their members. Nor could they draw upon a wide body of walking literature.

One of the first walking guides in Australia was HJ Tompkins' 1906 publication, *With Swag and Billy: Tramps by Bridle Paths and the Open Road*,

---

[8] Harper, *The Ways of the Bushwalker*, 5.
[9] Harper, *The Ways of the Bushwalker*, 5, 201–202. Note that the term had been used publicly in 1889, 1892 and 1911, but was formally used in naming an organisation in 1927.
[10] Don Garden, 2011. 'Growing Comfortable in the Land: Environmental History and Bushwalking'. Paper delivered to the Royal Historical Society of Victoria, 11 October 2011.
11 Alfred Hart, 1944. *History of the Wallaby Club*. Melbourne: Anderson, Gowan Pty Ltd.; Graeme Wheeler, 1991. *The Scroggin Eaters: A History of Bushwalking in Australia*. Melbourne: Federation of Walking Clubs (VicWalk) Inc., 14–18; Harper, *The Ways of the Bushwalker*, 128, 142.

and Andrew and his friends may have consulted this text for material advice, as noted below, as well as to shape their understanding of how bushwalking should be defined. While Tompkins' guidebook focused on walking trips in New South Wales – he was Secretary of the Warragamba Walking Club – the work also contained information which would have been relevant to Andrew and his friends as they embarked on their journey. Tompkins' text was a veritable how-to manual, providing advice on all types of necessities. In addition to his recommendation of the 'good old historic and poetic' swag for carrying food and equipment, Tompkins advised the carrying of a journal, which he recommended writing in for half an hour or so after the evening meal. This was not to be filled with 'secret musings' but rather with notes on the experiences of the day. In addition, he urged, a definite itinerary should be planned, with fixed stages for each leg of the trip.

Tompkins' recommendations on walking were 'the even, rhythmic, 3-miles-an-hour style' and ideally no more than 25 miles per day; the formation of a walking party, stating that 'two is the handiest number'; and he sang the praises of an early start and breakfasting on the track. Tompkins argued that the 'fixed rule should be to travel light', although his list of clothing to be taken included a 'collar and tie – these for state occasions'. Tompkins also espoused the health benefits of exercise, arguing that walking stimulated the mind and body: 'the blood is cleansed of impurities, the eyes become clearer, the complexion improved, and the flesh firmer'. A critical benefit of walking, however, was that in contrast to rushed, destination-oriented rail, bike or motor travel, pedestrian touring provided greater insights into the countryside and acted as a tonic to city life, enabling the walker to:

> obliterate his Me, and become part of the Australian bush: to thread his way through more or less scented woods, amid the music of the birds, with here and there traces of some devastating fire: possessed of a rapturous sense of irresponsibility; when there were no arbitrary divisions of time; and, filled with the spirit of revolt, he did not hesitate to pronounce civilization a desperate failure.

Providing concrete advice about physical activity, the structure of the walking tour, and the provisions to be carried seems to have been only part of Tompkins' design. He also promoted a particular kind of bushwalking, one which had no reminders of modern-day industrial life and which, through the extended walking tour, would allow the walker to reconnect with nature. He aimed to formalise bushwalking as a recreational pursuit as well as to insist on a precise conception of nature and the place of humans within it. Thus, even in the years before the 1920s and 1930s, when a view of the 'correct' way to walk began to form in the emerging bushwalking clubs, Tompkins' promotion of the idea that the walker should be 'at one' with nature might be seen as evangelical.[12]

Whatever their particular enthusiasms, however, walkers in the late nineteenth and early twentieth centuries were significantly hampered by their rudimentary equipment, which often prevented lengthy forays into the bush. The carrying of their kit posed particular difficulties. At the turn of the century there were few options apart from basic low-riding rucksacks and the humble-yet-romantic swag, though a number of bushwalkers, Tompkins among them, preferred the latter. Walkers noted the ways in which Australian swagmen had negotiated the difficulties of carrying supplies with this method and Tompkins, for instance, had complained of the rucksack's uncomfortable distribution of weight. Adjustable, high-riding knapsacks, he noted – which allowed for the flow of air behind the back – were not procurable in Australia, though they could be had at exorbitant prices from London.[13] But swagmen tended to obtain their food from farms and other places on their travels. For the bushwalker who intended a trip of several days or more, the limitations of the swag's carrying capacity was considerable.

Lack of adequate packs, however, was only one problem among several. Bill Waters, who began walking around 1917, recorded that at the time of his first walk, one-person tents had barely been heard of, though his party were 'dimly aware' of the existence of sleeping bags.[14] Waters' ignorance might be partly ascribed to the fact that he was not yet a member of a walking club, though it was not until the early 1900s that MAWTC club members began to use improvised sleeping bags, and not until 1909 that imported, ready-made bags were used by members.[15]

In addition, apart from Tompkins' 1906 text there was little written material to advise the amateur walker. RH Croll, an early member of the MAWTC, published guides to walking after the mid 1920s, and the *Weekly*

---

[12] H. J. Tompkins, 1906. *With Swag and Billy: Tramps by Bridle Paths and the Open Road.* Sydney: Govt. Tourist Bureau, Intelligence Dept., 2–14 passim, 80.

[13] Ibid., 8.

[14] Harry Stephenson (comp.), 1982. *W. F. 'Bill' Waters: A Biography. A Man of the Mountains.* Melbourne: Rover Section, the Scout Association, 6.

[15] Alan Budge, 1992. *No End to Walking: One Hundred Years of Walking by the Melbourne Walking Club.* Melbourne: H. H. Stephenson, 24. Note that Tompkins also suggested the improvisation of a sleeping bag in his 1906 publication.

Figure 4.1: 'We Three': Photograph taken on 14 December 1920, one week into Andrew's walking tour.
Image courtesy of the Andrew family.

Times published regular accounts of MAWTC activities, but a general lack of knowledge about how to prepare and undertake a long walking trip was characteristic of walking in the late nineteenth and early twentieth centuries.[16] As Waters wrote:

> no one seemed able to say what one should take on a tour; and equally important, what one could wisely leave behind… The gear necessary for camping out on a walking tour gave lots of exciting thought and

---

[16] Budge, *No End to Walking*, 115–116; Robert Henderson Croll, 1928. *The Open Road in Victoria: Being the Ways of Many Walkers*. Melbourne: Robertson and Mullens.

planning, and though we knew something of fixed camps and the weight of gear that one staggered under from railway to camp and return, no one seemed able to say just how one should be equipped for a walker's camp.[17]

Further to this lack of general knowledge was an absence of adequate maps. Alan Budge, in his history of the MAWTC, argues that word of mouth was sometimes the best available information and that leaders of club walks were encouraged to keep track notes which would be kept by the organisation as an aid to future walkers.[18] And, as Harper notes, even until World War II most published maps lacked the topographical detail that would be useful to bushwalkers.[19]

However, it would be wrong to think that walkers in this early period were generally concerned only to map their own paths through untamed and untrodden country. While many walkers shared Tompkins' enthusiasm for 'becoming part of the bush', others, such as Wallaby Club members, preferred walking in the 'pleasant company' of an 'assembly of good fellows' – and even then, not for an *entire* day.[20] Consequently, for many early walkers this lack of maps was a minor consideration. Some Australian walkers had been undertaking demanding and lengthy walking tours through the bush for decades.[21] Others (including Andrew and his companions) took advantage of existing country roads, where they shared the space with farmers, other rural workers and the occasional car, until the increase in motor transport made these roads unattractive. In addition, an extensive network of operational tram lines, used for timber extraction, were often utilised as walking trails.[22] Further to this, and until around 1910, it was not uncommon for early walkers to find accommodation in a series of pubs or hotels along their route.[23]

---

[17] Stephenson, *W. F. 'Bill' Waters: A Biography*, 6.

[18] Budge, *No End to Walking*, 17. See also Stephenson, *W. F. 'Bill' Waters: A Biography*, 8.

[19] Harper, *The Ways of the Bushwalker*, 253.

[20] Hart, *History of the Wallaby Club*, 11, 20.

[21] For instance, Harper notes that such walks had been carried out since the 1870s. Harper, *The Ways of the Bushwalker*, 59.

[22] Budge, *No End to Walking*, 54–56. For further details of early twentieth century tramlines in this region see Sally Symonds, 1982. *Healesville: History in the Hills*, Lilydale: Pioneer Design Studio, 85–86, 94.

[23] Budge, *No End to Walking*, 24, 89; Harper, *The Ways of the Bushwalker*, 150, 154.

Indeed, Harper argues that there was a sense in which travelling country roads and encountering farmers and 'authentic' bushies in a settled landscape, 'gave the city walker's bush experience a sense of authenticity'. While 'real' bushwalking would increasingly come to be associated with forging a path through untamed and untrodden country, 'in the late nineteenth and early twentieth centuries the walkers' landscape was one where the delights of nature easily co-existed with the settlers' bush'.[24]

Andrew's activities

Partly because of this lack of formalised bushwalking, at the time when Andrew, Burnet and MacKenzie set off for their walking tour there was not even a common descriptor for what they were doing. Andrew himself referred only to 'our trip', 'our tour' and 'the expedition', but he lacked a distinct terminology for the walking and camping components of their journey.[25] That their activity was relatively uncommon is evidenced in Andrew's diary where he describes their departure from Flinders Street Station amid, as he says – somewhat facetiously – 'the admiring gaze of curious sightseers'.[26] Later too, at Healesville, Andrew described '[walking] the main street length under gaze of curious eyes' and on their return noted in his diary that there had been 'remarks' from men on the train and that there had been 'curious crowds [because of] our disheveled appearances'.[27] Their improvised, bushman-like attire seems to have transgressed social standards, as discussed below, but it also points to the unusual nature of their activity.[28]

That Andrew kept a journal for each day of their journey was conceivably inspired by Tompkins' exhortation to prepare a travel itinerary and to make notes on the trip, or possibly by the MAWTC's advice to keep notes

---

[24] Harper, *The Ways of the Bushwalker*, 159.

[25] James Moore Andrew, 1920. 'Diary'. Copy of unpublished manuscript, J. M. Andrew and Andrew Family Collection, Old Gippstown, Moe, 7 December 1920, 1, 2, 26 December 1920, 28. Original diary owned by the Andrew family.

[26] Andrew, 'Diary', 6 December 1920, 1. Harper notes that leisure walking gained recognition as an acceptable form of recreation in the early twentieth century, and became popular by the 1920s. Harper, *The Ways of the Bushwalker*, 163.

[27] Andrew, 'Diary', 24 December 1920, 26; 26 December 1920, 28.

[28] As Harper notes, no ready-made clothing or equipment was available to walkers at this time. Harper, *The Ways of the Bushwalker*, 218, 220.

for the benefit of fellow walkers.[29] In any case, Andrew recorded carefully the miles travelled each day, the group's camping places and the meals they consumed. Burnet, too, kept a diary for the duration of their walking tour in which he recorded similar information, and had earlier calculated the number of calories per mile the men would require, the friends distributing their rations and equipment accordingly.[30] While Harper notes that a number of early walkers kept diaries and journals, and Andrew and Burnet were therefore not unique in this regard, Andrew's notes on the type of activities he undertook, as well as his reflections on these and the mood that we can sometimes glean from his narrative, deepen our understanding of the different ways in which the bush was appreciated during this period of walking.[31]

As we have seen, Andrew and his friends were not walking through isolated wilderness. Instead, the small party travelled through domesticated bushland. By the 1920s, the areas in which Andrew and his peers walked had experienced a long history of white settlement. A sheep run had been established in 1839 between Healesville and Yarra Glen, while cattle and timber industries were also set up to provide meat and timber to Melbourne. Gold was later discovered across the region, and new and improved roads, and the introduction of rail links and waterworks, also encouraged settlers to move into the district. An increased wartime demand for timber and the establishment of the Maroondah dam and Eildon weir led to a further clearing of the land, while tramways were extended to supply timber to the local mills.

It is through these settled areas and along established roads and tramlines that Andrew and his friends travelled. The first full day of walking was typical of the expedition with Andrew, MacKenzie and Burnet travelling along a road that was 'most part densely wooded', and traversing timber tracks which were 'very Rough & uneven'.[32] Rather than forging their own path, their days involved encounters with workmen, such as timber workers from a nearby sawmill, travelling through townships, observing horses pulling trolleys with loads of wood, and passing through cultivated areas with ponds, orchards and farm houses.

---

[29] According to the author of a history of the Melbourne Walking Club, the first records of food carried by the Melbourne Walking Club members are from a trip in December 1909. See Budge, *No End to Walking*, 89.

[30] Andrew, 'Diary', 7 December 1920, 3, Burnet, 'Diary'.

[31] See, for instance, Harper, *The Ways of the Bushwalker*, x, 11, 27, 77, 214.

[32] Andrew, 'Diary', 7 December 1920, 2.

Andrew and his friends camped each night of their three-week trip and, at least to some extent, carried their provisions. Some of the time, the friends made their camp in the bush. They dossed down in a combination of sleeping bags and blankets (due to the cold they were fully clothed much of the time) and, although it is not completely clear from Andrew's diary, they appeared to have slept only under an oilskin shelter or sheet rather than in a tent. On numerous occasions, however, their camp was made not in the bush but on the outskirts of town, often within view of houses. On one evening, in desperation because of incessant rain and the wet ground, they made their camp under a bridge at Thornton, where a train stopped overhead and offered them a ride to Alexandra.[33]

On such a lengthy journey, it was not possible that the men could be entirely self-sufficient. Although they carried food (Andrew's own pack weighed 30 pounds, or around 13.5 kilograms), without the benefits of freeze-dried or dehydrated foods available to campers from the 1940s the amount they could reasonably carry could sustain them for only a few days.[34] Andrew's diary records that, between them, the men carried bread, rice, flour, cheese, eggs, both butter and dripping, jam and meat. Such was the typical bushwalker's diet in the 1920s.[35] However, because the quantity and variety of these foods were insufficient to sustain Andrew and his friends for a full three weeks of trekking – as mentioned, Burnet had already established the number of calories per mile the men would need – they supplemented their provisions in a variety of ways, as discussed below.

The night that the three men spent under the bridge at Thornton parallels clearly the traditions of the Australian swagman, the itinerant traveller of the Australian bush who moved from town to town seeking temporary work. In some ways, and in the absence of a long tradition of sustained, self-sufficient walking, Andrew and his friends appeared to deliberately look to the swagman as a model. In this, they were not alone. Bill Waters, who joined the MAWTC in 1923, recalled the difficulties with carrying equipment and food during his first walking trip in 1917:

> We lived in a northern suburb and it was not unusual at times to see swagmen tramping northerly along Sydney Road heading towards

---

[33] Andrew, 'Diary', 16 December 1920, 13–14.
[34] Harper, *The Ways of the Bushwalker*, 224.
[35] In the late 1920s, for instance, Robert Croll, of MAWTC, recommended a combination of bread, biscuits, meat, butter, jam, sugar, tea and coffee. Budge, *No End to Walking*, 89–90. Cf. Harper, *The Ways of the Bushwalker*, 157, 223.

woolsheds or harvest fields… and odd occasions found me studying their strangely assorted equipment and ways of carrying it, to get some ideas, but I entirely failed to reach the point of halting one to ask his advice.[36]

Early walkers appear to have adopted the swaggie tradition with such enthusiasm that there were instances of mistaken identity. Robert Croll, from the MAWTC, recorded club members being offered jobs or small sums of money while on walking expeditions. Harper has also noted the extent to which the swagman offered the early walker not only a practical model for extended walking tours but an Australian nationalist identity, a 'symbol of the genuine bushman'.[37]

Although they had gained entry into an elite medical fraternity, the walk allowed the friends to connect with this figure of the 'true Australian' bushman. Andrew noted self-consciously the resemblance between his group and the swagman style. At the start of the trip Andrew had commented in his diary that the three friends 'looked a regular hobo brigade', with their canvas bag and swags swung over their shoulders.[38] Between them they carried a gun, billies and an axe, while their clothing consisted of military trousers, tennis shirts and heavy boots – though Burnet added a tie to this ensemble. (Burnet, Andrew noted privately in his diary, had 'quaint ideas of necessity of respectability'.[39]) Although Andrew reported that he was in 'fear and trembling lest my military trousers should be discovered' by his Ormond College peers, and the men had also taken their packs to the station in the early morning so they were 'less conspicuous', the diary also points to a valorisation of the hobo image.[40] While the stares from members of the public suggest that the men's clothing was improper, or even subversive, there is nonetheless a tone of pleasure in Andrew's noting of each

---

[36] Stephenson, *W. F. 'Bill' Waters: A Biography*, 7.
[37] Harper, *The Ways of the Bushwalker*, 158.
[38] Andrew, 'Diary', 6 December 1920, 1.
[39] Andrew, 'Diary', 24 December 1920, 26. Burnet was a member of the Wallaby Club from 1934 and president and life member from 1945. This club, more conservative than other walking clubs, placed emphasis on companionship and entertainment as much as walking, and limited its activities to day walks only. See Burnet's short biography on the Wallaby Club website, at http://www.wallabyclub.org.au/compendium/biogs/E000397b.htm, accessed 12 December 2013. We can speculate that Burnet's desire to wear a tie when setting off for his 1920 walk reflected a conservatism which was to later find expression in his membership of the Wallaby Club.
[40] Andrew, 'Diary', 6 December 1920, 1.

Figure 4.2: Campsite sketch by Andrew.
Image courtesy of the Andrew family.

of these episodes. Indeed, the swagman model reached its apotheosis when Andrew noted that the group had camped by a billabong – the local term for a waterhole – which recalled the famous Australian folk song, *Waltzing Matilda*.[41]

Their approaches to local farmhouses were also reminiscent of the practices of swagmen, and they frequently met kindly locals who gave them milk, cherry plums, apples and, at one house, steak and gravy. They also hunted rabbits, collected food from abandoned orchards and roadside bushes, or purchased supplies in towns, while workmen on the road also offered advice on the best routes to travel and the location of fresh water. On one occasion they returned these favours by giving a rabbit they had shot to a woman at one of the houses they passed.[42] In taking on aspects of a swagman identity, then, the group was able to exploit the ongoing rural customs and generosity that had their origin, in part, in the dependence on this itinerant workforce.[43] The friends thus benefited in ways which

---

[41] Andrew, 'Diary', 7 December 1920, 2.
[42] Andrew, 'Diary', 22 December 1920, 21.
[43] Russel Ward, 1958. *The Australian Legend*. Melbourne: Oxford University Press, 177, 186.

would not be accessible to walkers today. Like swagmen, however, they also encountered some suspicion and fear, and near Jamieson they were setting up their camp when a woman at a nearby house asked them to move on as there were no men at her place. When she stated she would have to go to a neighbour's house for the night they decided to leave, travelling on an additional mile to find a level spot with fresh water that was not too close to a farmhouse.

Andrew and his friends also engaged in activities that sought to test their powers of endurance, and these pursuits help us gain a more complete understanding of the variety of ways in which they approached bushwalking. Andrew's diary records numerous occasions of his grim determination – of the way, for instance, he 'trudged on manfully' through the pitch dark in search of a campsite, or the way he decided to 'force' himself onwards at the end of a long day, of the sufferings endured by blisters, of the cold, and of ill health in the final week.[44] Andrew dutifully recorded the distances they covered each day of their 205 mile walk (in one case it was at least 21 miles).[45] He also recorded in some detail his feats of masculine endurance in scaling a mountain at the steep and rocky Cathedral Ranges. Here, after some deliberation, he decided to go 'straight up the main peak', finding precarious toe holds only by scraping moss from the rocks. He was, in his words, 'panting along: sweat pouring out… [with] all feeling lost in the great nervous strain for self preservation: retreat impossible'.[46]

Andrew also saw bushwalking as a way to develop his hunting skills. On at least one occasion the men spent part of the day shooting at tins with a pea rifle and revolver, with Andrew recording: 'FMB not a bad shot: I am rotten on the trigger pull: with Revolver I got the mark fairly well'.[47] By the end of the second week MacKenzie had increased his average to three rabbits for nine shots. Burnet was averaging four for eight, while Andrew's average was reduced from four for five to four for six when he missed a running shot in the ferns. Andrew's peers seem to have been similarly willing to regard the bush as a material resource. Indeed, Myles Dunphy also practised shooting on his bushwalking expeditions even

---

[44] Andrew, 'Diary', 6 December 1920, 1; 7 December 1920, 2.
[45] Andrew, 'Diary', 22 December 1920, 22.
[46] Ibid., 17.
[47] Ibid., 21.

while he was adopting a conservationist agenda in the 1920s.[48] The killing of wildlife has fallen into disrepute today. Modern-day bushwalkers focus on minimising their impact on the landscape, while hunting is commonly regarded as an adversarial pursuit undertaken by a discrete group of hunters.

But there were also activities that the friends took part in which seem to have been less common among their contemporaries. While it was usual enough to engage in birdwatching and other forms of nature study – Andrew recorded sightings of robins, white-winged choughs, gang gangs and 'hundreds of tiny birds unnamed' – we are unaware of any other early bushwalkers who pursued entomology or dissection.[49] Each day the friends would pause to hunt for beetles, with Burnet, who had introduced Andrew to entomology, recording the specimens they had collected. On the third day of the trip Andrew also wrote that the friends had engaged in a 'nature study', which included a chase for a rabbit, the shooting and skinning of an echidna and the discovery of a larger one still. At the end of the day the friends dissected the echidna, recording its sex, and making reference to its 'ductless glands', 'heavy Patisma [sic] muscle' and to the size of its spines.[50] There is some evidence that Andrew intended to use the skin for a later taxidermy project.[51] This is more in keeping with the activities of naturalists, who combined hunting and collecting, although Tom Griffiths notes that this was being replaced with an ethic of reverence and non-interference by the early decades of the twentieth century.[52] It is also probable that the three medical students were aware of the work of orthopaedic surgeon William Colin MacKenzie, whose anatomical studies of monotremes and marsupials had informed understandings of human musculature around this time.[53]

In participating in these myriad activities, Andrew also revealed that he regarded the bush in diverse ways. He undoubtedly saw the beauty of the

---

[48] Harper, *The Ways of the Bushwalker*, 205–209, 256–258.

[49] Andrew, 'Diary', 8 December 1920, 3. See Harper, *The Ways of the Bushwalker*, 42–43 for a discussion of nature studies.

[50] Andrew, 'Diary', 8 December 1920, 3.

[51] Ibid., 4.

[52] Tom Griffiths, 2001. *Forests of Ash: An Environmental History*. Cambridge: Cambridge University Press, 107–108.

[53] Libby Robin, 2007. *How a Continent Created a Nation*. Sydney: University of New South Wales Press, 80.

Australian bush. On his first day, after the night on his sawdust heap, he exclaimed:

> Everywhere the scenery is so magnificent that it defies my powers of describing it, I can only leave it to imagination of readers who have seen. those who have not, we pity. everywhere a white flowering shrub of which we don't know the name, is bursting into bloom and lends colour to the eye, though the gums themselves show a hundred different tinges of reds, browns, yellow, greens and blues to our admiring eyes: Truly it is good to be alive and to be here.[54]

However, in ways that are possibly in tension with this understanding, Andrew both wrote and sketched his interpretation in terms that illustrate an appreciation for a more domesticated Australian bush. He drew cartoon-like sketches of soft landscapes, many of them depicting farmhouses, roads and fences, and also wrote of 'level ferneries', 'clematis bower[s]' and 'archways of ferns', using English gardening terms to describe the bush, and his journal stated explicitly that the area around Marysville Junction, with its 'tall beech trees with moss-covered trunks tower[ing] over the Spring' was 'quite English'.[55] We cannot know with certainty whether Andrew was acknowledging a distinction between the domestic and the wild, or between different kinds of beauty, or between the Australian and English countryside. As Harper has observed, walkers in this period could take pleasure in accessing settled countryside, although she maintains that from the late nineteenth century they also sought the creation of 'wilderness' areas which were free from the marks of human habitation.[56] However, Janice Newton insists that a preference for a 'tamed' environment was not uncommon for tourists in the interwar period. Indeed, Andrew's diary bears some resemblance to her description of the areas close to Melbourne being transformed from a threatening, wild frontier into a soft accessible environment that was characterised by waterfalls, ferny glades and panoramic views.[57]

---

[54] Andrew, 'Diary', 7 December 1920, 2.
[55] Andrew, 'Diary', 8 December 1920, 3; 9 December 1920, 4.
[56] Harper, *The Ways of the Bushwalker*, 5, 159.
[57] Janice Newton, 1996. 'Domesticating the Bush'. *Journal of Australian Studies* 20 (49), 76, 78.

Figure 4.3: Andrew sketched a number of scenes during the tour.
He depicted nature as a non-threatening environment.
Image courtesy of the Andrew family.

A distinct bushwalking identity

Although the fundamentals of 'bushwalking' in any era were evident in the walk from Warburton to Woori Yallock, it was not until later in the 1920s and into the 1930s that we can observe the development of a unique 'bushwalking' identity with particular ideas about the nature of acceptable practice. This can be attributed, in part, to the establishment of new bushwalking clubs from 1922. Among these the Melbourne Women's Walking Club was formed in 1922, the Sydney Bushwalkers in 1927, and the Hobart Walking Club in 1929, while other clubs were established in the following decade. These clubs differed from the older Wallaby Club, in particular, with its deliberate class-based, respectable walking attire, gendered notions of physical activity, leisurely half-day walks along well-known routes, and its view that club activities served as a means for intellectual discussion as much as for exercise.[58] While the Melbourne Amateur Walking Club and

---

[58] Harper, *The Ways of the Bushwalker*, 142–148.

Warragamba Walking Club memberships were also drawn from the middle and upper classes, and shared the Wallaby Club's view that walking was a masculine activity, these clubs favoured more vigorous activity, and by 1910 their members had adopted aspects of the swaggie tradition, as mentioned above.[59] The staid Wallaby Club is regarded by Harper as providing a model for the newer clubs to work against, and a 'bushwalking movement' emerged, with members coalescing around the idea of a specific bushwalking identity.[60] An assessment of Andrew's diary suggests some elements of this thinking even prior to the formal establishment of these clubs, although it is difficult to determine whether Andrew was ever fully converted to what would become the preferred bushwalking style.

Within these clubs a clear distinction was drawn between 'real' bushwalkers and those who were merely 'hikers'. While the terms 'bushwalking' and 'hiking' might be used interchangeably today, the term 'bushwalking' in the 1930s suggested a very different activity to hiking, which was a popular group activity that developed in the 1930s, and which was especially popular among the working classes. The Sydney Bushwalking Club summed up the distinction by claiming that 'The hiker is… the muddling inefficient; the bushwalker the expert'.[61] Even more than this, the *Sydney Bush Walker Annual* argued, hikers were people in 'gaudy slacks, who litter the bushland with their tins and rubbish, and who are in no way akin to the serous members of the Sydney Bush Walkers, with large packs on their backs and provision for a fortnight's sojourning away from civilization'.[62] According to Harper, for some members of these bushwalking clubs, bushwalking involved negotiating paths through unmapped bush, and camping out.[63] It was an activity undertaken by those who could carry their own gear and show powers of endurance and skills in bushcraft and, it might be said, those who could demonstrate a true appreciation of the bush.

These evolving ideas about the 'proper' way to experience the bush are also evident in Andrew's record of interwar tourist activities. A number of the areas which the small group visited on their trip through Warburton, Woods

---

[59] V. C. Routley, 1994. *Footsteps From the Past: A Centenary Publication of the Melbourne Walking Club Inc. 1894–1994*. Melbourne: Melbourne Walking Club, 16; Harper, *The Ways of the Bushwalker*, 154.

[60] Harper, *The Ways of the Bushwalker*, 149, 215–216.

[61] Melissa Harper, 1995. 'The Battle for the Bush: Bushwalking vs Hiking Between the Wars'. *Journal of Australian Studies* 19 (45), 44.

[62] Ibid., 43.

[63] Harper, 'The Battle for the Bush', 42; Harper, *The Ways of the Bushwalker*, 247–249.

Point, Jamieson, Eildon, Alexandra, Marysville, Healesville, Launching Place and Woori Yallock were regarded as tourist destinations, and through the 1920s between eight and ten thousand people visited Healesville during the Christmas and Easter holidays.[64] These townships were within easy reach of the city, with tourists taking advantage of affordable train fares to escape the bustle of city life and being attracted by the idea of a peaceful mountain retreat.[65] The diary of the 1920 walk provides evidence of the popularity of weekend bush retreats, with the entry for 12 December noting that 'crowds of people kept passing us all day, on foot, buggies, bikes, horseback and the good old gig… the [church] Services are mostly weekday ones so Sunday is a day of visiting, enjoyment etc. as well as of rest'.[66] His account for 19 December – which notes that 'dozens of foot, etc. passengers pass by up hill quite late at night & some wished us a cheery goodnight – others sung lustily' – implies some annoyance at being interrupted by these hordes.[67] Likewise his statement on Christmas day that 'Launching Place is alive with [campers], absolutely crawling' indicates some disdain for these sightseers.[68] Although the 'hiking craze' began in the 1930s, with thousands of amateur walkers participating in mystery hikes that were organised by the railways and which involved walking en masse along pre-arranged routes, bush excursions were also popular in the previous decade and these tourists seem to have some parallels with accounts of the 1930s 'hiker'.[69]

The interest in designing lightweight equipment and suitable walking attire also created a firm distinction between the bushwalker and other users of the bush. As noted earlier, without ready-made clothing or equipment for walkers in this period, walkers had to experiment or, from the 1920s, to import expensive gear from overseas.[70] Increasingly, walkers abandoned

---

[64] Symonds, *Healesville: History in the Hills*, 99.

[65] See Newton, 'Domesticating the Bush', 67; Rod Ritchie, 1989. *Seeing the Rainforests in the Nineteenth Century*. Sydney: Rainforest Publishing, 135, 140; and Symonds, *Healesville: History in the Hills*, 74 for a discussion of the accessibility of train travel and increasing ownership of house blocks in this area, as well as the burgeoning interesting in working class leisure.

[66] Andrew, 'Diary', 12 December 1920, 7.

[67] Andrew, 'Diary', 19 December 1920, 18.

[68] Andrew, 'Diary', 25 December 1920, 27.

[69] See, for instance, Newton, 'Domesticating the Bush', 67. Cf. Symonds, *Healesville: History in the Hills*, 107, which records a drop in rail travel in the 1930s.

[70] Harper, *The Ways of the Bushwalker*, 220. When the friends carried out their tour, the weight and practicality of every item needed to be considered, with walkers only determining the types of material to carry through trial and error. As Andrew noted,

what VC Routley described as 'normal street clothes' – including three piece suits, stiff collars and, at times, bowler hats and walking canes – in favour of specialist clothing.[71] Bill Waters introduced the MAWTC to shorts in 1924, although there was some resistance to this attire and the clothing was described in a club newsletter of 1934 as 'perfectly good trousers the legs of which have been deliberately and maliciously cut in halves in order to facilitate snake bite and to afford the wearer greater opportunities for having his legs picturesquely marked by blackberries and nettles'.[72] The Melbourne Women's Walking Club also ruled that members were not required to wear a dress or a skirt and blouse, although their riding breeches and long boots caused some laughter when they walked through a town.[73] Andrew also embraced this new garb and even wore his Rover shorts while practicing medicine during the war.[74] One can only imagine how the patients responded to his attire.

As commercial bushwalking companies were established to cater for the desire for outdoor equipment, bushwalkers also set aside the swag in favour of the rucksack, and Andrew was an early adopter of this newly available gear. His pack, purchased in the early 1930s, was made of canvas, with a cane A-frame and leather straps across the back to allow air to circulate between the walker and pack, while the label 'Camp Gear / Paddy Made / For Walkers' distinguished between bushwalker and hiker. Pallin had initially supported the hiking movement – his company's advertisements stated that he produced 'gear for hikers, made for hikers, by a hiker' – but his branding suggested that he quickly dissociated himself from this term.[75] The sleeping bag also gained in popularity, while Japara and 'campette' tents – weighing half a kilogram – were also created.[76]

'every detail of attire has to be considered in any attempt like this': clothing should be designed for comfort rather than respectability (although this *was* also a consideration), and foodstuffs and equipment reduced to essentials. Nonetheless, Andrew's pack still weighed 30 pounds, and by the second day he had reported that 'we ought to have had another blanket each at least with us, & a lot less tucker also'.

[71] Routley, *Footsteps From the Past*, 15–16.

[72] Ibid., 41.

[73] Amy Eastwood, Isabel Eastwood, and Hazel Merlo, 1985. *Uphill After Lunch: Melbourne Women's Walking Club, 1922–1985*. Melbourne: Melbourne Women's Walking Club, 12.

[74] Personal communication with Catherine Jerome, 18 February 2013.

[75] Harper, 'The Battle for the Bush', 43.

[76] Wheeler, *The Scroggin Eaters*, 34.

The importance of this new gear should not be underestimated. By the end of the 1930s walkers were completely self-sufficient, and could be readily distinguished from the swaggie or hiker. As Harper argues:

> Finding suitable clothing and equipment, and knowing how to carry it, along with perfecting the bushwalker's diet, were important practical issues for the bushwalker who also camped. As bushwalkers turned their hand to the task, innovation and experience became part of the armoury of expertise. But shorts, boots and rucksacks also carried a heavy symbolic weight. This more or less uniform set of accoutrements was worn with pride as powerful visible markers of the bushwalker's unique identity. No longer could the recreational walker feasibly be mistaken for a swagman. At the same time, in their announcement of the commitment and toughness of the walker who pushed through rugged bushland, they were a means of distinguishing the bushwalker from the tourer and the hiker.[77]

Andrew in later life

Although some bushwalking clubs might have decided that their approach to walking was superior, Andrew continued to follow his own interests and initiate trips on his own terms. The Andrew collection at Old Gippstown provides us with evidence that he was not wholly concerned with understandings of the 'correct' way to approach the Australian bush. Indeed, Andrew's activities in later years continue to problematise narrow understandings of bushwalking. On the one hand, as discussed below, Andrew made a number of trips to Gippsland's Alpine region. The evidence for these trips points not only to a desire to access wilderness areas, but suggests that Andrew took on a key role in opening up access for bushwalkers in these more remote regions. As the organiser of at least some of these trips, and by encouraging others to walk or ski in these areas, Andrew might be seen as an ardent convert to orthodox views about 'real' bushwalking. On the other hand, Andrew's diaries for other camping trips in the 1920s show that he carried on making trips to popular tourist destinations such as the south-west Victorian coast in 1921 and the northern Grampians region in 1925. Such locations were, and

---

[77] Harper, *The Ways of the Bushwalker*, 227.

continue to be, popular with recreational walkers.[78] These areas might have more closely resembled the 'big bush' or frontier wilderness, as described by Newton, but here too efforts were made by those encountering the bush to 'humanise' the landscape and recast it in recognisable terms.[79] The evidence also suggests that Andrew and his family undertook a number of sedentary camping and fishing trips throughout the 1940s and 1950s. These possibilities were enlarged by his purchase of a car, which is shown in a number of his early camping photographs, and by the use of packhorses for some expeditions. A homemade timber and flywire camp safe, which is housed with the Andrew family collection at Old Gippstown, provides further evidence of his interest in sedentary camping trips.

In addition, by the mid 1920s Andrew had begun to make packing lists for his trips away from home, and a number of these survive. These are valuable artefacts. Alan Budge's history of the Melbourne Walking Club notes that the club holds no records of worth that relate to the food carried by members before 1909, and while walkers such as Tompkins and Croll issued their recommendations for bread, biscuits, meat, butter and so on, they lack the specificity of Andrew's lists, which vary according to destination, method of transport, season, number of people and what might be assumed to be personal preferences over time.[80] This variety is evident in the surviving trip lists from his five-day trip to the popular Grampians region in 1925. Unlike the 1920 Warburton trip, Andrew and his companions – who included his brother Hugh and possibly Burnet – were entirely self-sufficient, carrying perishable items such as bread, eggs, butter, tinned food, dripping and a leg of cold meat, as well as items a modern bushwalker might discard in an attempt to remove even minimal amounts of extra weight from a pack, such as soap and towel, a nailbrush and comb. An estimated 15 kilograms of food was listed, with Andrew later recording that 'I've never in all my experiences slung such a pack for weight or bulkiness'.[81] Other items, however, such as tent fly and billies were not recorded and it is unclear whether the listed

---

[78] See for instance, Julia Horne, 2005. *The Pursuit of Wonder: How Australia's Landscape Was Explored, Nature Discovered and Tourism Unleashed*. Melbourne: The Miegunyah Press, 111.

[79] Newton, 'Domesticating the Bush', 76, 77.

[80] Budge, *No End to Walking*, 89.

[81] James Moore Andrew, 1925. '[Diary] Walking Tour Aug. 31st – Sept 5th '25'. Unpublished Manuscript, J.M. Andrew and Andrew Family Collection, Old Gippstown, Moe, 31 August, 1.

items only applied to Andrew; they may have been carried by him alone or shared between his companions.

While visiting popular destinations and making family-friendly trips, Andrew's move to Yallourn in 1926 also opened up new bushwalking possibilities for him in more isolated and relatively untouched regions. The packing lists also reveal that he attempted what are still considered to be difficult and strenuous walks in areas of Central Gippsland, such as Mount Wellington and Lake Tarli Karng. Like his earlier treks, Andrew saw these walking tours as an opportunity to test his physical abilities. Titles such as '3 men, 8 days, Dec 1935' or '2 men – 14 days – January 23rd, 1943', rather than the bland 'Mt Wellington tour', suggests that the diary entry from at least 10 years earlier – 'preparing for the "adventure", it is long since such a thrill as a camping and walking tour has been felt' – also rings true for the later trips.[82] Similarly, although it is unclear who coined the titled 'Operation TK' for the Melbourne Scouts' 28-mile walk, which Andrew directed from Licola to Tarli Karng, we can assume that the 'mission' or 'expedition' alluded to in this title would have pleased him.

Figure 4.4: Andrew's involvement in the Rover Scouts from the 1930s included skiing trips across Alpine regions.
Image courtesy of Dr Andrew and Andrew family collection,
Old Gippstown heritage park.

---

[82] Ibid., 1.

Andrew's packing lists, as well as photographs within his collection, reveal that he also undertook camping trips in the snow season, sometimes with the Yallourn Rover Scouts, an organisation he had joined in the mid 1930s. Andrew's skiing trips were, at least to some extent, encouraged and aided by Bill Waters, a fellow Gippsland resident. A number of items among Andrew's personal effects indicate that the two men had met by at least 1938 and that they had maintained their friendship until at least the late 1950s. During the early 1930s, Waters had encouraged the Yallourn Rovers to attempt skiing on the Baw Baw Plateau, an area that he considered to have been largely neglected for these purposes. Andrew assisted in organising working bees to cut ski runs and in building the Rover hut at Mount Erica, which opened in 1940. Several wooden skis and ski tips within the Andrew collection are testament to both his interest in skiing and to his early involvement in this activity. One of these ski tips is inscribed with the words 'Cope Hut' and is dated 1937. This hut was the base for the Bogong Rover crew, and the ski tip indicates that Andrew was one of the first walker-skiers to this area, the Rover crew having crossed the Bogong High Plains for the first time only five years earlier, in the winter of 1932.[83]

Andrew was consequently a walker who explored more isolated regions, including Bogong and Mount Wellington, and was amongst those early skiers who travelled across the Alpine regions in the winter season. However, he continued to place importance on accessing more popular regions. This suggests that Andrew, and likely many other walkers, continued to blur the ways in which they appreciated the bush, even after 'real' bushwalking had come to be associated with a narrower range of activities.

Conclusion

Through an assessment of the Andrew collection we can obtain a more complete picture of the nature of bushwalking in Australia in the first half of the twentieth century; of the unsuitable and rudimentary equipment, clothing and foodstuffs that were the lot of bushwalkers at this time. We can also see that Andrew's activities resist systematic classification. Never a part of a formal bushwalking club, Andrew carved out his own understanding of what it meant to be a bushwalker. While his activities display aspects of what might be regarded as 'bushwalking' he could also be said to have engaged in

---

[83] Stephenson, *W. F. 'Bill' Waters: A Biography*, 72.

'hiking'. While celebrating the swagman tradition, he also embraced more practical and innovative options. While humanising nature, he also sought isolated 'wilderness' areas. These contradictions, we argue, suggest that a fuller account of bushwalking in Australia must look beyond the activities of formalised groups to consider that people like Andrew had a diverse understanding of what it means to negotiate the Australian bush. It is by assessing the activities of individuals that we come to a more complete sense of how Andrew and other walkers might have engaged with nature. While bushwalking has now come to be associated with the quest for wilderness, walkers in the early twentieth century undertook a variety of activities and strode along a range of paths. A history of bushwalking in Australia must encompass all of these ideas.

Acknowledgements

We wish to thank the Andrew family for sharing their stories of Dr Andrew and providing permission to publish images from the 1920 diary. We are also grateful to Melissa Harper and the anonymous reviewer for commenting on a draft of the chapter, and to Deirdre Slattery for her generous guidance on a later draft.

References

Andrew, James Moore, 1920. 'Diary'. Copy of unpublished manuscript, J. M. Andrew and Andrew Family Collection, Old Gippstown, Moe. Original diary owned by Andrew family.

Andrew, James Moore, 1925. '[Diary] Walking Tour Aug. 31st – Sept 5th '25'. Unpublished Manuscript, J. M. Andrew and Andrew Family Collection, Old Gippstown, Moe.

Andrew, Margaret, 1976. 'The Life and Times of Dr and Mrs J. M. Andrew. A monograph produced for the opening of the "Andrew Memorial'" at the Gippsland Folk Museum, 7 March 1976'. Unpublished manuscript.

Budge, Alan, 1992. *No End to Walking: One Hundred Years of Walking by the Melbourne Walking Club*. Melbourne: H. H. Stephenson.

Burnet, Frank Macfarlane, 1920. 'Diary'. Frank Macfarlane Burnet Collection, 1989/0034 Series 2/5 University of Melbourne Archives.

Croll, Robert Henderson, 1928. *The Open Road in Victoria: Being the Ways of Many Walkers*. Melbourne: Robertson and Mullens.

Eastwood, Amy, Eastwood, Isabel and Merlo, Hazel, 1985. *Uphill After Lunch: Melbourne Women's Walking Club, 1922–1985*. Melbourne: Melbourne Women's Walking Club.

Fenley, Julie and Lothian, Kathy, 2012. Dr J. M. Andrew and Andrew Family Collection. Old Gippstown Significance Assessment. Heritage Victoria, Department of Community Planning and Development.

Garden, Don, 2011. 'Growing Comfortable in the Land: Environmental History and Bushwalking'. Paper delivered to the Royal Historical Society of Victoria, 11 October 2011.

Griffiths, Tom, 2001. *Forests of Ash: An Environmental History*. Cambridge: Cambridge University Press.

Harper, Melissa, 1995. 'The Battle for the Bush: Bushwalking vs Hiking Between the Wars'. *Journal of Australian Studies* 19 (45), 41–52.

Harper, Melissa, 2007. *The Ways of the Bushwalker: On Foot in Australia*. Sydney: UNSW Press.

Hart, Alfred, 1944. *History of the Wallaby Club*. Melbourne: Anderson, Gowan Pty Ltd.

Horne, Julia, 2005. *The Pursuit of Wonder: How Australia's Landscape was explored, Nature Discovered and Tourism Unleashed*. Melbourne: The Miegunyah Press.

Newton, Janice, 1996. 'Domesticating the Bush'. *Journal of Australian Studies* 20 (49), 67–80.

Ritchie, Rod, 1989. *Seeing the Rainforests in the Nineteenth Century*. Sydney: Rainforest Publishing.

Robin, Libby, 2007. *How a Continent Created a Nation*. Sydney: University of New South Wales Press.

Routley, V.C., 1994. *Footsteps From the Past: A Centenary Publication of the Melbourne Walking Club Inc. 1894–1994*. Melbourne: Melbourne Walking Club.

Slattery, Deirdre, 2009. 'Bushwalking and Access: The Kosciusko Primitive Area Debate 1943–6'. *Australian Journal of Outdoor Education*. 13 (2). 14–23.

Stephenson, Harry (comp.), 1982. *W. F. 'Bill' Waters: A Biography. A Man of the Mountains*. Melbourne: Rover Section, the Scout Association.

Symonds, Sally, 1982. *Healesville: History in the Hills*. Lilydale: Pioneer Design Studio.

Tompkins, H. J., 1906. *With Swag and Billy: Tramps by Bridle Paths and the Open Road*. Sydney: Govt. Tourist Bureau, Intelligence Dept.

Wallaby Club. 'Burnet, Frank Macfarlane'. Accessed 12 December 2013 at: http://www.wallabyclub.org.au/compendium/biogs/E000397b.htm.

Ward, Russel, 1958. *The Australian Legend*. Melbourne: Oxford University Press.

Wheeler, Graeme, 1991. *The Scroggin Eaters: A History of Bushwalking in Australia*. Melbourne: Federation of Walking Clubs (VicWalk) Inc.

Chapter 5

Port Albert and its strategic role, 1841–1860

Cheryl Glowrey

On Sunday, 14 February 1926, a crowd of 300 locals gathered at Toms Cap, above Yarram, to dedicate a memorial to Angus McMillan's arrival in the district in 1841. Sons and daughters of Gippsland's early European pioneers attended, despite smoke from nearby bushfires clouding the views of the surrounding coast.[1] Addressing the crowd, local historian the Reverend George Cox concluded that the fortunes of Gippsland were altered when influential trader Benjamin Boyd missed an auction held in Melbourne on 28 September 1843.[2] Cox told the assemblage that 'by this circumstance, the destiny of Gippsland was linked with Melbourne instead of New South Wales' (NSW), and that 'if Gippsland had been linked up with NSW we would have seen a railway from Sydney via Twofold Bay, East Gippsland, Sale and through to Welshpool as the Port for Gippsland'.[3] Instead, the land in question, Port Albert, came under the control of the Melbourne-based merchants and pastoralists, who used it to advantage their livestock trade to Van Diemen's Land between 1841 and 1860.

This chapter focuses on the presence of the mercantile traders, the way in which the physical resources of Gippsland were used, and the responses of the colonial government to the private development of Gippsland in the

---

[1] John Adams (ed.), 1990b. *Notes on Gippsland History by Rev. George Cox Vol. 1*. Port Albert: Port Albert Maritime Museum, 52.

[2] Rev. George Cox, the Anglican Minister of Yarram, published *Notes on Gippsland History* between 1911 and 1930, a series of articles on the early history of Gippsland printed in the Yarram newspaper, the *Gippsland Standard*. Rev. Cox's research drew on locally available documents and family knowledge of settlement history in the district, as well as the Sydney and Melbourne newspapers. Cox's much sought after *Notes* were reproduced in six volumes by the Yarram and District Historical Society in 1990 and since republished by the Port Albert Maritime Museum in 1995.

[3] Adams, 1990b. *Notes on Gippsland History*, 52.

1840s and 1850s. These were the decades of sail; of tall ships and coastal traders following the deep-time relationships between land, wind and sea along the Hobart to Sydney shipping route. In order to understand the strategic importance of Port Albert and Corner Inlet to the merchants we need to view it from the sea and realise that at this time Gippsland was more connected to the north and south than to the west, where natural barriers of mountains, forests, rivers and swamps limited overland routes.

Settlement of Gippsland from the north

The 'opening up' of Gippsland for settlement occurred from the north and the south at the same time in 1841, although a handful of pastoralists had ventured into the northern valleys by 1839. McMillan was told about the land to the south by Monaro Aborigines, and he followed traditional pathways used by the Gunai-Kurnai to travel through Omeo to Bruthen on their way to the gatherings for bogong moths and to the south coast of NSW.[4] Gunai-Kurnai trade and cultural ties extended along the coast and into NSW, and this was one of three main Aboriginal trading routes in Victoria.[5] The early pastoralists adopted the same trade routes and, as historian AW Greig notes, 'within three or four months a dray track stretched right across the hitherto wilderness from the Upper Tambo to the "Old Port"'.[6] By establishing markets in Hobart the pastoralists extended these routes south, with their ships following the ancient migratory route of the Bassian Rise, marked by the granite islands of Bass Strait.

The 'discovery' of Gippsland in 1840 is related to the quest for pastoral lands and over-writes a much longer history of European activity. The Bass Strait islands were known to American sealers and whalers before George Bass's exploratory journey in January 1798, and his subsequent voyage later that year with Matthew Flinders.[7] Bass's journey was itself prompted by

---

[4] A.W. Greig, 1912. 'The Beginnings of Gippsland Part 1.' The Victorian Historical Magazine 2 (1), 6; J. Flood, 1980. *The Moth Hunters of the Australian Capital Territory; Aboriginal Traditional Life in the Canberra Region*. Canberra: Australian Institute of Aboriginal Studies, 116; John Blay, 2005. *Bega Valley Path Ways and Trails Mapping Project*. Bega Valley Heritage Study, Public Version. Bega (NSW), NSW National Parks and Wild Life Service and Bega Valley Shire Council, 18.

[5] R. Broome, 2005. *Aboriginal Victorians: A History Since 1800*. Crows Nest: Allen & Unwin, 6.

[6] Greig, 1912, 'The Beginnings of Gippsland', 8.

[7] A.G.L. Shaw. 2003. *A History of Port Phillip District; Victoria Before Separation*. Melbourne, Melbourne University Press, 3.

the arrival of three survivors of the *Sydney Cove* who had walked to Sydney from the east coast of Gippsland.[8] Wilsons Promontory was viewed as an island by Europeans at this time, with a series of small mountain-backed coves facing the shipping route and offering shelter from the prevailing winds of Bass Strait. Before 1841, sealing and whaling establishments existed along the east coast of Wilsons Promontory, at Sealers Cove; Lady's Bay (now called Refuge Cove); and Bareback Cove (now named Little Waterloo Bay).[9] These were connected by shipping routes to other coastal colonies; for example, the Imlay Brothers were based at Twofold Bay and had whaling establishments at Lady's Bay on Wilsons Promontory and Launceston.[10]

The sea was a critical factor in the venture to settle Gippsland. Fears of the shifting sand bars at the entrance to Corner Inlet and strong tidal surges had previously deterred ships from sailing close to the coast, but curiosity about the land beyond prompted further exploration in 1839.[11] Camden (NSW) pastoralist James MacArthur was on board HMS *Pelorus* in mid 1839 when it was driven southwards down the coast from Cape Howe by strong winds towards the entrance to Corner Inlet.[12] It was MacArthur's proposal to the Polish explorer, the self-styled Count Paul Strzelecki, which led to their overland journey into Gippsland the following year.[13] Governor Gipps encouraged Strzelecki to explore undeveloped areas of the colony, laying the foundations for the later dispute over who discovered Gippsland, Strzelecki or McMillan.

Strzelecki, with MacArthur and their Aboriginal guide, Charlie Tarra, reached Westernport in May 1840 having followed McMillan's route for much of the journey. On arrival in Melbourne, Strzelecki promoted the good pastures to be had in Gippsland, despite having barely survived the westward expedition through the ranges that would later bear his name. His words captured the attention of William Brodribb and others looking

---

[8] M. Le Cheminant, 1988. 'Who Discovered the Gippsland Coast and Bass Strait?' *Gippsland Heritage Journal* 1 (3), 15.

[9] George Henry Haydon, 1846. *Five Years Experience in Australia Felix*. London, Hamilton Adams & Co., 50.

[10] Jane Lennon, 1975. 'Squatters, Merchants and Mariners; an Historical Geography of Gipps' Land 1841–1851'. Melbourne: MA Thesis, University of Melbourne, 60.

[11] Division of Crown Land Management, 1980. *Corner Inlet/Seaspray Coastal Study Vol. 1*. Resource Document, Melbourne: Department of Crown Lands and Survey, 71.

[12] Greig. 'The Beginnings of Gippsland', 9.

[13] Adams, 1990b. *Notes on Gippsland History*, 10.

for new land for their herds during a drought, but 'the great difficulty was to discover a practical route to Gippsland by which stock could travel'.[14]

It was the sea which provided a solution. Within days of the wreck of the paddle steamer *Clonmel*, at the entrance to Corner Inlet on the southern Gippsland coast in January 1841, a consortium of pastoralists formed the Gipps Land Company and took action to secure the land. Members of the Gipps Land Company included J. Hawden, J. Orr, A. Rankin, N. McLeod, W. Brodribb, A. Kinghorn, E. Kirsopp and Dr Stewart.[15] The company reformed with a change of membership as the Port Albert Company shortly after settlement under the leadership of the Melbourne merchant John Orr.

McMillan, in the employ of Monaro pastoralist Lachlan Macalister, arrived at Port Albert from the north on 14 February 1841 on his third overland attempt to reach the coast, coinciding with the landing of the Gipps Land Company and the *Singapore* entering from the south. Several pastoralists had preceded McMillan into Gippsland from the north, settling stations at Omeo, but McMillan's exploration to the coast, together with the identification of a site for a port by the men of the Gipps Land Company, provided the access to markets necessary to sustain a pastoralist industry.

Tension between private and public interest underpinned the European settlement of Gippsland from the beginning. The public acclaim extended to Strzelecki, the international explorer who had the sanction of Gipps on his arrival in Melbourne, contrasts with McMillan's early silence on his exploratory trips. As a result the official European naming of rivers, mountains and the region was adopted from Strzelecki's expedition rather than the earlier inscriptions chosen by McMillan. Despite the evidence that Gippsland was settled from the north by the Monaro squatters it was the speculative venture of the Melbourne-based Gipps Land Company on the *Singapore* that led to the founding of the port and trade networks that dominates accounts of this period.

The special surveys

If official claims of discovery in 1841 were about pastoralism and trade, then the coincidence of a decision made by the Colonial Office in London to set a

---

[14] W.A. Brodribb, 1978. *Recollections of an Australian Squatter 1835–1883.* 2nd edition, Sydney: John Ferguson, 24.

[15] John Adams (ed.), 1990c, *Notes on Gippsland History by Rev. George Cox Vol. 1.* Port Albert: Port Albert Maritime Museum, 1.

fixed, or uniform, price on land sales in the Port Phillip District, known as 'special surveys', provided the mechanism for private rather than government settlement to shape the character of early Gippsland. This short-lived scheme, introduced in February 1841 and withdrawn in August of the same year, was not favoured by either Governor Gipps or by the Superintendent of Port Phillip District, Lieutenant Charles La Trobe.[16]

Designed to encourage settlement away from major centres, special survey purchasers could apply for 5120 acres, or 8 square miles, for a fixed price of one pound per acre without an auction, on the condition they establish a town, complete with churches and a school.[17] Special surveys were to be located more than 5 miles from the coastal ports and settlements in the Port Phillip District, listed as Melbourne, Geelong and Portland. In applying this regulation, Gipps asserted his power to retain control over shipping and trade in the public interests of the colony. By this means, Gipps sought to limit the influence of land speculators using the special survey legislation to take up prime land at very low cost.

The settlement at Port Albert, of little more than a few huts, was founded after the special survey legislation was drafted and therefore not listed by Gipps as a shipping port to be protected from sale. It was this omission that the Gipps Land Company hoped to exploit by securing the land on the waterfront, including the port site, and which Gipps acted unsuccessfully to forestall. His letter instructing La Trobe to reserve a township site at Port Albert of 'not less than 20 or 25 square miles', despite the protests of the 'claimants', failed to reach Melbourne before Brodribb, of the Gipps Land Company, departed for Port Albert on board the *Isabella* with the surveyor, TS Townsend.[18] La Trobe had already acquiesced to the Company's request for the coastal site where they had landed, on the condition that a much smaller area of 640 acres was set aside for a government town.[19] On his return to Melbourne aboard the *Singapore* in April 1841, John Orr had applied to La Trobe for a special survey on behalf of shareholders, including most members of the Gipps Land Company.[20] John Reeve, a Monaro pastoralist, also applied for land at Port Albert with a plan to subdivide

---

[16] A.G.L. Shaw, 1989. *Gipps-La Trobe Correspondence 1839–1846*. Melbourne: Miegunyah Press, 77.

[17] Adams, 1990c, *Notes on Gippsland History*, 15.

[18] Shaw, *Gipps-La Trobe Correspondence 1839–1846*, 77.

[19] Brodribb, *Recollections of an Australian Squatter*, 45.

[20] Adams, 1990c, *Notes on Gippsland History*, 15.

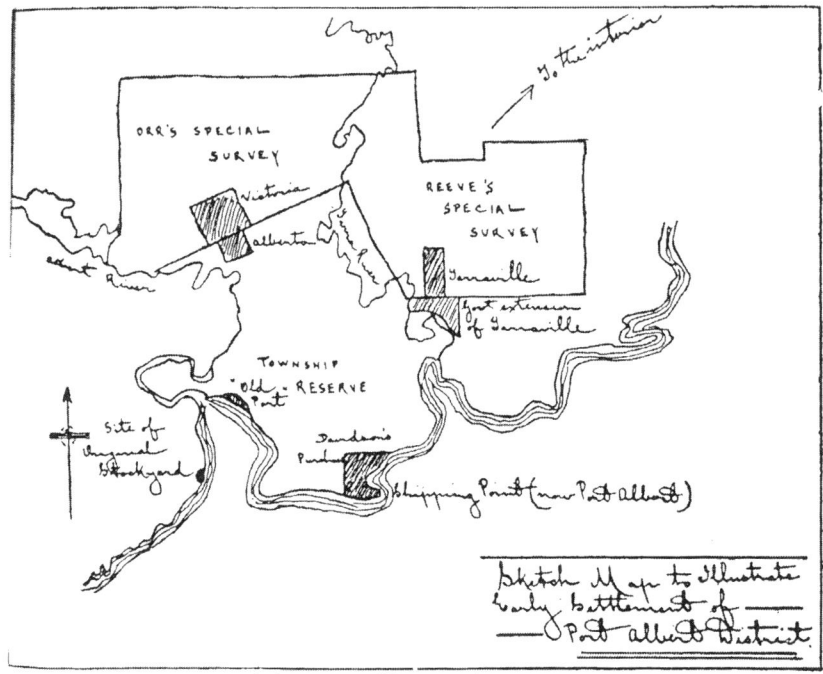

Figure 5.1: Map of Port Albert.
Port Albert Maritime Museum.

it into smaller blocks. A third application by William Rutledge was later exchanged for land at Warrnambool.[21]

Gipps despatched urgent directions to Townsend via *The Brothers*, which sailed from Sydney in July 1841, directing the surveyor to refrain from selecting the government township site, or the special surveys, and instead to map the rivers and coastline of Corner Inlet. Gipps had asserted his right to reserve land for ports and townships, causing considerable delays for the pastoralists and land speculators arriving at Port Albert. His actions forced Orr and Reeve to select land 5 miles inland from the coast, adjacent to the Albert and Tarra Rivers, inadvertently granting them the best agricultural land near the port.[22] Figure 5.1 shows the special surveys in relation to the original huts erected by the Gipps Land Company at Old Port and what would become the contestable land at Shipping Point, now Port Albert.

---

[21] Adams,1990a. *From These Beginnings: History of the Shire of Alberton (Victoria)*. Yarram: Alberton Shire Council, 10.

[22] *Gippsland Times*, 23 November 1872.

In 1842, Townsend returned to the settlement to identify a site for the township and survey the road to the north as far as Omeo. Lack of fresh water on the coast itself led Townsend to propose a township on an inland site, now Alberton. As a result, the swampy foreshore land, including Shipping Point, was made available for sale, ostensibly to Benjamin Boyd who had requested the 180-acre block. Boyd was negotiating with the NSW Government, which had oversight of land sales in the Port Phillip District, for the purchase of the land at auction to expand his proposed coastal trading enterprise.[23] The former London stockbroker had convinced the Colonial Office to support his scheme to develop the steamship trade along the east coast of Australia and he was 'promised every facility of assistance' by them to purchase five or six coastal properties.[24]

As a founding director of the London-based Royal Bank of Australia, Boyd raised investment capital to provide the financial backing for his venture. By the time of his arrival in Sydney, in July 1842, the trading routes along the east coast were already taken, but an opening was created by the NSW Government on the southern route from Port Jackson to Port Phillip via Launceston.[25] Boyd's paddle steamer, *Seahorse,* operated from August 1842 until damaged irreparably when she hit a rock on the Tamar River in June 1843. Twofold Bay, located 560 kilometres south of Sydney, had become a regular stopping place for the *Seahorse* by March 1843, and Boyd began to develop Boydtown and his port.[26] Port Albert was to be the second site along the coastal trade route of his dreams.

Despite negotiations between Boyd and the NSW Government, who put the land at Port Albert up for auction specifically for him, the sale scheduled for August 1843 was cancelled as a result of the London-based decision to locate Gippsland under the administration of Port Phillip District and La Trobe. Boyd's agent failed to arrive at the auction on time when it was relocated from Sydney to Melbourne at short notice in September 1843 and held quietly at the Survey Office in Melbourne, the only land for Gippsland auctioned on the day. A brief letter from WH Buckley to the Surveyor General in Sydney later expressed 'the far from satisfactory' outcome of the auction held in the absence of Surveyor Hoddle, and cited the location

---

[23] Adams, 1990d, *Notes on Gippsland History by Rev. George Cox Vol. 1.* Port Albert: Port Albert Maritime Museum, 41.

[24] G.P. Walsh, 1968. 'Benjamin Boyd'. In *Australian Dictionary of Biography 1788–1850* edited by Pike, Douglas. Melbourne: Melbourne University Press, 140.

[25] H.P. Wellings, 1950. *Benjamin Boyd in Australia (1842–1849).* Sydney: s.n., 4.

[26] Ibid., 36.

as a factor, 'this office being situated at the extreme end of the town the attendance of purchasers is much smaller than if the Sales [sic] were held as formerly at the Government's Auctioneer's Rooms'.[27]

While the NSW government favored Boyd and, by extension, ongoing links between Gippsland and NSW, the actions of La Trobe and the Gipps Land Company had effectively stymied these plans. Victoria's new status as a self-governing colony enabled the influential Melbourne merchants involved in the Gipps Land Company to seize the opportunity to influence land sales.

In 1843, Boyd was at the beginning of his rise to power, and his interest in Port Albert as a base for his coastal trading enterprise was due to two factors. Firstly, in 1842, the commissariat responsible for supplying the convict settlements in Van Diemen's Land was unable to source sufficient fresh meat locally and began to import livestock from the other colonies. Secondly, proximity gave Port Albert an economic advantage compared to the ports of Twofold Bay, Melbourne and Geelong, which were all trading into Van Diemen's Land at this time.[28] Port Albert was only three days' sailing to Hobart, allowing livestock to arrive in good condition still fat from the grasslands of central Gippsland which were proving suitable for breeding the much sought after animals. In 1842, an influx of settlers from NSW moved into the northern Gippsland settlements to take advantage of the Hobart trade, strengthening the political and economic links along the eastern coast.[29]

Boyd later rose to prominence in NSW through his expansion into pastoralism, which began from the time his steamships were operating in 1842.[30] He amassed more than a million hectares of land between the Riverina and the south coast of NSW, centred on the Monaro region where he held at least fourteen pastoral leases. By 1844 he had turned his attention to the whaling industry, operating several whaling boats successfully from Twofold Bay, and his former plans to build a coastal trading empire waned.

The failed plan to provide Boyd with the land at Port Albert, which would have resulted in a stronger north–south trade link to NSW, was to lead to the commercial acquisition of the region by the merchants from Melbourne. Major Alexander Davidson, newly arrived from military service in India,

---

[27] Adams, 1990d, *Notes on Gippsland History*, 27.

[28] Wayne Caldow, 2003. 'The Commercial Inn, Tarraville; Maritime Trade and Early Settlement of Gippsland'. *Victorian Historical Journal* 74 (1), 26; Lennon, 'Squatters, merchants and mariners', 32.

[29] Adams, 1990d. *Notes on Gippsland History*, 32.

[30] Wellings, *Benjamin Boyd*, 5.

proved to be the only bidder, purchasing the swampy land for a fixed price of one pound per acre. Davidson, a 'dummy' for Turnbull, Orr & Co., was living with Orr at the time.[31] Orr, who operated his mercantile business from Market Street in Melbourne, claimed to a Select Inquiry into Port Albert in 1857 that he 'bought it of [sic] the Government at a pound an acre', confirming on whose authority Davidson acted.[32] Davidson's purchase ensured the landlocked special surveys of Orr and Reeve retained their privileged access to the sea for shipping livestock.[33]

It seems the Sydney-based Gipps, alerted to the creation of this monopoly, attempted to stop the sale from proceeding by withdrawing the land after the auction, but was unable to intervene.[34] There can be no certainty about the reasons behind Davidson's successful purchase, although several factors may have contributed. The difficulties of communication between the colonies played a part. Equally, the economic depression may also have contributed to the lack of interest from speculators, as 'money was scarce in Melbourne that year'.[35] Tensions between settlers in the Port Phillip District and the NSW Government over the belief they were being starved of rightful funds drove separatist agendas at the time.[36] The Melbourne-based merchants, having won access to the wealth of land in Gippsland with its addition to Port Phillip District, now acted to secure the prime site at the port.

Does Gippsland belong to NSW or Port Phillip?

The competition between Boyd (with his Sydney ties) and the Gippsland Company (with their Melbourne base) reflected tensions between the two cities as the political movement for separation of the colonies grew in urgency. This became more complex as decisions made by the Colonial Office in London about administrative boundaries were often at odds with the growing knowledge of local officials such as Gipps, La Trobe and the surveyors. Moves towards separation of Port Phillip District from NSW, and debates about boundaries during the 1840s, saw Gippsland shift between

---

[31] Adams, 1990d. *Notes on Gippsland History*, 27.
[32] Victoria Parliament Legislative Assembly, 1857. *Select Committee Upon Shipping Point, Port Albert*. Victorian Parliamentary Papers, Melbourne: Victorian Parliament, 18.
[33] Adams, 1990d. *Notes on Gippsland History*, 27.
[34] *Gippsland Times*, 23 November 1872.
[35] Greig, 'The Beginnings of Gippsland', 63.
[36] Shaw, *A History of Port Phillip District*, 153.

the two administrations. In 1840, Gippsland was a part of Monaro, as was the south coast of NSW, which was in keeping with the vision of Governor Bourke in the mid 1830s to link 'the eastern ports with the western pastures'.[37] By August 1841 the tensions between Port Phillip and Sydney over the potential wealth of the Gippsland livestock trade were reported in the press, with the *Port Phillip Herald* arguing 'If the Melbournites do not secure this country to themselves now, the Maneroo settlers will most assuredly take possession of it for fattening their superabundant stock'.[38]

In September 1841 the Colonial Office in London, intent upon dividing the colony of NSW into manageable districts, included 'the whole of the coast as far as Twofold Bay in the district of Port Phillip', setting the boundary along the Murrumbidgee River with a border line straight across to Cape Howe.[39] Gipps continued to argue that Gippsland belonged in the 'Middle' or Sydney District, and thus revenue raised from land sales must go to Sydney.[40]

In February 1842 the Melbourne-based colonists looked to nature to define the boundary between Port Phillip and NSW, with the *Port Phillip Patriot* arguing, 'if the great Worrogong Chain, or Australian Alps, had extended in one unbroken line to Wilsons Promontory' it would form a 'proper boundary', but 'its continuity is interrupted' by 'a comparatively level tract of country, although thickly covered to some extent with an impenetrable scrub, from Gipps Land to Melbourne'.[41] This early description of the Latrobe Valley route as 'traversable' was disputed by the pastoralists attempting to reach the Melbourne market. William Brodribb reflected on his overland journey in March 1841: 'we had entirely failed to find a good overland route by which stock could be taken to the new country'.[42] He claimed that a government and several private parties had all 'returned unsuccessful'. The advocates for separation sought to secure the wealth of trade in Gippsland for Melbourne at a time when travelling north into Monaro was the only viable overland route.

Despite the difficulties in accessing the region from the west, Gippsland was gazetted as a part of Port Phillip District on 1 July 1843 and proclaimed one of four districts for the administration of Land Laws in

---

[37] W.K. Hancock, 1972. *Discovering Monaro; A Study of Man's Impact on his Environment.* Cambridge: Cambridge University Press, 7.

[38] Greig, 'The Beginnings of Gippsland', 56.

[39] Adams, 1990e, *Notes on Gippsland History by Rev. George Cox Vol. 1.* Port Albert: Port Albert Maritime Museum, 45.

[40] Ibid.

[41] Ibid.

[42] Brodribb. *Recollections of an Australian Squatter,* 44–45.

August.[43] In September, La Trobe appointed Charles Joseph Tyers as Crown Lands Commissioner for the region. The three-year delay between the first settlers arriving at Port Albert in January 1841 and the presence of government officials in December 1843 allowed the pastoralists and merchants to establish their networks and authority, a factor which would make the role of government more difficult.

Tyers was initially given the task of opening a dray track from Melbourne to Alberton, but after several attempts he was driven back by the environment; the swamps, forests and mountains.[44] He eventually travelled to Port Albert by sea, arriving in January 1844. Within months, two expeditions – a government party led by GA Robinson, the Chief Protector of Aborigines of Port Phillip District and the other by the pastoralist, Hobson, with 600 head of cattle – opened a dray and cattle route along the coast from Westernport.

La Trobe continued to seek a route for overland communication to Port Albert, noting that Tyers was required to send mail on ships sailing via Van Diemen's Land, taking three to four weeks to reach Melbourne.[45] La Trobe journeyed into Gippsland following the coastal route, arriving at Corner Inlet on 4 March 1845. On his return, La Trobe advised Walsh, in charge of police in Gippsland, to find an alternative road, and between 1845 and 1847 several attempts, including one by Captain Dana, were made to construct what would eventually become the Latrobe Valley route.[46] La Trobe's report to Gipps in December 1847 expressed his concern that 'every day the necessity of a certain mode of communication with Gipps Land becomes more and more evident'.[47] His sense of frustration at the difficulty of governing Gippsland was tangible.

Establishing the link from Melbourne remained a problem, and historian Patrick Morgan suggests that it never became strong.[48] The decisions made elsewhere to determine political boundaries cut across an older connected landscape along the coast.

---

[43] Adams, 1990d, *Notes on Gippsland History*, 42; Adams, 1990a, *From These Beginnings*, 14–15.

[44] Adams, 1990f, *Notes on Gippsland History by Rev. George Cox Vol. 1*. Port Albert: Port Albert Maritime Museum, 15–16.

[45] Cuthill, 1991. 'The Gipps Land Road 1836–1848 Part 2'. *The Gippsland Heritage Journal* 10, 22–23.

[46] Ibid., 23.

[47] Ibid., 26.

[48] Patrick Morgan, 1997. *The Settling of Gippsland: A Regional History*. Leongatha: Gippsland Municipalities Association, 47.

The Turnbull brothers and the private ownership of Port Albert

The role of the government in Gippsland was hampered from the outset by the private control of the port, the main point of entry and trade for the region. Within months of Davidson's dummy purchase, the Turnbull's shifted their buildings from the Old Port, or original settlement, onto the private land at Shipping Point, a move which coincided with the arrival of Commissioner Tyers in 1844. The dissolution of the Turnbull, Orr & Co. partnership in 1845 resulted in the ownership of Port Albert shifting to the Turnbull brothers, Robert and David, trading as Turnbull & Co.[49] According to Rev. George Cox these were 'the extraordinary transactions by which the only convenient port at that time in the whole of Gippsland passed out of the hands of the government and into the possession of private individuals'.[50] One consequence was that rather than cementing ties with Sydney through Boyd and via Twofold Bay, Gippsland would increasingly be linked to Van Diemen's Land and Melbourne through the operation of the Melbourne-based company of the Turnbull brothers.

There is little doubt that the Turnbull brothers ventured into Gippsland as merchants. John Fowler Turnbull migrated from Scotland to New Zealand before moving to Van Diemen's Land and then Port Phillip in 1839. There he was joined by younger brother, Robert, and later by George, Patrick, David and Phipps, all of whom participated in various business partnerships and pastoral leases in Gippsland.[51] Robert Turnbull, who sailed from Leith (UK) to Hobart with a cargo of goods, changed his plans on arrival and moved to Melbourne in January 1840, where he had formed a partnership with John Orr in a mercantile business.[52] John Turnbull was a founding partner of the Port Albert Company, which had been re-formed under Orr from the Gipps Land Company. In 1842, John Turnbull was already 'building several large stores' to receive wool from Gippsland at Old Port, the original settlement of Port Albert, located a kilometre west of the current township. When he returned to pastoralism in 1846, taking up the *Loyang* and *River Tyers* runs at Traralgon, the management of the port and shipping trade was handed to Robert, then 25 years old.[53]

---

[49] Adams, 1990a. *From These Beginnings*, 14.
[50] Adams, 1990e. *Notes on Gippsland History*, 1.
[51] Gwen O'Callaghan, 2006. *Clonmel to Federation: Guide to People in the Port Albert Area 1841–1901*. Port Albert: self-published, 694–697.
[52] *The Argus*, 22 November 1872.
[53] R.V. Billis and A.S. Kenyon, 1974. *The Pastoral Pioneers of Port Phillip*. Melbourne: Stockland Press, 152.

Figure 5.2: Robert Turnbull.
Port Albert Maritime Museum.

The Turnbull brothers gained control of an immense resource base. Between 1843 and 1857, no land was sold at Port Albert and no competition encouraged and, according to Don Watson, 'not a man in Gippsland was not on the company books'.[54] The Turnbull's owned the hotel, bond store and stockyard, leasing them to managers like John Campbell, a pastoralist who arrived at Old Port with the Gipps Land Company, or operating them in partnership with men like John Hawden, a shareholder on Orr's survey.[55]

---

[54] Don Watson, 1984. *Caledonia Australis; Scottish Highlanders on the Frontier of Australia.* Sydney: Collins, 191.

[55] Adams, 1990c. *Notes on Gippsland History*, 15, 43; Adams 1990e, *Notes on Gippsland History*, 14.

The Turnbulls charged pastoralists for the use of their stockyards and wharf, offering half price for those using a ship owned by their company. While taking advantage of their monopoly, these measures increasingly angered the locals.[56]

For a period before 1850 the Turnbulls leased all the land lying on either side of the entrances to Port Albert and Corner Inlet. Robert Turnbull leased land on Wilsons Promontory, including Sealers Cove and the Singapore Peninsula. They held onto the lease that included Refuge Cove until the 1870s, ensuring their ships could shelter from the strong Bass Strait winds in the only cove that was fully protected from all directions. George and Patrick Turnbull took out the first pastoral lease on Snake Island and Sunday Island from 1847 when Land Acts offered some tenure to pastoralists, although the Turnbulls most likely squatted before this.[57]

The Turnbulls became the central agents for an interconnected pastoral industry which depended on cheap land and labour, coastal trading ships for transport and, above all, an ability to navigate the shallow waters and shifting sands of Corner Inlet. The coastal environment had proved challenging for shipping from the beginning and ships' captains were not always prepared to cross the bar at the Port Albert entrance. Although initially a high number of ships ran aground or were wrecked on the banks, the introduction of a steam-driven tug in 1854 made it possible for pastoralists to continue their established practices.

In practice, pastoralists drove livestock, both sheep and cattle, from the central plains of Gippsland and onto coastal holding properties, including land on Orr's or Reeve's special survey, while waiting for ships to Hobart.[58] The road from Gippsland ran through this survey and the private settlement was a haven for labourers, ex-convicts and others. In response to the initial attempt by Tyers to exercise authority over the settlers at Old Port and collect unpaid land taxes, most of the residents relocated to the village on Reeve's survey, an area that would become known as St Giles, after the neighbourhood in London known for the notoriety of the people living there.[59]

An understanding of the magnitude of the livestock market is evident in the report that Angus McMillan alone shipped 900 head of cattle to

---

[56] Victoria Parliament Legislative Assembly 1857, 36.
[57] Adams, 1990e. *Notes on Gippsland History*, 10.
[58] Caldow. 'The Commercial Inn, Tarraville', 28.
[59] Adams, 1990a, *From These Beginnings*, 13.

Hobart in 1842.[60] Between March 1842 and the end of February 1843, 25 ships traded into Port Albert.[61] In 1843, 76 vessels were cleared out of Port Albert, two-thirds destined for Van Diemen's Land.[62] Bass Strait became 'a highway' between the colonies of Port Phillip and Van Diemen's Land as Gippsland pastoralists became an integral part of the latter's convict economy.[63] Gippsland regional historian Patrick Morgan refers to this as the *Vandemonian Trail*, along which ex-convicts, bringing with them the hardened attitudes of their experiences, secured work with Hobart merchants holding runs in Gippsland, such as John Johnson and John Foster.[64]

The livestock trade peaked in 1847 and continued with some strength during the 1850s. In terms of the value of trade from Port Albert during the 1840s, Hobart was the most important destination, with three-quarters of total exports being of livestock sent to Van Diemen's Land, while wool shipped to Sydney was the second-most important.[65] The initial commercial connection to Sydney held by wool-producing Monaro pastoralists soon weakened.[66] The discovery of gold in central Victoria opened new markets through Melbourne and, with a burgeoning demand for timber, the Turnbulls turned their attention to the forests on Wilsons Promontory and Corner Inlet. They were shipping blackwood staves and building materials from Sealers Cove to Melbourne from 1849 in partnership with William Buchanan.[67] In 1856 they purchased established mills on Corner Inlet and consolidated these into a single enterprise at Muddy Creek (Toora), using their own ships to transport timber to Melbourne for major infrastructure projects across the state.

It is easy to underestimate the anticipated role Port Albert, as the pivotal site for Gippsland, was expected to play in the future of the colony at this time. An indication may be found in the 1859 refusal by the Victoria Government to grant Joseph Thomson, the lessee of Snake Island, a pre-

---

[60] John Wells, 1980, *More Colourful Tales of Old Gippsland*. Melbourne: Rigby Publishers Limited, 128.
[61] Lennon. 'Squatters, Merchants and Mariners', 102.
[62] Ibid., 103.
[63] James Boyce. 2008. *Van Diemen's Land*. Melbourne: Black Inc., 244; Caldow. 'The Commercial Inn, Tarraville', 28.
[64] Morgan. *The Settling of Gippsland*, 75.
[65] Lennon. 'Squatters, Merchants and Mariners', 138.
[66] Ibid., 138–139.
[67] Adams. 1990a, *From These Beginnings*, 49.

emptive right to 640 acres of freehold land on the island on the basis that it may be required for fortifications in the future.[68] The number of ships entering Corner Inlet prompted the appointment by the Victorian Government of a pilot and customs officer, Charles Peterson, to Biddys Cove on Wilsons Promontory in 1859. A lighthouse was constructed on Snake Island in 1859. A short-lived market with New Zealand in 1864 saw large ships at Port Welshpool, but the halcyon days of the pastoralists and their port were ending.

Environmental problems associated with shifting sands at the entrance contributed to the decline of Port Albert. By 1864, the Snake Island lighthouse was moved back from the encroaching coastline, while on the other side of the entrance, on Clonmel Island, plans for a lifeboat and boatshed were complicated as the island was rapidly washing away. In October 1865 the signal station was blown down in a storm.[69] Further along the coast, the opening of the Gippsland Lakes to shipping in 1864, a closer port to the central Gippsland pastoralists, and the improved Latrobe Valley road, curtailed livestock shipping from Port Albert.[70] The development of an east–west axis for trade and transport resulted in the relative isolation of Port Albert, and the district declined.

The influence of the Turnbulls was also weakening. The inquiry conducted by the Victorian Parliament in 1856 responded to pressure from pastoralists to establish a free port at Welshpool and break the monopoly on Gippsland trade held by the Turnbulls.[71] It was evident that the pastoralists preferred Port Albert because of the closer proximity to their stations further north. As a result of the inquiry, Robert Turnbull sold a small portion of the waterfront back to the government in 1857 for £2500, compared to the £2 he paid in 1843. On Turnbull's nomination by the pastoralists as the only candidate for election to the Victorian Legislative Council for Gipps Land Province in 1851, the Melbourne *Argus* commented:

> Here is a person who has stood in the gap of our rights with a whole host of mercenaries and hangers-on at his back to oppose us in the acquisition of a free Gippsland port, now coming forward as our

---

[68] PROV Lands Department VPRS, Unit 1157, Pastoral Run Papers, Fiche 726, Snake Island Run.
[69] *Gippsland Guardian,* 13 October 1865.
[70] Morgan. *The Settling of Gippsland*, 43.
[71] Adams, 1990f, *Notes on Gippsland History*, 36.

representative under the banner of free trade and at the same time holding the sole monopoly of our Port.[72]

On news of his death in 1872, the *Gippsland Times* also commented on the way in which the private ownership of Port Albert had caused the decline of the district. It observed that 'the connection was an unfortunate one for the district in many respects, no one can deny but it was mistaken policy not the man which experience condemns'.[73] As a result of the land ownership by Reeve and Orr through the special surveys and the Turnbulls at the port, three townships developed on private land: Tarraville, Victoria and Port Albert. Alongside these, the surveyed township of Alberton and the administrative centre adjacent to the port, Palmerston, were built. According to Cox, this fracturing of the economic and social wealth of the district due to private ownership was to impact on the prosperity of the district for future decades. Robert Turnbull and the private port were major factors in the changing relationships between people and their use of the environment.

It seems unlikely, despite the Rev. Cox's belief, that Boyd's influence at the time that Gippsland was included in the Port Phillip District would have been great enough to have overcome the growing influence of the Melbourne market. Improved access to Melbourne and the ability of Europeans to apply technology to overcome the natural barriers at the same time as they opened the shipping entrance to the lakes served to isolate the Port Albert district. As successive Land Acts stripped back the size of their leases after 1862, many of the pastoralists selected their central Gippsland properties for permanent residency, letting go of property at Port Albert. The township at Sale became the commercial centre for Gippsland trade and the withdrawal of capital investment and influential citizens from Port Albert contributed to its decline. The shift in the economic and political axis from north–south to east–west signalled a change in the deep-time relationships between people and the environment. The role of the Melbourne-based merchants and their ability to capitalise on the shipping trade in the early years of settlement was a significant factor in the early development of Gippsland.

---

[72] *The Argus*, 30 August 1851.
[73] *Gippsland Times*, 23 November 1872.

References

Adams, John., 1990a. *From These Beginnings: History of the Shire of Alberton (Victoria)*. Yarram: Alberton Shire Council.
Adams, John (ed.), 1990b. *Notes on Gippsland History by Rev. George Cox Vol. 1*. Port Albert: Port Albert Maritime Museum.
Adams, John (ed.), 1990c. *Notes on Gippsland History by Rev. George Cox, Vol.II*. Port Albert: Port Albert Maritime Museum.
Adams, John (ed.), 1990d. *Notes on Gippsland History by Rev. George Cox Vol III*. Port Albert: Port Albert Maritime Museum.
Adams, John (ed.), 1990e. *Notes on Gippsland History by Rev. George Cox Vol. IV*. Port Albert: Port Albert Maritime Museum.
Adams, John (ed.), 1990f. *Notes on Gippsland History by Rev. George Cox Vol. V*. Port Albert: Port Albert Maritime Museum.
The Argus, 1872.
Billis, R.V. and Kenyon, A.S., 1974. *The Pastoral Pioneers of Port Phillip*. Melbourne: Stockland Press.
Blay, John, 2005. *Bega Valley Path Ways and Trails Mapping Project, Bega Valley Heritage Study, Public Version*, Bega: NSW National Parks and Wild Life Service and Bega Valley Shire Council.
Boyce, James, 2008. *Van Diemen's Land*. Melbourne: Black Inc.
Brodribb, W.A., 1978. *Recollections of an Australian Squatter 1835–1883*. 2nd Edition, Sydney: John Ferguson.
Broome, Richard, 2005. *Aboriginal Victorians: A History Since 1800*. Crows Nest: Allen & Unwin.
Caldow, Wayne, 2003. 'The Commercial Inn, Tarraville: Maritime Trade and Early Settlement of Gippsland'. *Victorian Historical Journal* 74 (1), 23–45.
Cuthill, W.J., 1991. 'The Gipps Land Road 1836–1848 Part 2', *The Gippsland Heritage Journal* 10, 21–27.
Division of Crown Land Management, 1980. *Corner Inlet/Seaspray Coastal Study Vol. 1*. Resource Document, Melbourne: Department of Crown Lands and Survey.
Flood, J., 1980. *The Moth Hunters of the Australian Capital Territory: Aboriginal Traditional Life in the Canberra Region*. Canberra: Australian Institute of Aboriginal Studies.
Gippsland Guardian, 1865.
Gippsland Times, 1872.
Greig, A.W., 1912. 'The Beginnings of Gippsland Part 1'. *The Victorian Historical Magazine* 2 (1), 1–14.
Hancock, W.K., 1972. *Discovering Monaro: A Study of Man's Impact on his Environment*. Cambridge: Cambridge University Press.
Haydon, G.H., 1846. *Five Years Experience in Australia Felix*. London: Hamilton Adams & Co.
Le Cheminant, Marion, 1988. 'Who Discovered the Gippsland Coast and Bass Strait?' *Gippsland Heritage Journal* 1 (3), 14–17.
Lennon, Jane, 1975. 'Squatters, Merchants and Mariners: An Historical Geography of Gipps' Land 1841–1851'. Melbourne: MA Thesis, University of Melbourne.
Morgan, Patrick, 1997. *The Settling of Gippsland: A Regional History*. Leongatha: Gippsland Municipalities Association.
O'Callaghan, Gwen (ed.), 2006. *Clonmel to Federation: Guide to People in the Port Albert Area 1841–1901*. Port Albert: self-published.
PROV Lands Department VPRS, Unit 1157, Pastoral Run Papers, Fiche 726, Snake Island Run.

Shaw, A.G.L., 2003. *A History of the Port Phillip District: Victoria Before Separation*. Melbourne: Melbourne University Press.
Shaw, A.G.L. (ed.), 1989. *Gipps-La Trobe Correspondence 1839–1846*. Melbourne: Miegunyah Press.
Victoria Parliament Legislative Assembly, 1857. *Select Committee Upon Shipping Point, Port Albert*. Victorian Parliamentary Papers, Melbourne: Victorian Parliament.
Walsh, G.P., 1968. 'Benjamin Boyd'. In *Australian Dictionary of Biography 1788–1850*, by Douglas Pike, editor. 140–142. Melbourne: Melbourne University Press.
Watson, Don, 1984. *Caledonia Australis: Scottish Highlanders on the Frontier of Australia*. Sydney: Collins.
Wellings, H.P., 1950. *Benjamin Boyd in Australia (1842–1849)*. Sydney: s.n.
Wells, John, 1980. *More Colourful Tales of Old Gippsland*. Melbourne: Rigby Publishers Limited.

Chapter 6

Northern Gippsland Aboriginal associations with country
The frontier period, 1834–1859

Ruth E. Lawrence

The Aborigines of northern Gippsland

The traditional Aboriginal occupants of the Omeo region of northern Gippsland have been variously named as 'the Kunora alias Gundanora tribe', 'the Yaymirttong', 'the Thed-dora', 'the 'Gundanora', and 'Ya-it-ma-thang'.[1] In 1904, anthropologist Alfred Howitt proposed that the Kandangora-mittung were a sub-set of the Yaitmathang, and in 1974 anthropologist Norman

---

[1] Today, the Omeo region of eastern Victoria is geographically linked to Gippsland. Defined by landscape characteristics as an area of undulating plains topography, the Omeo Plains range in elevation from 600 to 900 metres and are surrounded by higher topography to the east and north and the Bogong High Plains (1500–1900 metres) to the west. The absence of a major mountain range to the south is the main reason for Omeo's geographical connection to Gippsland. The headwaters of several rivers flowing into both the Gippsland Lakes to the south and the Murray River to the north rise in the Omeo region. The present-day township of Omeo is located in the south-western corner of the Omeo region and the township of Benambra is adjacent to the normally dry Lake Omeo that approximates the centre of the Omeo Plains.

A.E.J. Andrews, (ed.) 1979. *A Journey from Sydney to the Australian Alps Undertaken in the Months of February and March 1834 by Dr. John Lhotsky*. Hobart: Blubber Head Press; Robinson in I.D. Clark, 1998a, *The Journals of George Augustus Robinson, Chief Protector, Port Phillip Aboriginal Protectorate*, Volume 4: 1 January 1844 – 24 October 1845. Melbourne: Heritage Matters, 109; J Buntine, undated, Letter to A.W. Howitt, Sale, from John Buntine, Toongabbie, entitled 'Omeo tribe'. MS 9356, Box 1054/2, Manuscript Collection, State Library of Victoria; Wills in Victoria, Legislative Council, 1859, Report of the Select Committee of the Legislative Council on the Aborigines; Together with the Proceedings of Committee, Minutes of Evidence, and Appendices, Parliamentary Paper D19, John Ferres, Melbourne, 26; Notes from John Bulmer in Howitt manuscripts, MS 9356, Box 1054/2 (b), La Trobe Library, Melbourne.

Tindale mapped the Aborigines of the Omeo and adjacent areas to be the Jaimathang.[2] These names might reflect different ways in which Aboriginal people understood themselves, or may point to European difficulties in pronunciation and spelling. For the purposes of this chapter, the Omeo Aborigines of northern Gippsland are referred to as the Yaitmathang, in alignment with geographer Sue Wesson's mapping of Aboriginal groupings in south-eastern Australia and Ian Clark's language and social grouping reconstruction.[3]

According to Howitt and Clark the Yaitmathang Aborigines were closely associated with the Ngarigo Aborigines of the Monaro region of New South Wales, and the Walgalu people who lived in what is now the Canberra region.[4] In 1895 naturalist Richard Helms maintained that the Yaitmathang, Ngarigo and Walgulu were all friendly towards each other and had similar habits and customs, such as annual pilgrimages to feast on the bogong moth (*Agrotis infusa*).[5] The Gippsland region to the south of the Yaitmathang was occupied by the Gunaikurnai people, which consisted of many named groups and about which much is known.[6] There was a strong antagonism between the Gunaikurnai and Yaitmathang that was reported by many authors, which apparently represent traditional rivalries.[7] Alfred Howitt recorded that 'the old road from Omeo to Bruthen followed the

---

[2] A.W. Howitt, 1904. *The Native Tribes of South East Australia*. London: Macmillan, 77; N.B. Tindale, 1974, *Aboriginal Tribes of Australia: Their Terrain, Environmental Controls, Distribution, Limits, and Proper Names*. Canberra: Australian National University Press.

[3] S. Wesson, 2000. *An Historical Atlas of the Aborigines of Eastern Victoria and Far South-eastern New South Wales*. Monash Publications in Geography and Environmental Science No. 53, 62; I.D. Clark, 1996. *Aboriginal Language Areas in Victoria*. Melbourne: Victorian Aboriginal Corporation for Languages; I.D. Clark, 2009. 'Dhudhuroa and Yaithmathang Languages and Social Groups in North-east Victoria – A Reconstruction', *Aboriginal History*, 33, 201-229.

[4] Howitt, *The Native Tribes of South East Australia*, 78; Clark, 'Dhudhuroa and Yaithmathang Languages'.

[5] R. Helms, 1895. 'Anthropological notes', *Proceedings of the Linnean Society of New South Wales*, 20, 388; J. Flood, 1980. *The Moth Hunters: Aboriginal Prehistory of the Australian Alps*. Canberra: Australian Institute of Aboriginal Studies, 107 ff.

[6] Wesson, *An Historical Atlas*, 17 ff.; L. Fison and A.W. Howitt, 1880. *Kamilaroi and Kurnai*. Melbourne: George Robertson; Pepper, P. and DeAraugo, T. 1985. *What did Happen to the Aborigines of Victoria: Volume I: The Kurnai of Gippsland*. Melbourne: Hyland House; P.D. Gardner, 1993. *Gippsland Massacres*. Ensay: Ngarak Press; A. Campbell, and R. Vanderwal, 1994. *Victorian Aborigines: John Bulmer's recollections: 1855 – 1908*, Museum of Victoria.

[7] For example, see G.A. Robinson 1844, Journal, MLms A1236, Mitchell Library, Sydney, 711; Helms, 'Anthropological Notes', 388.

trail by which the Gippsland and Omeo blacks made hostile incursions into each other's countries'.[8] There were some major land disputes between the Yaitmathang and the Gunaikurnai, and a strip of debatable land in northern Gippsland was claimed by both the Yaitmathang and Gunaikurnai.[9]

This chapter provides an insight into northern Gippsland Aboriginal connections to their country during the first quarter century of European settlement. Firstly, associations of the Yaitmathang with explorers, squatters, miners and police of northern Gippsland are outlined. In contrast to common perceptions that Aborigines kept to the periphery of European settlement and watched in bewilderment or anger as their lands were occupied and transformed, there is evidence that the Yaitmathang accommodated European arrivals and attempted to incorporate the newcomers into their traditional societal structures. Secondly, the chapter examines Yaitmathang and European connections to country, and reveals vast differences in attitudes towards the north Gippsland environment.

Source information for this chapter is entirely European documentation, with all their associated and recognised inadequacies.[10] Particular emphasis has been placed on sourcing information from authors who had a professional interest in understanding traditional Aboriginal cultures, such as Victorian Chief Protector of Aborigines George Augustus Robinson. Reference is also made to materials produced by those who lived during or immediately after the years under examination, such as naturalists Alfred Howitt and Richard Helms. It is recognised that there are many gaps in the accessible knowledge base, and that no Yaitmathang voices have been found.

Yaitmathang land affinities during early European settlement

Guiding the 'explorers'

The Omeo region was one of the first areas in Gippsland to be explored and settled by Europeans. Pastoralists from the Monaro region in south-eastern New South Wales began to move south-west into the area in the

---

[8] Howitt, *The Native Tribes of South East Australia*, 78.

[9] P.D. Gardner, 1992. 'Aboriginal History in the Victorian Alpine Region', in *Cultural Heritage of the Australian Alps*, ed. B. Scougall, Proceedings of a symposium held at Jindabyne, New South Wales, 16-18 October 1991, Australian Alps Liaison Committee, 91.

[10] B. Attwood, 2011. 'Aboriginal History, Minority Histories and Historical Wounds: The Postcolonial Condition, Historical Knowledge and the Public Life of History in Australia', *Postcolonial Studies*, 14 (2), 171-186.

mid 1830s. A notable feature of this exploration phase was the involvement of Aborigines in European 'discoveries', as was the case in other parts of Australia. Johann Lhotsky was the first European to attempt to reach Omeo from the Monaro. It is not certain that he actually arrived in Omeo, but he did record in 1834 that the word 'Omeo' was an Aboriginal term describing a very extensive plain.[11] This information may have been gained from either Monaro or Omeo Aborigines and records the first Yaitmathang association of landscape features with place nomenclature.

The first Europeans to have definitively reached Omeo were graziers from the Monaro region looking for new pastures on which to establish grazing leases. In early 1835, George MacKillop travelled from Monaro to Omeo in the company of four men: an unnamed Aboriginal from the Monaro region, and John Pendergast, James McFarlane and George Livingstone. MacKillop recorded that the Aboriginal guide told him the climate at Omeo was bland all the year round.[12] Familiarity with the environment of the Omeo area by the unnamed Monaro Aborigine is consistent with the tribal liaisons between the Yaitmathang and Ngarigo.

In the late 1830s, more European graziers sought new pastures in the Omeo area. Two such men were Cody Buckley and Lachlan MacAlister who arrived there in 1839 and who 'first visited it [Omeo] by information of natives'.[13] Again, familiarity of the local or neighbouring Aborigines with the Omeo region assisted the explorers and graziers. There was a meaningful language exchange between the two cultures for this to happen, and the language exchange included spatial information relating to quality grazing country, suggesting a Yaitmathang appreciation of European economic values.

In 1839, Angus McMillan was sent from Monaro to Omeo by his employer Lachlan Macalister to find new grazing lands.[14] Because the Omeo region was already occupied, McMillan attempted to pass through Omeo to Gippsland in May 1839, but was thwarted in his efforts. He was accompanied by Jemmy Gibber – described as the chief of the Monaro people – but was forced to return to Omeo due to Gibber's fear of encountering

---

[11] J. Lhotsky, 1834. 'Australian Alps Expedition', *Sydney Gazette*, 15 April 1834, 3.
[12] G. MacKillop, 1836. 'Australian Colonies', *Quarterly Journal of Agriculture, Edinburgh*, 70, 162.
[13] Robinson in Clark, 1998a. *The Journals of George Augustus Robinson*, 105.
[14] Macleod in A. Skene and R. Brough Smyth, 1874. *Report on the Physical Character and Resources of Gippsland*. Melbourne: Government Printer, 7.

the Gunaikurnai Aborigines.[15] This was an example of the willingness of Aboriginal guides to assist the newcomers only into country over which they held some sovereignty.[16] McMillan established a temporary base at 'Numblamunjie' south of Omeo, but set out for Gippsland again in January 1840. On his second attempt to reach Gippsland, McMillan was accompanied by two Yaitmathang Aborigines: Cobbon Johnny (also known as Jargair or Hingebira) and Boy Friday (Mitinggong, alias Friday); three settlers from Omeo: Mathew Macalister, Angus Cameron and Thomas Bath; and an unnamed stockman.[17] The Yaitmathang successfully navigated their way into Gunaikurnai territory along what may have been an old war path, and must have felt safe penetrating hostile territory under cover of the protection of European guns. McMillan subsequently claimed discovery of Gippsland.[18]

Paul Edmund de Strzelecki followed McMillan from Monaro to Omeo to Gippsland in March 1840, and then found his way to the Port Phillip Bay region. The *Port Phillip Herald* reported 'the party consisted of Count Strzelecki, Messers McArthur and Riley, with servants, etc'.[19] One of the members of Strzelecki's party, who was obviously regarded as a servant, was an Aboriginal from Sydney named Charley Tarra.[20] As Strzelecki followed the same route to Gippsland as McMillan, Charley Tarra must have had the right to, (or acquired) both directional knowledge from, local Aborigines and safe passage courtesy of European firearm cover.

Once the explorers had established routes to and through Omeo with the help of the Aborigines, it was not long before the squatter settlers followed suit. John Pendergast established 'Omeo A' and James McFarlane

---

[15] McMillan in Thomas Bride, 1898. *Letters from Victorian Pioneers: Being a Series of Papers on the Early Occupation of the Colony, the Aborigines, etc., Addressed by Victorian Pioneers to His Excellency Charles Joseph LaTrobe, Esq., Lieutenant-Governor of the Colony of Victoria*, Government Printer, Melbourne, 203.

[16] F. Cahir, 2012. *Black Gold: Aboriginal People on the Goldfields of Victoria, 1850-1870*. Aboriginal History Monograph 25. ANU E Press and Aboriginal History Incorporated.

[17] Robinson in I.D. Clark, 2000. *The Papers of George Augustus Robinson, Chief Protector, Port Phillip Aboriginal Protectorate*, Volume 2: Aboriginal Vocabularies: South East Australia, 1839 – 1852. Melbourne: Heritage Matters, 205; McMillan in Bride, *Letters from Victorian Pioneers*, 204.

[18] Bride, *Letters from Victorian Pioneers*.

[19] P. Strzelecki, 1841. 'Report by Count Strzelecki', in *England, Parliamentary Papers Respecting New South Wales, Despatch from Sir G Gipps, Governor of New South Wales, to the Right Honourable Lord John Russell, &c, &c, Dated 28-9-1840*, Appendix C, Colonial Office, London, 2.

[20] Macleod in Skene & Brough Smyth, *Report on the Physical Character*, 8.

founded 'Omeo B' squatting runs in 1835. In 1839, Patrick Cody Buckley moved from Monaro to Omeo, and took up both the 'Benambra' and 'Tongiomunjie' runs. Angus McMillan founded the 'Numblamunjie' run (later 'Ensay') for Lachlan Macalister in late 1839.[21] In the early 1840s, two further stations were established in the Omeo region: Edward Crooke took up 'Hinnomunjie' station in 1841, and the 'Bindi' run was officially occupied in 1845. After the bushfire in 1851, the Grey brothers from Wangaratta took out four leases in the Cobungra area: 'Bundaramunjie', 'Cobungra', 'Bynomunjie' and 'Darbarlary'.[22]

Between 1835 and 1851, the pastoral country in the Omeo region was fully occupied by squatter settlers, allowing for the exploration of the surrounding country. An Aborigine from Cobungra, west of Omeo, again related spatial information to the Europeans which enabled the 'discovery' of the Bogong High Plains. Two stockmen named James Brown and John Wells, both employed by George Gray of Wangaratta, had been sent to find grass for their stock immediately after the 1851 bushfires. An Aboriginal man named Cor-vo-mung, alias Slarney or Larnie, directed them to the subalpine pastures of the spurs extending south-east from mounts Hotham and Cope, and Brown and Wells subsequently discovered the Bogong High Plains on their return journey to Wangaratta.[23] Cor-vo-mung effectively communicated his knowledge of country and its attributes to the Europeans, thus facilitating the economic progress of the newcomers.

It is evident that the European explorers were dependent on Aboriginal guides to find 'new' lands. In assisting the Europeans in their exploration and settlement of northern Gippsland, all Aborigines demonstrated a keen sense of place, intimately knew the many routes to and through the Omeo area, and were willing to relate that information to the Europeans who sought new lands. In turn, Aboriginal guides acquired status by their association with the newcomers. By travelling under European protection beyond their traditional boundaries, they gained access to enemy country thus extending both their traditional land associations and their sphere of influence in their own societies.

---

[21] A.M. Pearson, 1969. *Echoes from the Mountains and History of the Omeo Shire Council*. Omeo: Omeo Shire Council, 25.

[22] R. Spreadborough and H. Anderson, 1983. *Victorian Squatters*. Ascot Vale: Red Rooster Press, 40-45.

[23] S.G.M. Carr, 1962. 'The Discovery of the Bogong High Plains', *Proceedings of the Royal Society of Victoria*, 75, 285.

Assisting the squatters

As the European settlers moved into the Omeo region, it is interesting to note the nomenclature they used for the squatting runs they established. Without exception, the names of all the squatting runs in the Omeo region were of Aboriginal origin and, where the meaning is known, the nomenclature was always associated with place (Table 6.1). The suffix *mungie* was frequently used in place names, being a reference to the presence of fish or codfish in the streams.[24] It is noteworthy that the Yaitmathang shared their knowledge of local food resources with the settlers, thus transferring information vital to the survival of the newcomers. It is apparent that the initial European settlers interacted amicably with the Aborigines, learnt the Yaitmathang names and features of specific locations, and thought those terms appropriate names for their runs.

Table 6.1: Details of nomenclature of squatting runs in the Omeo area

Name of squatting run	First year of squatting run	Aboriginal name	Aboriginal meaning	Source of information on Aboriginal names
Omeo A	1835	Omeo or Karengo	Very extensive plains Unknown	Lhotsky 1834; Robinson in Clark 2000, 205
Omeo B	1835	Omeo Nabbimunjie	Very extensive plains Unknown	Lhotsky 1834; Robinson in Clark 2000, 205
Benambra	1839	Benambra	Unknown	
Tungie Mungee	1839	Tongiomungie	Place where fish are found in streams	Dawson 1858;1 Robinson in Clark 2000, 205
Numblamungie (later Ensay)	1839	Numblamungie	Place where fish are found in streams	Dawson 1858; Robinson in Clark 2000, 205
Hinnomunjie	1841	Enieomungie	Place where fish are found in streams	Dawson 1858
Bindi	1845	Bendi	Home of the 'Bindi-mittung' (*mittung* = a number or many)	Dawson 1858; Howitt manuscripts
Cobungra	1851	Karbungerer	Named for a chief living in the area	Robinson 1844 journal
Bundaramunjee	1857	Bundaramunjee	No fish in river	Neumayer 1869, 82
Bynomunjie (Binnomungie)	1860	Binnomungie	Place where fish are found in streams	Dawson 1858; Robinson in Clark 2000, 205
Darbarlary	1861	Darbarlary	Unknown	

---

[24] Robinson in I.D. Clark, 1998b. *Place Names and Land Tenure – Windows into Aboriginal Landscapes: Essays in Victorian Aboriginal History*. Melbourne: Heritage Matters, 109; George Neumayer, 1869. *Results of the Magnetic Survey of the Colony of Victoria Executed During the Years 1858-1864*, J. Schneider, Manheim, 82.

It is likely that the Yaitmathang Aborigines previously knew of the ways of the squatter settlers through their close association with the Ngarigo Aborigines. The Monaro area had been settled in the 1820s, so the Ngarigo Aborigines had observed the ways of the settlers for a decade prior to settlement at Omeo. Some of the early interactions between the Monaro Aborigines and the Europeans had not been friendly.[25] It is possible that European settlement in the Monaro region prepared the Aborigines for that which was to come at Omeo.

From a European settler perspective, the Yaitmathang did not appear overawed by the incursion of settlers into their territory. Cody Buckley reported that the 'natives were there peaceable' when they established the squatting runs in the 1830s.[26] Descendants of the Pendergast family also recorded that 'the first white settlers [on the Omeo Plains] found the blacks very friendly and cooperative'.[27] At least one early settler and a Yaitmathang Aboriginal woman had children together, but descendants are loath to acknowledge that Aboriginal association today. Initial European settlement at Omeo does not appear to have produced the bloody confrontations between the Europeans and Aborigines that occurred elsewhere. By contrast, the Yaitmathang were inclusive of the newcomer's arrival into their country.

During the 1840s, the settlers employed some Yaitmathang in the management of their squatting runs. In 1844, the Pendergasts at 'Omeo A' employed 'a black boy from Omeo' and Archibald Macleod at Ensay employed an Omeo Aboriginal named Joe.[28] In 1846, the Chief Commissioner of Lands in Gippsland, Charles Tyres, reported the following about the Aborigines of Omeo and Gippsland:

> During the shearing season some of these assist in washing sheep, receiving for their services about a shilling a day, which they generally lay out at the stores in tea, sugar, &c. A few of the more civilised, or such whose intercourse with Europeans has been of a longer duration, are found useful as stock-keepers and bullock-drivers, and receive

---

[25] Peterson in K. Hueneke, 1994. *People of the Australian High Country*. Canberra: Tabletop Press, 207, 265; J. Lingard, 1846. *A Narrative of the Journey to and from New South Wales, Including a Seven Year Residence in that Country*. Chapel-en-le Frith: J. Taylor & Co.

[26] Cited by Robinson in Clark, 1998a. *The Journals of George Augustus Robinson*, 105.

[27] J.V. Pendergast, 1968. *Pioneers of the Omeo District*. Melbourne: Riall Print, 2.

[28] Robinson in Clark, 1998a. *The Journals of George Augustus Robinson*, 108, 99.

for their labour rations, and wages somewhat less than are paid to European labourers.[29]

Within a decade of the arrival of the settlers, the Yaitmathang demonstrated an understanding of the economic basis of European culture. Not only did the Yaitmathang serve the Europeans on their squatting runs, they also demonstrated good handling skills of introduced stock and animal transport systems, and were rewarded with wages accordingly. It would appear that the Yaitmathang adapted their diets to accommodate European commodities acquired via the newcomer's economic system.

After initial European settlement, the Yaitmathang quickly learned English. It was recorded that by 1844 the 'Omeo Blacks' 'speak English fluently but depraved stockmen have taught them to blaspheme'.[30] They soon acquired European clothing and were particularly impartial to wearing boots.[31] They also became acquainted with horses and horse riding; for example, an Omeo Aboriginal named Bit-to-cort alias Billy Blue accompanied George Robinson to the Monaro region in 1844 and they 'put him on [the] police horse Punch'.[32] By 1845 at least two Yaitmathang had joined the Native Police Corps and rode their own horses.[33] The Yaitmathang generally made very good riders, as an O'Rourke family member noted in his memoirs:

> Jim Brindle the half caste was recognised the best rider of them all. He worked for the O'Rourkes most of the time. Some people recognise him as the man from the Snowy River.[34]

Within a decade of European arrivals, the Yaitmathang had learned their language, found useful employment, and accessed their modes of travel, suggesting a substantial level of Indigenous adaptation to the new

---

[29] Anonymous, 1875. 'The South Gippsland of the Past', *Gippsland Mercury*, 18 March 1875, 3.
[30] Robinson in George Mackaness, 1941. 'George Augustus Robinson's Journey into South-eastern Australia, 1844', *Journal of the Proceedings of the Royal Australian Historical Society*, 27, 326.
[31] Macleod in Skene and Brough Smyth, *Report on the Physical Character*, 9; Robinson in Clark, 1998a. *The Journals of George Augustus Robinson*, 110.
[32] Robinson in Clark, 1998a. *The Journals of George Augustus Robinson*, 111.
[33] *Native Police Corps, Day Book of the Native Police Corps, Narre Narre Warren, 1845–1853*, VPRS 90, Public Records Office, Melbourne; M.H. Fels, 1988, *Good Men and True: The Aboriginal Police of the Port Phillip District 1837 – 1853*, Carlton: Melbourne University Press.
[34] O'Rourke family miscellaneous papers and photographs, State Library of Victoria, Manuscripts Collection, MS 12608, Box 3435/7.

and pervasive culture. It seems that the Yaitmathang added European items as status symbols to their own value system. Those Yaitmathang who participated in European authority structures by becoming members of the Native Police were extending their sphere of influence and importance in the non-Indigenous culture; however, it is not known how this impacted on their standing within the Aboriginal community. The latter was predicated on their acquired riding skills, which were acknowledged by whites, although one of the legends of Australian folklore – the Man from Snowy River – denied this Aboriginal identity.

Living with the gold miners

Gold was officially discovered at Omeo in November 1851.[35] Initially, alluvial gold was worked in the upper Mitta Mitta River, Morass Creek, Livingstone Creek and tributaries. In July 1853, sluicing operations began in the elevated riverbeds, and in 1858 several quartz reefs in the vicinity of Omeo were discovered and worked. This activity saw an influx of Europeans and Chinese to the Omeo district, although the remoteness of Omeo from the major ports kept the numbers of miners lower than in other mining centres.[36] The population of miners at Omeo grew steadily throughout the 1850s. Aboriginal people assisted the miners navigationally and also helped settlers and miners experiencing hardship. For instance, assistance was provided in negotiating difficult terrain, while as an anonymous journalist recalled:

> On one trip coming back from Port Albert we [a mining party including John T. Reid] escaped [the steep and treacherous] Tongio Hill by coming up Swift's Creek and had a black fellow as a guide. Blacks were very numerous at Omeo then [mid 1850s].[37]

Other miners accessed Aboriginal pathways in their pursuit of new goldfields.[38]

---

[35] J. Flett, 1970. *The History of Gold Discovery in Victoria*. Melbourne: Hawthorn Press, 168.

[36] Pearson, *Echoes from the Mountains*, 48.

[37] Anonymous, 1908. 'Gippsland's First Gold: How it was Won', *Bairnsdale Advertiser and Tambo and Omeo Chronicle*, 25 July 1908, 4. For an example of the help given in navigating some difficult terrain in the Omeo area see G.C. Fead, c. 1880. *Notes on an Unsettled Life, Memoires of Life on the Victorian and New South Wales Goldfields, 1853-1860, Written c1880*, Manuscript held by Margaret Fead, 11 Esperance Street, Red Hill, ACT 2603.

[38] A. Massola, 1969. *Journey to Aboriginal Victoria: A Guide and History to the State's Aboriginal Antiquities*. Adelaide: Rigby, 148.

The discovery and mining of gold in the Omeo area resulted in a profound change to the Aboriginal population. Glimpses into Aboriginal interaction with the miners can be gleaned from the following stories. Omeo miner George Fead recorded interactions between Aborigines and the miners in the 1850s as follows.

> Travelling down the long sloping track which led to the diggings, I overtook Metoaka, King of the Omeo Blacks, returning from hunting. He was a frank, manly old fellow, much liked by the whites, over whom he claimed some kind of sovereignty as well, and welcomed me as one of his former subjects. As we walked along, his hand resting on the mare's neck, he told me, in his own way, of the changes that had lately taken place in his little world – of the erection of a bakery, a restaurant, and a public house, and with a merry laugh – what I already knew – that a number of white gin immigrants, candidates for domestic service, having arrived at Port Albert, a party of diggers and others had gone down and secured wives a few minutes or, at most, a few hours, after they had met them for the first time in their lives. Such marriages were, at that time, not uncommon…
>
> Shortly afterwards, the first local race meeting was held… Metoaka had been holding horses all day and was proud of it; his honest cheery voice could be heard during the races, urging on his favourites. A grog seller, who had made much money that day by far less reputable means, gave him that night a long drink, and in the morning the good old king was dead.[39]

This account illustrates a number of interesting points about this Yaitmathang elder. The mutual respect that Metoaka and the Europeans displayed towards each other is clearly evident. Metoaka was very friendly towards the Europeans, including those he had not seen for some time. He was portrayed as being fully aware of current affairs and accepting changes in a good-natured manner. It seems that he was not a fringe dweller to European society but participated in the lifestyle of the settlers. He spoke English well and had a good sense of humour. The fact that he 'claimed some kind of sovereignty' over the Europeans is indicative of his own identity with the Omeo area, his strong association with inhabitants of the Omeo area, and his people's inclusion of the settlers at Omeo. It also indicates an ongoing

---

[39] Fead, *Notes on an Unsettled Life*.

sense of Aboriginal ownership and authority that was not diminished by European settlement and which involved some form of reciprocity and exchange: he was recognised by the white settlers as a 'king'.[40]

Another miner who came to the Omeo gold fields in the mid to late 1850s was Frank Shellard. He told the story of 'a fancy dress ball on a grand scale' that was held in Omeo in the early days of mining (mid to late 1850s). The new European wives of the miners referred to by Meteoka hosted the ball. Some Aborigines attended the ball, as Shellard recalled:

> The ball was a great success: all the ladies were handsomely dressed in striking contrast with their partners, who were only 'cleaned up' for the occasion. However it was a grand sight for that part of the world, the coloured ladies enjoyed it amazingly. All the windows and doors were thrown open and the spaces filled with shining black faces. King Billy of Kabungara was in full dress with a red night cap, and a red woollen comforter around his neck, and a short white shirt with his brass breast plate, but that was as far as he was dressed, he had stopped at that. His Queen had on her fancy costume but she did not create such a sensation as her consort. The chief feature of the ball was when the King and Queen, Princesses and Princes gave a nature dance in the ball room, when they very much frightened their white sisters, for they had never seen anything like it in their lives. The great noise, and the scantiness of their costumes, with the absence of any shyness on the part of the dancers in exposing their limbs, fairly astounded their white sisters, and as no one could get out of the hall, as all the openings were jammed, the poor young wives had to sit out terror stricken and it will never be forgotten by them.[41]

It is evident that parts of the traditional Aboriginal lifestyle were still operating in the 1850s. At that time, there was a family of Aborigines still living at Cobungra: 'King' Billy, his wife, and sons and daughters. This family was seen by the miners to be associated with a specific place: Cobungra. The group had partially absorbed European ways, as they wore some clothing, but had not disbanded their traditional ways as they performed a traditional dance. The European reaction to the Aboriginal dance was multifaceted: on the one hand there was a fascination with the exotic nature of the act, and

---

[40] Cahir, *Black Gold*.
[41] F. Shellard, c. 1850. *Reminiscences of an Old Gold Digger in the Early Fifties*, National Library, Canberra, MS 1890.

on the other hand embarrassment over the 'otherness' of the performance.[42] Shellard's use of royal nomenclature reflected the frontier practice of bestowing selected Aborigines with leadership status irrespective of tribal leadership choices. Once bestowed, the Aborigines usually accepted their acquired 'royal' status by proudly wearing the brass plates given to them by colonists, thus reinforcing their importance in both cultures, as was the case with 'King' Billy.[43]

Assisting the police as trackers

One of the consequences of the wealth generated from gold mining in Victoria was the increased number of men who became bushrangers and outlaws. Gold was transported from regional centres, such as Beechworth and Omeo, to Melbourne by police escort. These escorts usually kept to a tight timetable, which made their interception a relatively easy operation. The rugged mountain ranges and vast forests of north-eastern Victoria provided an ideal setting for bushranging activities.[44] Although police were present at Omeo in 1854, there was a lot of territory for them to cover to try to bring offenders to justice. The Victorian police were not equipped to apprehend offenders because they did not know their way around the country, so they turned to the one group of people who were intimately acquainted with the bush: the Aborigines. In January 1859, a Yaitmathang Aboriginal tracker was employed by the Victorian police to catch two outlaws, as noted below. His pay rates are unknown. By cooperating with the police, trackers made themselves useful to Europeans and acquired status within both European and their own societies.

An incident reported extensively in *The Constitution and Ovens Mining Intelligencer*, told of the murder of Cornelius Green, a storekeeper from the Livingstone Creek, whilst conveying gold from Omeo to Sale on 5 January 1859. The following day, the murderers were named as George Chamberlain and William Armstrong and a reward posted for their capture. A week later, local identity Tom Toke informed police that the two men had visited his shanty at the Gibbo Creek, after which he shadowed them until he saw where they were camped. The police assembled a team to apprehend the

---

[42] Per Cahir, *Black Gold*.

[43] J. Troy, 1993. *King Plates: A History of Aboriginal Gorgets*. Canberra: The National Museum of Australia.

[44] G. Jones, 1991. *Bushrangers of the North East: The Golden Years of Bushranging*. Wangaratta: Charquin Hill Publishing, 11.

suspects 'taking some black fellows with them for tracking'. The skill of the trackers was evident as 'the trackers were scenting the trail like bloodhounds, over rocks and dead timber that made it almost impossible to follow, and the police bring up the rear'. After the trackers found two horses tethered behind some huge rocks, they increased their pace:

> ... when the leading tracker suddenly stopped and sat down at the foot of a large gum tree as if he was exhausted, so when the others came up the sergeant asked: What was up? The blackfellow then pointed upwards saying 'White fellow a long at tree all same possum'. The police on looking up saw a man partly concealed with leaves and boughs they immediately covered him with their rifles calling him to come down or they would fire, so he had to come down and surrender.[45]

The trackers had tracked the men over 50 kilometres of very rough country in two days, which was a noteworthy effort considering the difficulties they encountered, as the two men were good bushmen who understood tracking and did everything they could to obliterate their own tracks.

It was later recorded that the name of one Omeo Aborigine tracker who located the offenders was 'Omeo Tommy'.[46] Local residents who remembered the incident spoke of Omeo Tommy's skill as a tracker. He was able to tell that one horse had a broken shoe and that the other horse had no shoes.[47] The names of the other tracker/s to accompany Omeo Tommy have not been recorded. Omeo Tommy demonstrated an intimate knowledge of his country and was able to navigate and trace the movements of people through his territory in a manner far superior to the Europeans. The settlers and police obviously recognised the skill of the Aboriginal trackers from Omeo and were happy to utilise those skills in the pursuit of justice. In turn, Omeo Tommy and the other trackers obtained recognition and value within European society.

These three incidents during the mining era of the 1850s indicate the flexibility of the Yaitmathang to new circumstances. As was the case during

---

[45] Anonymous, 1859. 'Murder and Attempted Highway Robbery at the Omeo', *The Constitution and Ovens Mining Intelligencer*, 14 January 1859, 2-3.

[46] Haughton in Victoria, Legislative Assembly 1861, *First Report of the Central Board Appointed to Watch over the Interests of the Aborigines in the Colony of Victoria*, Parliamentary Paper 39, John Ferres, Melbourne, 20. The Chief Commissioner of Police of the time verified the above story. Newspaper journalists recorded that 'thanks to the native black fellow, they [Chamberlain and Armstrong] were speedily secured, and marched back to Omeo'. See J. Sadlier, 1913. *Recollections of a Victorian Police Officer*. Melbourne: George Robertson & Company, 88; 'The Late Murder at the Omeo', 2.

[47] M. Dyer, 1997. *Yarns and Characters from the Mountains*. Bairnsdale: Max Dyer, 79.

the settlement era, the Yaitmathang adapted to the influx of miners to their territory. The Aboriginal 'chief' of the Omeo area, Metoaka, was apparently happy to accommodate the miners into his territory. The family of 'King' Billy from Cobungra readily participated in a fancy dress ball hosted by the miners' wives. Omeo Tommy was happy to assist the police in the capture of outlaws. Those three Yaitmathang Aborigines all demonstrated a strong sense of belonging to the Omeo region and shared part of that knowledge with Europeans. In so doing, each of those men demonstrated loyalty to the Europeans and were rewarded with either royalty status or monetary payment, though it is not clear whether this enhanced their importance within their own society in all cases.

However, the reaction of the settlers and miners to the Aborigines was not always accommodating. The Europeans generally showed little appreciation for the habits and customs of the Yaitmathang. This lack of respect by the Europeans was evident in their embarrassment at the fancy dress ball. While prepared to invite Aborigines to social occasions and to provide employment, it seems there was an uneasy acceptance of the Yaitmathang on the part of the Europeans.

Contrasted connections to country: Yaitmathang v European

For a quarter of a century after Europeans arrived in northern Gippsland, the Yaitmathang and newcomers demonstrated an uneven level of understanding and adaptation to each other's ways. Having outlined the progress of settlement of Europeans onto traditional Yaitmathang country, and Yaitmathang adaptations to the new European lifestyle and culture, what follows is a discussion of how the Yaitmathang simultaneously maintained their traditional connection to country whilst incorporating European attributes, thus demonstrating a high level of lifestyle flexibility. There is little evidence to suggest the Europeans adopted the cultural values and environmental knowledge of the Aborigines of northern Gippsland in their pursuit of survival in a new land, preferring to transpose European values and knowledge on the north Gippsland landscape. The contrast between the two groups is noteworthy when considering patterns of movement, food sources harvested, technological resources, and the use of fire.

That the Yaitmathang were intimately familiar with their land was evidenced by the many tracks they established for varying purposes throughout their country. There were at least two tracks used to travel from the Omeo area to Ngarigo country: the first was along the northern scarp of the Bowen

Mountains and Limestone Creek to the Kosciusko Plateau, and the second along the Omeo Creek to Tom Groggin, and along the Murray River to the Snowy Mountains.[48] Tracks linking Yaitmathang country to others' territories included tracks along tributary valleys to the Murray River and the reported war path from Omeo into Gunaikurnai country. Within their own territory, the Yaitmathang developed many tracks along the river valleys and spurs leading to such destinations as bogong moth aestivation sites.

Many of these tracks were subsequently used by explorers and miners, and have formed the basis of the present-day road network in the Omeo region. For example, as noted above, an Aboriginal track was utilised in the 'discovery' of the Bogong High Plains.[49] Likewise, Aldo Massola recorded that 'in April 1854, gold was discovered at Omeo, and the lonely track between Beechworth and Omeo, used by the Aborigines over the centuries, became a highway for innumerable parties of miners of all nationalities'.[50] In 1861, it was recorded that a group of prospectors took a mining warden from Omeo to the newly discovered goldfield at Wombat Creek along what the prospectors called the Blacks Trail.[51] Each of these traditional Yaitmathang routes had a gentle grade and was upgraded as a designated mining track or road during the gold era, and some remain in use by cars or off-road vehicles today. Europeans were happy to utilise and modify the existing Aboriginal track network to facilitate their aim of accessing new lands for grazing and mining purposes.

The food resources available to, and utilised by, the Omeo Aborigines were abundant. Alfred Wills recorded that 'game was formerly so abundant in this [Omeo] district, that by half a day's hunting, a native could support himself for four days together'.[52] Helms recorded that the food of the Yaitmathang 'consisted of all kinds of game, birds and birds' eggs, reptiles, fishes, and insects' and that they 'usually chose a cleared space for their camps, in the neighbourhood of water, as fish and birds were their principle articles of food', although 'the opossum furnished probably the most frequent meal, because it occurred very abundantly'.[53] The men hunted for game every few days and the women collected edible plant matter daily.

---

[48] Massola, *Journey to Aboriginal Victoria*, 150.
[49] Carr, 'The Discovery of the Bogong High Plains', 285.
[50] Massola, *Journey to Aboriginal Victoria*, 148.
[51] Flett, *The History of Gold Discovery in Victoria*, 159.
[52] Wills in Victoria, Legislative Council, 1859, 37.
[53] Helms, 'Anthropological Notes', 394.

The Yaitmathang and associated friendly tribes engaged in an annual trek to the alpine area during the summer months to collect and feast on the bogong moth (*Agrotis infusa*). In 1845, McMillan recorded the technique used to obtain the moths and reported that they were 'very good' to eat.[54] George Robinson recorded that the Yaitmathang Aborigines visited the Monaro area to feast on the bogong moths in the company of the Ngarigo people, and that the bogong moth was 'called "Cori" by the Omeo, and "Boogong" by the Yass blacks'.[55] Helms added:

> As early as October, as soon as the snow had melted on the lower ranges, small parties of natives would start during the fine weather for some of the frost-riven rocks and procure 'Bugongs' for food. A great gathering usually took place about Christmas on the highest ranges, when sometimes from 500 to 700 aborigines belonging to different friendly tribes would assemble almost solely for the purpose of feasting upon roasted moths. Sometimes these natives had to come great distances to enjoy this food.[56]

Josephine Flood documented the ceremonial procedures associated with hunting the bogong moth.[57] She suggested that the Aborigines commenced their annual migration to the mountains soon after the moths started aestivating there about 4000 years ago, indicating an acquired knowledge of food sources transcending many generations.[58] Whilst in the mountains collecting moths, certain birds were also targeted for consumption. In 1895 Helms recorded that 'the crows fattened rapidly on the moths and were also highly prized as food: they were consequently much pursued by the natives during their bugonging pic-nics'.[59]

The food sources accessed by the Yaitmathang contributed to a balanced diet. Protein was sourced from a variety of native mammal, reptile, avian, aquatic and insectivorous sources, while carbohydrates and green vegetables were readily available in native roots and leafy plants.[60] By contrast, the early European settlers typically consumed a diet of imported foodstuffs such as

---

[54] Per Robinson in Clark, 1998a. *The Journals of George Augustus Robinson*, 88.
[55] Robinson in Clark, 1998a. *The Journals of George Augustus Robinson*, 716.
[56] Helms, 'Anthropological Notes', 398.
[57] Flood, *The Moth Hunters*, 61 ff.
[58] Ibid., 279.
[59] Helms, 'Anthropological Notes', 396.
[60] B. Gott, 2008. 'Indigenous use of Plants of South-eastern Australia', *Telopea* 12(2), 215-226.

beef, damper made from flour, and tea. No European acknowledgement of the abundance of native sources of green produce in north Gippsland has been found, suggesting the newcomers unnecessarily consumed an unbalanced diet. Ironically, introduced stock were not slow to avail themselves of palatable plants such as kangaroo grass (*Themeda triandra*) and the yam daisy (*Microseris* spp.).

The medicinal value of some plants was well known to the Yaitmathang. Helms recorded that 'as a cure [from a scurvy-type disease] the natives ate a kind of yam cooked in hot ashes or roasted on stones, as well as other vegetable food and certain herbs'.[61] Beth Gott has catalogued many other native plants with medicinal value that were known to the Aborigines.[62] No records have been found suggesting Omeo settlers or miners availed themselves of such bush medicines, albeit there is also an absence of records suggesting ailing or dying Europeans rejected such potential cures.

The stock the Europeans introduced had lasting effects on the local environment. Fawcett discussed the impacts of cattle grazing on the soils of the Omeo area, and tracked the demise of many native fauna species to the deliberate and accidental introduction of several domestic and feral species.[63] She provided multiple examples of reduced vegetation cover, altered stream characteristics, gully initiation, widespread erosion and sedimentation resulting from European pastoral practices. Within 100 years of settlement, the grazing-based economy imposed by the early settlers on the Omeo landscape had become unsustainable and necessitated remediation.[64] While the Aboriginal people may also have altered the environment, as discussed below, it seems they did not have such a deleterious impact of the landscape.

It also seems that Europeans were reluctant to adapt Aboriginal tools for their own purposes, although there is some evidence of an appreciation of the skill required to produce these items. The manufacture of Yaitmathang technology required knowledge of the land and its resources. Yaitmathang tool usage included yam sticks, stone axes, water containers, fibre nets, carry bags, boomerangs, spears, clubs, shields and canoes. Specific knowledge about geological and floral properties was applied to the manufacture of the tools. The construction of 'umigongs', or axe heads, necessitated obtaining

---

[61] Helms, 'Anthropological Notes', 398.
[62] Gott, 'Indigenous use of Plants'.
[63] S.G.M. Fawcett, 1955. 'Upper Hume Catchment: Ecological Report', in *A Book for Maisie*, ed. D.J. Carr. Fyshwick: Pirion Printers, 248-250.
[64] Ibid.

appropriate materials that were then transported to, and worked at, various localities around Omeo. One location that seems to have been a training ground devoted to the manufacture of stone implements was Horsehair Plain, where approximately 40,000 stone artefacts were uncovered during the construction of the Mount Hotham airport.[65] Selected in situ rocks were used to sharpen tools, and rocks 'well worn from the rubbing of spears and axes' have also been found in the Omeo area.[66]

Specific ecological knowledge was required for the manufacture of spears for different purposes, as the Yaitmathang manufactured 'three or four kinds of spears, which were made of reeds, seedstalks of the grasstree, boxtree, or if procurable, ironbark'.[67] Likewise, nets made for the collection of bogong moths utilised specific wetland plant species, as Helms described:

> The fine nets made of kurrajong fibre… seem to have been especially designed for the purpose of collecting the 'Bugong'. They had very fine meshes and were manufactured with great care, and being attached to a couple of poles they could be readily folded up when they had to be withdrawn from the crevices. A shrub (*Pimelia* sp.) growing abundantly in places by the river sides to a height of three or four feet, furnished the fibre. The bark of this bush was stripped and allowed to dry, was then placed in water, and weighted down with some stones for several days till the non-fibrous portions were partly rotted. It was then taken out of the water and spread in the sun to dry till it was quite crisp, after which the fibre was freed by beating with sticks or flat stones. All this was the women's work, and they managed to produce a tenacious material from it that could be spun into the finest threads… The Omeo blacks called the bush as well as the fibre kurrajong.[68]

Most Yaitmathang tools were developed for the purposes of food acquisition – others were for warfare and travel purposes – and all were derived from local materials and manufactured for specific purposes. After the 1830s, the Yaitmathang incorporated European commodities in the manufacture of their tools, suggesting an adaptable attitude to tool manufacture. For

---

[65] See paper Anonymous author, 2000. 'The Aboriginal People of the Mountains: An Exhibition of Archaeological Finds Made During the Construction of this Airport' June.

[66] Pendergast, *Pioneers of the Omeo District*, 3.

[67] Helms, 'Anthropological Notes', 400.

[68] Ibid., 396.

example, clubs hosting patterns made by hot wire and spears with embedded glass have been lodged with the Omeo Historical Society.

The Europeans had their own suite of equivalent tools (the plough instead of the yam stick, the gun in place of the boomerang or spear, etc.) which required high energy usage to manufacture and were imported from great distances at great cost. No evidence has been found that the early settlers of north Gippsland engaged in any Aboriginal technology manufacture, although the raw material for such things as net making was readily available. The superiority of European firearms over Aboriginal boomerangs and spears was a great attraction for the Yaitmathang, despite the non-renewable nature of the imported products, and the fact that their ongoing use (purchase of bullets, etc.) by Aborigines required participation in the European trade economies.

In common with many other Aboriginal groups around Australia, the Yaitmathang utilised fire as a management tool for signalling, hunting bogong moths, procuring small game, or for cooking game.[69] Of the Omeo Aborigines, Helms recorded:

> To make a signal, a fire was lit by the side of a dry tree and green bushes were heaped upon the flames when these had made a good start. The smoke would then rise alongside of the tree as if it were forced from a furnace.[70]

Whilst catching Bogong moths, Helms recorded that fire was used to 'stifle the thickly congregated moths... and make them tumble to the bottom of the cleft' where they were collected in kangaroo skins or fine nets. He stressed that the 'process required some care and attention in order to prevent the bodies of the moths getting scorched' whereupon the insect 'generally shrivelled to the size of a grain of wheat'. Helms noted that the Yaitmathang 'cooked their food either on the fire, or when they had a great deal of it and were not in a hurry, in a kind of oven in the ground'.[71]

The use of fire by the Yaitmathang to procure game was recalled by Omeo settler Harry Witham:

> Kangaroos and wallabies were stalked and speared but when these larger animals were scarce fires were lit for the purpose of obtaining small game.

---

[69] R. Jones, 1969. 'Fire-stick Farming', *Australian Natural History*, 16, 224 ff; J.L. Koehn, 1995. *Aboriginal Environmental Impacts*. Sydney: University of New South Wales Press, 35 ff.

[70] Helms, 'Anthropological Notes', 397.

[71] Ibid., 393, 394.

The area burnt at any one time was limited to one or two acres [less than one hectare]. As the grass burned, the blacks, who had taken up positions around the fire, pounced on any lizard or small animal unfortunate enough to come into view in its efforts to escape. He emphasised that the blacks showed skill and judgement in the management of the fires and took great care not to let them get out of hand.[72]

This is an example of the use of 'fire-stick farming' practiced by the Yaitmathang on the Omeo Plains. Historian Bill Gammage claimed that Aboriginal people across Australia practiced an extraordinarily complex system of land management using fire. Gammage's thesis is based on early explorer/settler writings describing vast park-like or treeless landscapes, including the Omeo Plains, which he attributes to careful land management strategies by Aborigines to minimise the effort required in maximising food production, and stands in contrast to the labour-intensive system of pastoralism practiced by Europeans.[73]

Once the miners arrived, the frequency and extent of fire in the Omeo region changed, as Maisie Fawcett recorded:

> The long grass which covered the country in those days was a nuisance to the miner looking for outcropping reefs or trying to locate them by 'loaming'. Fire was a simple way of removing the dense ground cover and it was used extensively.[74]

Miners across the mountains of eastern Victoria used fire indiscriminately to remove vegetation impediments in their search for gold.[75] However, the Yaitmathang had a strong reaction to the widespread use of fire by gold miners in the 1850s. Settler Edward McNamara recalled 'that the blacks were not only frightened of bushfires lit by white men, but protested that their hunting grounds were being ruined'.[76] An Omeo settler recalled that when the Yaitmathang came to Omeo in about 1854, 'they made bitter complaints about the number of fires burning in the mountains and said

---

[72] Cited in Fawcett, 'Upper Hume Catchment', 275.
[73] B. Gammage, 2011. *The Biggest Estate on Earth: How Aborigines made Australia*. Sydney: Allen & Unwin, 7-8.
[74] Fawcett, 'Upper Hume Catchment', 280.
[75] B.K. Smart, 1883. 'Beechworth Mining District: Buckland Division', in *Reports of the Mining Surveyors and Registrars. Quarter Ending 30th September 1883*. Melbourne: Mines Department, 32, 33.
[76] Edward McNamara in Fawcett, 'Upper Hume Catchment', 275.

that if the white man continued to burn the bush it would soon not be worth their while coming there to hunt'.[77] The Yaitmathang ethic of careful use of fire for ecological and food-procuring purposes was replaced in the 1850s by indiscriminate use of fire by Europeans as they exploited mineral wealth to the detriment of the north Gippsland environment.

In conclusion, the Yaitmathang Aborigines had a multi-faceted and complex cultural association with both their country and newcomers to the Omeo region of northern Gippsland. Indeed, the Yaitmathang actively participated in the exploration, squatting, mining and police tracking activities as northern Gippsland Europeans relied on the Yaitmathangs' intimate knowledge of their country for their own survival and development. The Yaitmathang sought to both incorporate Europeans into their traditional social structures and individually extend their sphere of influence by adopting European commodities and favour. At the same time, the Yaitmathang maintained traditional connections to country and expressed strong displeasure that their traditional food sources were adversely affected by activities such as European-imposed fire regimes. These strong Yaitmathang associations with country are today preserved in contemporary road network patterns and place name nomenclature. But the Europeans otherwise demonstrated little willingness to learn from and adopt the Yaitmathang relationship with their environment.

References

Andrews, A.E.J. (ed.) 1979. *A Journey from Sydney to the Australian Alps Undertaken in the Months of February and March 1834 by Dr. John Lhotsky*. Hobart: Blubber Head Press.
Anonymous, 2000. 'The Aboriginal People of the Mountains: An Exhibition of Archaeological Finds Made During the Construction of this Airport', June.
Attwood, B. 2011. 'Aboriginal History, Minority Histories and Historical Wounds: The Postcolonial Condition, Historical Knowledge and the Public Life of History in Australia', *Postcolonial Studies*, 14 (2), 171–186.
Bairnsdale Advertiser and Tambo and Omeo Chronicle, 25 July 1908.
Bride, Thomas 1898. *Letters from Victorian Pioneers: Being a Series of Papers on the Early Occupation of the Colony, the Aborigines, etc., Addressed by Victorian Pioneers to His Excellency Charles Joseph LaTrobe, Esq., Lieutenant-Governor of the Colony of Victoria*. Melbourne: Government Printer.
Bulmer, John. Undated. Notes from John Bulmer in Howitt manuscripts, MS 9356, Box 1054/2 (b), La Trobe Library, Melbourne.
Buntine, J. Undated. Letter to A.W. Howitt, Sale, from John Buntine, Toongabbie, entitled 'Omeo tribe'. MS 9356, Box 1054/2, Manuscript Collection, State Library of Victoria.
Cahir, F. 2012. *Black Gold: Aboriginal People on the Goldfields of Victoria, 1850–1870*. Aboriginal History Monograph 25. ANU E Press and Aboriginal History

---

[77] Fawcett, 'Upper Hume Catchment', 280.

Incorporated.
Campbell, A. and Vanderwal, R. 1994. *Victorian Aborigines: John Bulmer's Recollections: 1855 – 1908*, Museum of Victoria.
Carr, S.G.M. 1962. 'The Discovery of the Bogong High Plains', *Proceedings of the Royal Society of Victoria*, 75, 285–289.
Clark, I.D. 1996. *Aboriginal Language Areas in Victoria*. Melbourne: Victorian Aboriginal Corporation for Languages.
Clark, I.D. 1998a. *The Journals of George Augustus Robinson, Chief Protector, Port Phillip Aboriginal Protectorate, Volume 4: 1 January 1844 – 24 October 1845*. Melbourne: Heritage Matters.
Clark, I.D. 1998b. *Place Names and Land Tenure – Windows into Aboriginal Landscapes: Essays in Victorian Aboriginal History*. Melbourne: Heritage Matters.
Clark, I.D. 2000. *The Papers of George Augustus Robinson, Chief Protector, Port Phillip Aboriginal Protectorate, Volume 2: Aboriginal Vocabularies: South East Australia, 1839 – 1852*. Melbourne: Heritage Matters.
Clark, I.D. 2009. 'Dhudhuroa and Yaithmathang Languages and Social Groups in North-east Victoria – A Reconstruction', *Aboriginal History*, 33, 201–229.
Dawson, W.T. Monthly reports to Andrew Clarke, Surveyor General, Melbourne, 1855–56, from W.T. Dawson, Surveyor, Sale, Gippsland. PROV: VPRS 1328, Unit 1, Public Records Office Victoria.
Dyer, M. 1997. *Yarns and Characters from the Mountains*. Bairnsdale: Max Dyer.
Fawcett, S.G.M. 1955. 'Upper Hume Catchment: Ecological Report', in *A Book for Maisie*, ed. D.J. Carr. Fyshwick: Pirion Printers, 239–307.
Fead, G.C. c. 1880. *Notes on an Unsettled Life, Memoires of Life on the Victorian and New South Wales Goldfields, 1853–1860, Written c1880*, Manuscript held by Margaret Fead, 11 Esperance Street, Red Hill, ACT 2603.
Fels, M.H. 1988. *Good Men and True: The Aboriginal Police of the Port Phillip District 1837 – 1853*, Carlton: Melbourne University Press.
Fison, L. and Howitt, A.W. 1880. *Kamilaroi and Kurnai*. Melbourne: George Robertson.
Flett, J. 1970. *The History of Gold Discovery in Victoria*. Melbourne: Hawthorn Press.
Flood, J. 1980. *The Moth Hunters: Aboriginal Prehistory of the Australian Alps*. Canberra: Australian Institute of Aboriginal Studies.
Gammage, B. 2011. *The Biggest Estate on Earth: How Aborigines made Australia*. Sydney: Allen & Unwin.
Gardner, P.D. 1992. 'Aboriginal History in the Victorian Alpine Region', in *Cultural Heritage of the Australian Alps*, ed. B. Scougall, Proceedings of a symposium held at Jindabyne, New South Wales, 16–18 October 1991, Australian Alps Liaison Committee, 89–99.
Gardner, P.D. 1993. *Gippsland massacres*. Ensay: Ngarak Press.
Gippsland Mercury, 18 March 1875.
Gott, B. 2008. 'Indigenous Use of Plants of South-eastern Australia', *Telopea* 12(2), 215–226.
Helms, R. 1895. 'Anthropological notes', *Proceedings of the Linnean Society of New South Wales*, 20, 387–407.
Hueneke, K. 1994. *People of the Australian High Country*. Canberra: Tabletop Press.
Howitt, A.W. 1904. *The Native Tribes of South East Australia*. London: Macmillan.
Jones, R. 1969. 'Fire-stick Farming', *Australian Natural History*, 16, 224–228.
Jones, G. 1991. *Bushrangers of the North East: The Golden Years of Bushranging*, Wangaratta: Charquin Hill Publishing.
Koehn, J.L. 1995. *Aboriginal Environmental Impacts*. Sydney: University of New South Wales Press.

Lingard, J. 1846. *A Narrative of the Journey to and from New South Wales, Including a Seven Year Residence in that Country.* Chapel-en-le-Frith: J. Taylor & Co.

Mackaness, George 1941. 'George Augustus Robinson's Journey into South-eastern Australia, 1844', *Journal of the Proceedings of the Royal Australian Historical Society*, 27, 318–349.

MacKillop, G. 1836. 'Australian Colonies', *Quarterly Journal of Agriculture, Edinburgh*, 70, 156–169.

Massola, A. 1969. *Journey to Aboriginal Victoria: A Guide and History to the State's Aboriginal Antiquities.* Adelaide: Rigby.

Native Police Corps, Day Book of the Native Police Corps, Narre Narre Warren, 1845 – 1853, VPRS 90, Public Records Office, Melbourne.

Neumayer, George 1869. *Results of the Magnetic Survey of the Colony of Victoria Executed During the Years 1858-1864*, J. Schneider, Manheim.

O'Rourke family miscellaneous papers and photographs, State Library of Victoria, Manuscripts Collection, MS 12608, Box 3435/7.

Pendergast, J.V. 1968. *Pioneers of the Omeo District.* Melbourne: Riall Print.

Pearson, A.M. 1969. *Echoes from the Mountains and History of the Omeo Shire Council.* Omeo: Omeo Shire Council.

Pepper, P. and DeAraugo, T. 1985. *What did Happen to the Aborigines of Victoria: Volume I: The Kurnai of Gippsland.* Melbourne: Hyland House.

Robinson, G.A. 1844, Journal, MLms A1236, Mitchell Library, Sydney, 711.

Sadlier, J. 1913. *Recollections of a Victorian Police Officer.* Melbourne: George Robertson & Company.

Shellard, F. c. 1850. *Reminiscences of an Old Gold Digger in the Early Fifties*, National Library, Canberra, MS 1890.

Skene, A. and Brough Smyth, R. 1874. *Report on the Physical Character and Resources of Gippsland.* Melbourne: Government Printer.

Smart, B.K. 1883. 'Beechworth Mining district: Buckland Division', in *Reports of the Mining Surveyors and Registrars. Quarter ending 30th September 1883*. Mines Department, Melbourne, 32, 33.

Spreadborough, R. and Anderson, H. 1983. *Victorian Squatters.* Ascot Vale: Red Rooster Press.

Strzelecki, P. 1841. 'Report by Count Strzelecki', in *England, Parliamentary Papers Respecting New South Wales, Despatch from Sir G Gipps, Governor of New South Wales, to the Right Honourable Lord John Russell, &c, &c, dated 28–9–1840, Appendix C*, Colonial Office, London.

Sydney Gazette, 15 April 1834, 3.

Tindale, N.B. 1974. *Aboriginal Tribes of Australia: Their Terrain, Environmental Controls, Distribution, Limits, and Proper Names.* Canberra: Australian National University Press.

The Constitution and Ovens Mining Intelligencer, 14 January 1859.

Troy, J. 1993. *King Plates: A History of Aboriginal Gorgets.* Canberra: The National Museum of Australia.

Victoria, Legislative Council, 1859. *Report of the Select Committee of the Legislative Council on the Aborigines; Together with the Proceedings of Committee, Minutes of Evidence, and Appendices*, Parliamentary Paper D19, John Ferres, Melbourne.

Victoria, Legislative Assembly 1861. *First Report of the Central Board Appointed to Watch over the Interests of the Aborigines in the Colony of Victoria*, Parliamentary Paper 39, Melbourne: John Ferres.

Wesson, S. 2000. *An Historical Atlas of the Aborigines of Eastern Victoria and Far South-Eastern New South Wales*, Monash Publications in Geography and Environmental Science No. 53.

Chapter 7

'It's just fly fishing against net fishing...'
Commercial fishing, angling and the Gippsland Lakes, 1870s to the 1880s

David Harris

Introduction

The Gippsland Lakes share with the larger coastal and marine environments of North America and Europe a common story of environmental degradation and declining fish numbers. Where wild fish are harvested for sport and for profit there are similar histories of contested shorelines, government regulations, and vested interests intertwined with politics, economics, culture and competing views of nature.[1] On the Gippsland Lakes, in the late nineteenth century, an added dimension was the environmental transformation that accompanied rapid colonisation and nation building.[2]

---

[1] C.Y. Chiang, 2008. *Shaping the Shoreline: Fisheries and Tourism on the Monterey Coast*. Seattle: University of Washington Press; A. Garner, 2005. *A Shifting Shore: Locals, Outsiders, and the Transformation of a French Fishing Town, 1823–2000*. Ithaca: Cornell University Press; J.E. Taylor, 1999. *Making Salmon: An Environmental History of the Northwest Fisheries Crisis*. Seattle: University of Washington Press; A.F. McEvoy, 1986. *The Fisherman's Problem: Ecology and Law in the California Fisheries, 1850–1980*. Cambridge: Cambridge University Press; M. McKenzie, 2011. *Clearing the Coastline: The Nineteenth-Century Ecological & Cultural Transformations of Cape Cod*. Lebanon: University Press of New England; T.D. Smith, 1994. *Scaling Fisheries: The Science of Measuring the Effects of Fishing, 1855–1955*. Cambridge: Cambridge University Press.

[2] J. Adams, 1981. *The Tambo Shire Centenary History*. Bruthen: Tambo Shire Council; C. Dow, 2004. 'Tatungalung Country: An Environmental History of the Gippsland Lakes.' Melbourne: PhD thesis, Monash University; E.C.F. Bird and J. Lennon, 1989. *Making an Entrance: The Story of the Artificial Entrance to the Gippsland Lakes*. Bairnsdale: James Yeates & Sons; P. Synan, 1989. *Highways of Water: How Shipping on the Lakes Shaped Gippsland*. Drouin: Landmark Press; K.L. Hotchin, 1990. 'Environmental and Cultural Change in the Gippsland Lakes Region, Victoria, Australia.' Canberra: PhD thesis, Australian National University.

This chapter deals with the period after 1878 when the Gippsland Lakes experienced a boom in commercial fishing and angling made possible by the opening of the rail connection between Sale and Melbourne. Opened in March 1878, the railway multiplied the economic value of the fish in the Gippsland Lakes by providing the means to deliver the catch in a fresh condition to the Melbourne Fish Market and more distant regional centres at Bendigo and Ballarat.[3] At the same time, it increased the economic potential of the Lakes for the holiday trade as it brought anglers and shooters from Melbourne in search of fish and game.[4]

In May 1879, three colonial politicians took evidence and heard opinions at public meetings in Sale, Bairnsdale, Rosherville and Toonallook about the operation of the newly expanded fresh fish trade on the Lakes. The meetings received detailed coverage in city and local Gippsland newspapers that was possibly a measure of the public interest in the potential benefits that the railway would bring to this corner of the colony.[5] Unlike previous fishing inquiries, there was no report produced following the Gippsland visit. In a very brief statement to parliament, one of the politicians who took part noted that, as all the grievances presented during the meetings had been resolved, a report was unnecessary.[6]

The fact-finding excursion produced few tangible outcomes but the public meetings provided a forum for the broader discussion of contentious issues that went beyond the economics of commercial fishing to include competing views about the Lakes' ecology and the regulation of commercial fishing. Although concerns about the impact of commercial fishing on Victorian inshore fisheries had been raised previously, the growth of the Gippsland Lakes as a site for commercial fishing and angling seemed to amplify these long-standing tensions.

The debates resonated throughout the 1880s as claims and counterclaims were made about the environmental impact of commercial fishing and use of seine fishing nets.[7] Yet, given the lack of knowledge about the breeding habits of most

---

[3] S.J. Evans, 2003. *Fins, Scales, and Sails: The History of Fishing At Port Fairy 1845 to 1945.* Daylesford: Jim Crow Press, 90–91; Synan, *Highways of Water*, 160–161.

[4] C. Dow, 2008. '"A Sportsman's Paradise": The Effects of Hunting on the Avifauna of the Gippsland Lakes', *Environment and History* 14 (2): 145–164.

[5] *Gippsland Times*, 25 August 1879, 2. Parliamentary representative, Benjamin Davies was on the Board of The Gippsland Lakes Fisheries Company that advertised an offer of shares to the public a few months after the visit in May. See also *Argus*, 15 May 1879, 4, for the recommendation by the government for the colony's Chief Fisheries Inspector, Captain Payne, to accompany the local members.

[6] Victorian Parliamentary Debates vol. XXXI Session 1879–80, 1060–1061.

[7] The seine was a type of fishing net that had been used since ancient times. The net is particularly adapted to shallow water on gradually shelving beaches and is usually worked

fish species in the Lakes or the ecosystem they inhabited, commercial fishers and the angling groups were engaged in a contest between their competing views of nature. Although the debate about commercial fishing with nets on the Gippsland Lakes was framed in terms of whether or not it was destructive, the conflict between the different groups had more to do with their specific social and economic interests.

Local historians have referred to the year 1878 as the beginning of commercial fishing on the Gippsland Lakes, but rather than a 'beginning' it was more a transition from an earlier phase, dating from the 1860s, when Indigenous and non-Indigenous fishers sold fresh and preserved fish to a variety of markets.[8] While the discussion in this chapter commences with 1878, it is in the context of a shift from a mixed artisanal fish trade to a combination of family-based businesses of small capitalists and groups of more inexperienced commercial fishers supplementing other incomes.[9] Recent research into the Chinese fish-curing industry at Port Albert suggests the need for further investigation of the contribution to the colonial economy by the Chinese fishing community at Metung.[10] Equally, accounts of post-conquest Indigenous fishing on the south coast of New South Wales and on the Murray River indicate the need for further consideration of how the Indigenous population engaged with the European economy through the sale and barter of fish from the Gippsland Lakes.[11]

---

from the shore.' T.C. Roughley, 1951. *Fish and Fisheries of Australia*. Completely rev. and enl. ed. Sydney: Angus & Robertson, 190.

[8] Synan, *Highways of Water*, 160; Dow, 'Tatungalung Country', 84–110.

[9] There is no accurate indication of how many commercial fishers – whether European or non-European – were earning a livelihood on the Lakes in the period under discussion. The Parish Plans and Shire of Tambo Rate Books have individuals with Chinese and European names, some of whom appear as 'fishermen' in the Rate Books for Metung and at the eastern end of the Lakes from the early 1870s. Estimates based on the number of fishermen's licences taken out under section 47 of the Land Act only account for the licence holder, while estimates provided by European fishers to journalists do not usually include Chinese or Indigenous fishers. For example, the 18 boats referred to in a *Gippsland Times* article in 1878 do not include boats owned by Chinese commercial fishers who had been working the Lakes since the late 1860s. See PROV VPRS 1310 License Register Land Act 1862, Bairnsdale, 36 and VPRS 4303, Units 1–3 Shire of Tambo Rate Book 1885–1889; *Gippsland Times*, 1 May 1878, 3; Township of Metung, Parish of Bumberah, County of Tambo. Accessed 11 November 2010 at http://handle.slv.vic.gov.au/10381/137649.

[10] A.M. Bowen, 2007. '"A Power of Money": The Chinese Involvement in Victoria's Early Fishing Industry'. Melbourne: PhD thesis, La Trobe University; Synan, *Highways of Water*, 161–162.

[11] M. Bennett, 2007. 'The Economics of Fishing: Sustainable Living in Colonial New South Wales.' *Aboriginal History* 31: 85–102; G. Blainey, 2004 *Black Kettle and Full Moon: Daily Life in a Vanished Australia*. Camberwell: Penguin Books Australia, 300.

After 1878 the Gippsland Lakes appeared to be a contested space where commercial fishing interests faced the angling clubs and the influential fish acclimatisation societies. At the same time, individual anglers and commercial fishers developed interdependent relationships at holiday times when anglers paid commercial fishers to take them fishing. The focus of this chapter is with these entwined and contradictory relationships that were evident during the politicians' expedition to Gippsland in 1879.

The Paynesville visit of May 1879

At 10 o'clock in the morning on Saturday, 17 May 1879, a meeting was held aboard the steamer *Kangaroo* between a group of fishers, representing the local fishing community on the Gippsland Lakes, and the three Victorian parliamentarians. Beneath an awning erected to provide shelter from the constant drizzle, and seated on the raised platform of the *Kangaroo's* wheel deck, sat the colonial politicians and Captain Charles Payne.[12] Parliamentary representatives Francis Mason, Benjamin Davies and Frederick Smyth met for two hours with Duncan Alexander from Geelong, Henry Watson from Queenscliff and Point Nepean, William Robertson from Port Albert and William Carstairs from Western Port. The four had been appointed at a meeting the previous evening to represent the views of commercial fishers to the parliamentarians.[13]

While most who attended the meeting with the parliamentary delegation on the *Kangaroo* seemed to be recent arrivals on the Lakes, they appeared intent on creating a permanent settlement there. Toonallook was the official name of the place where they had gathered but, by August 1879, when applying for Fishermen's Licenses, the fishers were calling it Paynesville – after Captain Payne, the Chief Fishery Inspector.[14] Payne was known to have a good relationship with the fishers but whether the naming was sincere or ironic flattery of a government official, it served the purpose of defining the area as a fishing settlement. Much of the waterfront land was taken up by commercial fishers under section 47 of the Land Act 1869 under which they could take out licences to occupy Crown

---

[12] *Victorian Government Gazette* 8 January 1878, 72 Accessed 3 January 2013 at http://www.austlii.edu.au/au/other/vic_gazette/.

[13] *Gippsland Times*, 21 May 1879, 3–4.

[14] The first attempts to spell the word by the clerk at the Bairnsdale land's office resulted in 'Pains Ville'. PROV VPRS 13206 Register of Applications Bairnsdale Section 47 Land Act 1869, 13.

Figure 7.1: 'The Fishing Station, Paynesville, Raymond Island.'
Illustrated Australian News, 7 June 1879.
State Library of Victoria.

lands for 'fishermen's residences' or for other purposes such as net drying. Licence fees cost five shillings, 10 shillings, one pound or three pounds a year depending on the distance from Melbourne and whether the site was in a borough or a town.[15] Not all fishers took out a licence at Toonallook, as some had taken up land as selectors, while others set up temporary camps along the shoreline.[16]

The new wave of fishers who, with their families, settled at Toonallook brought a keen understanding of how the railway would benefit the growth of commercial fishing in the colony. They were aware of how similar circumstances had occurred in Britain where the railway provided quick access to metropolitan markets.[17] Experienced commercial fishers, such as William Carstairs, looked to the future based on the experience of fishing in Britain but there was a lingering perception in other sectors of the colony that commercial fishing was a pre-industrial livelihood. Throughout the 1870s,

---

[15] Land was limited to 20 perches at a distance of 100 feet from the high water mark. *Victorian Government Gazette* 1 March 1872, 460 Accessed 3 January 2013 at http://www.austlii.edu.au/au/other/vic_gazette/.

[16] *The Australasian Sketcher*, 3 August 1878, 31 January 1880.

[17] R. Robinson, 1996. *Trawling: The Rise and Fall of the British Trawl Industry*. Exeter: University of Exeter Press, 23–33.

concern was also voiced in the press that fishers lacked the necessary modern equipment and skills that were considered integral to the success of the fishing industry at 'home'.[18] The colonial fish trade was thought to be old-fashioned, and this view was evident in Samuel Calvert's illustration of the fishing station at Paynesville.

Employed as an artist for the *Illustrated Australian News*, Calvert accompanied the politicians on their visit to the Lakes in 1879. His engraving of the fishing station suggested a pre-industrial scene. Fishermen – with similar beards, hats and boots – lounge against their boats, or against their drying nets, idly engaged in conversation. Their huts, surrounded by piles of fishing baskets and stacked sawn timber, contribute to the sense of a pioneer setting overwhelmed by a primeval forest of tall trees.[19]

The sketch suggests that these people were simple fishers, eking out a subsistence livelihood. Absent from the scene were the women and children who had been amongst the 80 people welcoming the visitors the previous day. One report had counted a quarter of the throng as children between the ages of five and 12 years.[20] While the scene reached into the past, it was the present and the future that held the attention of these fishing families. Their children would be the first generation to benefit from the 1872 Education Act and one of their requests to the visiting politicians was for a school that the community offered to build if the government provided a teacher.[21] One report observed:

> There are portions of thirty or forty families living at this place engaged in the fishing industry; together with those of a few selectors, portions we say for nearly all the wives and younger children have been sent away because there is no school in the place, although it is said that there are now thirty children of school age, and were a school established the number would be soon doubled.[22]

For commercial fishers, an investment in fishing as a livelihood was an investment in the future for their families. They expected colonial government assistance through the provision of services for the community;

---

[18] *Argus*, 1 February 1873, 4–5. This was one of several articles run by the *Argus* dealing with the supply of fish in Melbourne. See also editions for 3, 15 February; 27, 28 and 31 January.
[19] *Illustrated Australian News*, 7 June 1879, 90.
[20] *Gippsland Times*, 21 May 1879, 3–4.
[21] *Gippsland Mercury*, 20 May 1879, 3.
[22] Ibid.

Figure 7.2: 'Gippsland Fisheries: The fishermen state their grievances.'
Illustrated Australian News, 7 June 1879.
State Library of Victoria.

restrictions on the creation of monopolies in the fish trade; a school and post office; government support through the lowering of rail transport costs; a restriction on the monopoly of the Melbourne Corporation Fish Market through the construction of additional metropolitan markets for the sale of their fish; and controls on fish hawkers who they alleged created a monopoly on the sale of fish and charged exorbitant prices to the public.[23] Earlier inquiries had not canvassed such a range of views. These were the concerns of individuals with a sense of how their livelihoods would be part of a future colonial economy.

Calvert framed the meeting between the fishers and the politicians as a dramatic tableau.[24] Journalists often portrayed 'the fishermen' as romantic

---

[23] *Gippsland Times*, 21 May 1879, 3–4; *Gippsland Mercury*, 20 May 1879, 3; Agnes Carstairs, Alice Mentiplay (nee Gettens) and Margaret Mentiplay (nee Carstairs) '1891 Women's Suffrage Petition' Accessed 20 November 2012 at http://wiki.prov.vic.gov.au.

[24] *Illustrated Australian News*, 7 June 1879, 90. Francis Mason, the youngest of the three politicians, is on the upper far left of the etching. Addressing the meeting, and seated

figures existing just outside the new industrial society; one referred to fishing as 'a natural, and time honored industry' and this sense of it was apparent in the description of the fishing settlement at Paynesville:

> The place is rustic, and of course a little primitive, the huts of the fishermen being originally canvas, more or less enclosed and built over with the ti-tree boughs or cane and thatched with reeds, the material for which is abundant. Many of these huts have good chimney stacks constructed of poles and inter woven with sticks, hurdle fashion, and more or less faced with mud where found necessary.[25]

The evidence that the fishers presented to the inquiry was in stark contrast to this romantic notion of fishing. William Carstairs' account, in particular, suggested he had a clear sense that fishing was an industry with economic potential. His contribution to the meeting, as reported in the *Gippsland Times* and the *Argus*, concentrated on a detailed breakdown of the costs of running his business, while the other three fishers generally confined themselves to particular grievances. His evidence to a later inquiry in 1892 contained similar precise observations about costs, weights and quantities.[26]

While commercial fishers were preoccupied with the economic potential of the fish trade on the Lakes, other locals were opposed to the growth in commercial net fishing. At a meeting on the Friday evening in Bairnsdale, they petitioned the visiting parliamentarians with concerns about the use of seine fishing nets by commercial fishers. Frederick Smyth's provocative reading of the petition to the meeting on the *Kangaroo* produced a predictable uproar of derisive and mocking laughter amongst the assembled group of fishers. One fisher called out that 'It's just fly fishing against net fishing, that's what the petition means, sir; any fisherman would laugh at it'.[27] As Roughley explains, a seine net 'consists of a long shallow strip of cotton net, the upper edge attached to a rope... with corks arranged at intervals to give it buoyancy, and the lower edge secured to a line... weighted at intervals with small pieces of lead'.[28] The concern was that these nets caught under-sized fish. Repeating criticisms that had appeared in local newspapers during

---

next to Cole on the right, is Frederick Smyth. Seated in the centre, to the right of Smyth, is Benjamin Davies. The identities of the four fisher representatives are less certain.

[25] *Illustrated Australian News*, 7 June 1879, 90.

[26] 'Progress Report From the Select Committee Upon the Fishing Industry of Victoria', *Victorian Parliamentary Papers* vol. 1 1892 q 3788–3871, 132–134.

[27] Ibid.

[28] Roughley, *Fish and Fisheries of Australia*, 190.

the previous year, the petition alleged that these fish were left by the 'cart load' to rot 'on the lake shore… polluting the atmosphere to such an extent as to be almost unbearable'.[29] The continued use of seine nets, the petition alleged, would potentially destroy the fishing industry on the Lakes. Being particularly adapted to use in shallow water, such as the waters of the Lakes, and as it was a traditional style of net, the suggestion to not use seine netting was asking commercial fishers to stop using a key tool of their trade.[30] For older fishers, the comment possibly revived earlier debates in the colony about fishing regulations protecting the class interests of a privileged few.

The comment about net and fly-fishing was also an expression of two divergent views of nature amongst European colonisers. For one group, the Lakes were a work place – a site where fishers earned their livelihoods – while for others it was where they came to play at the sport of angling. Even the original choice of the site for a settlement at the entrance to the Lakes was chosen on the basis of its views and the opinion that it would 'become a fashionable watering-place'.[31] Connie Chiang identified similar patterns at Monterey in California where 'working landscapes were often tourist attractions; tourist attractions were also working landscapes'.[32] But a difference between the Gippsland Lakes and other fisheries in North American or Europe was that commercial fishing and the holiday angling trade developed at the same time due to the railroad. Accordingly, the character of the working landscape on the Lakes appeared to shift at different times of the year because some fishers could benefit from the arrival of the sporting anglers. Government fishing regulations on the Lakes limited the times and places where net fishing was permitted, so anglers and commercial fishers were unexpectedly brought together in a mutually beneficial relationship. Tourism would become an alternative stream of income for some commercial fishing families during the times of the year that netting was banned on the Lakes. Fishers charged to take anglers out for a day's fishing on the water and, in many fishing communities, the front room of the family's house was rented out to summer visitors.[33] Although some commercial fishers benefitted financially from the holiday anglers

---

[29] *Gippsland Times*, 21 May 1879, 4.

[30] Roughley, *Fish and Fisheries of Australia*, 190.

[31] *Gippsland Times*, 22 November 1865, 3.

[32] Chiang, *Shaping the Shoreline*, 7.

[33] T. Lee and J. Ellis, 2002. *Casting the Net: An Oral History Project of the Lakes Entrance Family History Resource Centre*. Lakes Entrance: The Centre, 29; B. Duncan, 2006. 'The Maritime Archaeology and Maritime Cultural Landscapes of Queenscliffe: A

on the Lakes, tensions with particular angling clubs and fish acclimatisers continued throughout the 1880s. In addition, the increase in the numbers of commercial fishers challenged the capacity of the government to enforce its fishery laws and regulations on the Lakes.

The fish trade in transition

The meetings held in 1879 were a departure from previous official inquiries and foreshadowed a new direction in how the colonial government would respond to commercial fishing over the next decade. Increasingly, the management of the fishery became a question of balancing the types of concerns about commercial fishing referred to in the petition handed to the politicians in Bairnsdale. Previous fishing inquiries held in 1861–62 and 1873 only took evidence in Melbourne and they were more concerned with restricting the use of nets and regulating the minimum size of fish caught than with the operation of the fish trade.[34] The 1873 inquiry gathered evidence from beyond Port Phillip Bay, but the only response dealing with the Gippsland Lakes came from William Foster, the Police Magistrate at Sale. Foster was a gentleman angler who argued for an extension of the ban on the use of nets in rivers to include the shores of the Lakes.[35]

By 1879, the expansion of the colony's railways shifted the economic potential of the colony's fisheries. More than changes in political attitudes or perceptions of commercial fishing, the railway transformed the possible economic uses of the Gippsland Lakes for the tourist industry and for the fish trade. To the east and west of Melbourne, fishing towns along the coast benefited from the expansion of the railway network as they now had more efficient transport to supply the Melbourne market with fresh fish and, by the 1890s, fish were arriving from South Australia for sale at the Melbourne market. Portland fishers were sending their catch via the line to Hamilton from 1878 but it was not until 1889 that Port Fairy could supply Melbourne

---

Nineteenth-Century Australian Coastal Community.' Townsville: PhD thesis, James Cook University, 172–173.

[34] 'Report From the Select Committee Upon the Fisheries' Acts Together With the Proceedings of the Committee and Minutes of Evidence,' *Victorian Parliamentary Papers* vol. 2 1861–1862; 'Report of Board to Prepare Amended Fisheries Bill 1873 No 48 in Victoria,' *Victorian Parliamentary Papers* vol. 3 1873.

[35] 'Progress Report of Board to Prepare Amended Fisheries Bill 1873 No 48 in Victoria', *Victorian Parliamentary Papers* vol. 3 1873, 9.

and Ballarat with barracouta.[36] Queenscliff was connected by rail to Geelong in 1879 and, as Brad Duncan has observed, 'the opening of the Geelong to Queenscliff railway in 1879 dramatically expanded the scope and viability of the Queenscliff fishing industry to include Melbourne, Ballarat, Bendigo and Portland'.[37]

Although no official statistical account of the colonial fishing industry was taken until 1889, the expansion of these fishing grounds – particularly in Gippsland – increased the amount of fresh fish sent for sale at the Melbourne Market.[38] Local newspapers often carried reports about the number of baskets of fish that were sent to the Melbourne Fish Market by rail. For example, in 1878 the *Gippsland Times* reported:

> The present export, which is in its infancy… is approximately 250 baskets per week of four days (each basket weighing 56lb), and 8 or 10 cwt. of eels every night. These quantities, it is assumed, will be trebled when the present draw-backs are removed.[39]

A further indication of growth in the trade was the amount of unsold fish at the Market, which became a matter of increasing concern in the 1880s. During a three-month period in 1886 the Inspector of Fisheries and Game reported 65 tons of fish were destroyed due to the inability to sell it and, in the space of one morning, 80 baskets of fish that could not find a buyer were given to charity.[40]

Increasingly, there were individuals – other than the fisher or the fishmonger – who had an interest in the fish trade. Fish companies formed and boats were built for transporting the fish to the rail head, bringing increasing numbers of individuals into the trade who made an income from it without hauling a net or seeing a fish from the moment of capture to the point of

---

[36] Evans, *Fins, Scales, and Sails*, 90–91. Evans suggests that crayfish were the only catch that could arrive fresh at the Melbourne market from Portland because the rail connection went via Hamilton.

[37] Duncan, 'Maritime Archaeology', 180–181.

[38] PROV VPRS 16182/P0001 Outward Letter Book, Inspector of Fisheries and Game, Unit 1, 1885–1894, Annual Report of the Inspector of Fisheries on the Fisheries of Victoria 1889, 278.

[39] *Gippsland Times*, 1 May 1878, 3. 56 lbs is approximately 25 kilos and 10 cwt is about 508 kilos.

[40] PROV VPRS 16182/P0001, Report to the Consul for Sweden and Norway 29 June 1886, 150–153.

sale.⁴¹ As the trade increased and more livelihoods depended on it, colonial politicians in constituencies close to the fishing grounds began to take more notice of the local fishing communities, as discussed below.

Acclimatisers, anglers and commercial fishers

In 1878, a Bairnsdale publican, Frederick Gingell warned of the threat posed by seine nets to the survival of hatchery-bred salmon that had been released in the rivers that flowed into the Lakes:

> I beg to remind you that it is not many weeks since Sir S. Wilson at considerable trouble and expense placed a large number of young salmon in several of our Gippsland Rivers. What chance they will have if the objectionable practices alluded to… are allowed to continue I cannot conceive.⁴²

Gingell's perspective about the dangers of commercial fishing nets to hatchery-bred fish had been previously used as an argument during the 1873 inquiry into commercial fishing. On that occasion, William Foster referred to the English trout that had been imported from Tasmania and released in local rivers in 1870 at a 'considerable cost'. He advocated increasing the restrictions on the use of nets on the Lakes but particularly at the mouths of the rivers that flowed into them.⁴³

Gingell and Foster were repeating arguments against the use of nets by commercial fishers, particularly in rivers, that were presented by the fish acclimatisers in the Victorian Acclimatisation Society.⁴⁴ Established in Melbourne in 1861 by Edward Wilson, the editor and, by the 1860s, part owner of the *Argus*, the Acclimatisation Society was part of a wider movement, a popular response by European colonisers to the exotic lands they had conquered.⁴⁵ The Society did not oppose commercial fishing, but neither did it want commercial fishers netting the foreign fish that it was

---

⁴¹ For example, Benjamin Davies MLA had joined the delegation to Gippsland because he had an interest in the proposed in the proposed Melbourne Fishing Company. *Argus*, 8 May 1879, 6.

⁴² *Gippsland Times*, 26 April 1878, 4.

⁴³ 'Report of Board to Prepare Amended Fisheries Bill', 9.

⁴⁴ R.O'Brien, 1988. *Ballarat Fish Hatchery, a History: The First 117 Years*. Ballarat: Ballarat Fish Acclimatisation Society.

⁴⁵ T.R. Dunlap, 1997. 'Remaking the Land: The Acclimatization Movement and Anglo Ideas of Nature' *Journal of World History* 8: 304.

introducing, for the sport of anglers and the 'improvement' of nature, into Victorian rivers.[46] Its image as a scientific movement appealed to educated men, in particular, amongst the middle to upper class in Melbourne but also in regional centres across the colony where local acclimatisation societies were formed, usually as a way of organising the distribution of foreign fish into local streams and rivers. They lobbied parliamentarians, some of whom held official positions within the societies, and assisted police as honorary fishing inspectors, enforcing regulations on commercial fishers who were caught breaking the law. The membership of the societies in country towns mirrored the class character of the Melbourne group, usually with local pastoralists, professional men or politicians holding official positions.[47]

The movement also tapped into growing environmental concerns. In the 1850s, Victoria's commercial fisheries had been plundered with the same urgency that motivated the gold seekers who hoped to turn a quick profit. As in the California rush, oysters and shrimps were a popular catch because they were easily harvested, realising a speedy return when sold on the streets or in public houses. Local oyster beds were soon destroyed to the extent that by the end of the 1850s the oyster beds of Westernport Bay were barren while fish numbers in Port Phillip Bay were also thought to be declining.[48]

Attempts to control the exploitation of the fisheries met resistance in the early generation of colonists. For many gold rush immigrants, controls on netting were perceived as a way 'to prevent a set of hard working men from earning a livelihood' through the introduction of 'the infernal system of the English game laws'.[49] The link between fishing regulations and the protection of class interests and privilege were evident to many of this early

---

[46] I. Tyrrell, 2005. 'Peripheral Visions: Californian-Australian Environmental Contacts, 1850s–1910.' *Journal of World History* 8 (2): 297–298; Dunlap, 'Remaking the Land', 316. Aside from birds, animals and plants they introduced carp, roach, tench and goldfish into Victorian rivers; they were importing salmon ova from acclimatisation societies in California; they released Gippsland Perch at the junction of the Yarra and Saltwater Rivers in Melbourne and Murray Cod into Gippsland Rivers. See *The Yeoman and the Australian Acclimatiser* 1862; *Argus*, 12 August 1872, 4–5; The Acclimatisation Society, *Report of the Acclimatisation Society* 1863.

[47] A.E. Dingle, 1984. *Settling*. Sydney: Fairfax, Syme and Weldon Associates, 142; *Gippsland Times*, 22 November 1865, 3.

[48] *Mercury*, 11 September 1862, 6. The article refers to James Putwain, Inspector of Fisheries and Oyster Beds, who noted the decline of whiting, mullet, guard fish and bream; PROV VPRS 16182/P0001, Mandeville to Commissioner of Trade and Customs 21 August 1885 85/60.

[49] *Argus*, 30 July 1850, 4.

generation of colonists. One writer observed at the beginning of the 1850s that:

> With reference to the [English] Game Laws, the introduction of anything approaching to their nature, would lead, in all probability, to results which, unfortunately, too many in these colonies have reason to regret.[50]

These laws in England had restricted hunting to gentleman landowners and, as Bolton has suggested, the first generation of white Australians carried a deep resentment about any legislative controls on hunting in the colonies. Even in the 1860s, the spectre of the English game laws exerted an influence on debates about fauna protection legislation in the New South Wales and Victorian parliaments.[51]

Colonial observers also pointed to the local exploitation of native wild life. Naturalist and professional hunter Horace Wheelwright, writing of his experiences in Melbourne of the 1850s, observed that sea fishing like much else in the colony was 'now overdone'.[52] By the early 1860s there were reports on the disappearance of shrimps and other types of marine life and, according to James Putwain, the Inspector of Fisheries and Oyster Beds, the depletion of 'whiting, mullet, guardfish and bream' in the Bay and the careless destruction of the natural oyster beds in Westernport Bay.[53] The extent to which fish supplies were dwindling was a regular topic for discussion in the Melbourne papers, with blame linked to the small net-mesh size used or, sometimes, to Chinese fishers.[54]

After the environmental ravages of the gold rush, an aspect of the debate about net fishing that developed was a concern with the destructive impact of humans on the environment. City newspapers like the *Argus* carried articles drawing attention to the environmental destruction around Melbourne. The use of nets in rivers to catch fish not only deprived anglers

---

[50] Ibid.
[51] G.C. Bolton, 1981. *Spoils and Spoilers: Australians Make Their Environment 1788–1980*. Sydney: George Allen & Unwin, 15, 98.
[52] H.W. Wheelwright, 1979. *Bush Wanderings of a Naturalist*. Melbourne: Oxford University Press, 248.
[53] *Argus*, 3 September 1862, 4.
[54] See for example PROV VPRS 6605/P0000 Chief Commissioner of Crown Lands – Inwards Correspondence Unit 22 'Complaint Chinese fishermen Melbourne' 58/1298. Chinese fishers were accused of catching undersized fish. The petitioners claimed if it continues 'the legitimate purveyors of the fish market will not be able to gain a livelihood [sic]'.

of sport but was no different, the *Argus* suggested, from the shooters who had waged 'a war of extermination' against birds in Melbourne. There was a link, the paper warned, between the 'indiscriminate slaughter of small birds' by colonists and the devastation that had occurred in the summer of 1861–62 to 'gardens, fields and orchards' in the colony by 'aphids, grubs and beetles of various kinds'. Observing that Tasmania had introduced legislation to ban the use of nets in rivers, the paper supported similar bans in Victoria and a complete ban on fishing during the spawning season. The *Argus* looked to the precedents of European laws protecting swallows and storks – wildlife that was considered 'useful'.[55]

Other views identified the inexhaustible riches of the ocean. In 1864, GS Lang addressed the Acclimatisation Society on the untapped riches of the ocean:

> There is… every reason to believe that we have under the waters as extensive a field for the profitable exertion of our energies as we have on the land, though hitherto left as utterly useless and unprofitable as were our pastures before a white man trod upon them.[56]

Lang's dismissal of Indigenous industry and economy was typical but there were other attitudes expressing a common belief in the inexhaustible bounty of the oceans that were separate from notions of cultural entitlement.[57] One letter writer to the *Argus* in 1873 was optimistic about the resources of the ocean, observing that, 'the crop of the earth is mostly once a year, but the crop of the sea is to be had at all times and can not be exhausted'.[58]

By the 1870s, Dingle argues, the Victorian Acclimatisation Society's influence was waning.[59] Nonetheless, groups such as the Fish Acclimatisation Society, the Anglers' Society, the Geelong Fish Acclimatisation Society, the Ballarat Fish Acclimatisation Society and the Fish Protection Association sprang up, often with members of parliament holding executive positions.[60] Many of the fish acclimatisers were also involved with fish

---

[55] *Argus*, 7 January 1861, 4–5.
[56] The Acclimatisation Society of Victoria, *Third Annual Report of the Acclimatisation Society of Victoria* (1864), 59.
[57] *Argus*, 15 February 1873, 1.
[58] *Argus*, 15 February 1873.
[59] Dingle, *Settling*.
[60] *Gippsland Times*, 27 March 1882, 3; *Argus*, 19 August 1873, 6; *Victorian Parliamentary Debates* second session 1883, Vol XLIV, 1696; For example, David Gaunson MLA

hatcheries that enjoyed a measure of scientific authority during the period under discussion. Fish propagation was not a new idea but in the context of European colonisation it gained significance as a civilising measure of nature's improvement by the colonisers. Their assumed knowledge about the colonial fishery gave the fish culturists and acclimatisers a degree of political influence that sustained tensions over commercial fishing throughout the 1880s.[61]

Managing the fishery

The character of commercial fishing on the Gippsland Lakes also changed after 1878 as more fishers arrived to harvest the bounty. New legislation introduced by the Victorian parliament in 1873 had lifted many of the restrictions on the use of nets and, as discussed above, the spread of the rail network in the late 1870s opened more-distant fisheries to the Melbourne Fish Market. The *Argus* also raised public awareness of commercial fishing through a series of newspaper articles about the colony's fishing communities.[62] Moreover, the growth in the fishing industry in England and the various government inquiries and royal commissions that took evidence there during the 1860s and 1870s encouraged colonials to make comparisons with the local fishing industry.[63] Australia followed the English example with the first of the large official inquiries in New South Wales in December 1879.[64]

Although commercial fishing would face opposition from the angling clubs and fish acclimatisers throughout the 1880s, government administration of the fishery was generally supportive of commercial fishers in this period. Unlike larger foreign fisheries – such as in the United States – colonial fishers did not face vast faceless bureaucracies, or different levels of administration, where the competing political interests of state and national

---

was president of the Victorian Fisheries Protection Society *Victorian Parliamentary Debates* 1886, Vol. LIII, 2385; Lieutenant-Colonel Champ MLA was president of the Melbourne Anglers' Protective Society *Argus*, 23 August 1870, 6.

[61] *Gippsland Times*, 6 July 1883, 3; *Victorian Parliamentary Debates* 1883, Vol. XLIV, 1696.

[62] *Argus*, 18 March 6, 21 March 6, 25 March 6, 15 April 7, 26 April 6 and 3 June 1873, 1.

[63] Smith, *Scaling Fisheries*, 33–38, 51–59.

[64] 'Report of the Royal Commission to inquire into and report upon the actual state and prospect of the Fisheries of this Colony; together with Minutes of Evidence, and Appendix' *New South Wales Parliamentary Papers* vol. 3 1879–80; *Sydney Morning Herald*, 19 December 1879, 2.

authorities compounded a complex political landscape.⁶⁵ To some extent, the familiarity between fishers and administrators may have assisted the growth of commercial fishing in this early period.⁶⁶ Indeed, the seine net continued to be used with some restrictions on fishing in specific locations and at certain times. But commercial fishing on the Gippsland Lakes remained contentious throughout the 1880s. By 1886, Captain Mandeville, the Fisheries Commissioner, was recommending to the Minister for Trade and Customs that the Lakes should be entirely closed to netting because it was impossible to police the areas where commercial fishing had been banned.⁶⁷ A preferred alternative site for a fishing industry, he recommended, was at Port Albert.⁶⁸

For the period under discussion, the Fisheries Department remained a small addition to the Victorian Department of Customs and Trade, with only one paid officer who, amongst a host of other maritime duties, was also the Inspector of Fisheries.⁶⁹ Fishers were not dealing with an impersonal bureaucracy where there was little understanding or appreciation of maritime work. Captain Mandeville, who succeeded Captain Payne as Inspector of Fisheries, came to the position from a naval background.⁷⁰ The inspectors regularly visited the 'fishing stations' along the coast, engaging with commercial fishers from the common basis of a shared understanding of the maritime world.⁷¹ Following his appointment as Inspector of Fisheries in 1878, Captain Payne took advice from the fishers about the most effective way to regulate the industry on the Lakes, and Mandeville was similarly sympathetic to the fishing communities with whom he worked.⁷² For example, in correspondence to the Minister for Trade and Customs in 1885, Mandeville wrote at length about the quality of the fishing community at Port Albert:

---

[65] McKenzie, *Clearing the Coastline*, 111–136.

[66] *Argus*, 19 July 1888, 13; *Argus*, 9 December 1887, 8.

[67] PROV VPRS 16182/P0001, Mandeville to Minister for Trade and Customs 11 February 1886 786/3, 118.

[68] Ibid 9 November 1885 785/56, 82.

[69] *Argus*, 31 March 1890, 8. There were numerous voluntary positions of assistant fishery inspectors created across the colony but Captain Payne as the Inspector also had other significant duties as the Chief Harbour Master, Immigration Agent, Pilot Board Chairman and Steam Navigation Board President. Payne was also Secretary of the Royal Humane Society and of the St John Ambulance Association.

[70] *Argus*, 4 July 1887, 6.

[71] *Gippsland Times*, 30 August 1878, 3.

[72] Ibid.

The fishermen are a hard working well conducted set of men, who will I believe, bring up their children to their own calling, this being so, they take an interest in the fishing, and I am of the opinion that majority wish to have law and regulations and wish that they be enforced. They are bringing up a fine race of people, in fact the fine physique of the young men and boys strike the visitor... For instance one man has thirteen children, nice sons all about six feet and one they say is six feet six inches.[73]

During the decade after 1879, commercial fishers continued to pursue their economic concerns with the Victorian Government. As the industry grew and economic circumstances deteriorated, political influence became more significant in earning a livelihood and the formation of a representative body for the industry became essential. Increasingly, fishery management in the colony was influenced by changes overseas and, in the late 1880s, the government employed visiting experts to advise on fishing regulations. On the Gippsland Lakes, commercial fishing and angling continued to grow, though with frequent disputes arising from a continuation of the tensions that were evident at the 1879 meeting.

Conclusion

The visit by the parliamentary delegation to Gippsland in 1879 occurred within the context of growth brought about by the expansion of the colony's railway network. Fresh fish could now be supplied regularly to urban and regional markets. At the same time, it was the beginning of an intense period of discussion about commercial fishing both in the Australian colonies and overseas. These changes found expression through government inquiries into colonial fisheries and in the holding of the First International Fisheries Exhibition in London in 1883.

Environmental concerns about the growth in commercial fishing that had been part of the old debates about class and privilege were, for many in the colony, out of step with the optimism of economic development in the boom of the 1880s. Awareness about the environmental consequences of the uncontrolled exploitation of natural resources that was apparent and broadly discussed in the 1860s seemed to undergo a redefinition.

---

[73] PROV VPRS 16182/P0001, Mandeville to Minister for Trade and Customs 9 November 1885 785/56, 94.

The angling clubs and the fish acclimatisers continued to lobby effectively for regulations on commercial fishing on the Lakes but they were also increasingly perceived as pursuing their own interests at the expense of the commercial fishers' livelihood. On the Gippsland Lakes, tensions between commercial fishing and sectors of the tourist industry continued to simmer at the end of the 1880s yet it was not apparent amongst individual anglers who paid commercial fishers to take them out fishing during the summer holidays.

Commercial fishing was encouraged by the expansion of the railway to more-distant fishing grounds and governments offered rewards for the development of offshore fishing. Commercial fishers on the Lakes worked with the Inspector of Fisheries to regulate where and when fishing occurred but the size of the Lakes, combined with the varying levels of skill and expertise amongst commercial fishers, made the enforcement of regulations difficult.

Lack of financial resources limited the government regulation of commercial fishing but the small size of the bureaucracy meant that fishers' main concerns were less with politics than with the economics of the industry. They were part of an economy that was becoming more complex as the market for their product grew. The period between 1878 and the early 1890s was one of transition, where a traditional pre-industrial livelihood found itself becoming part of an expanding capitalist economy. By the 1890s, the political economy of commercial fishing was driving significant changes in the operation of the industry. This next stage would occur within the context of profound environmental change on the Lakes, as well as a more complex political and economic landscape brought about by Federation, economic growth, technology and the development of offshore trawling.

Acknowledgements

I wish to thank Katie Holmes, Richard Broome, Alice Garner and Nicole Curby for their comments on early drafts of this paper. I am also grateful to the two anonymous reviewers who gave particularly useful guidance on the chapter.

References

Adams, J. 1981. *The Tambo Shire Centenary History*. Bruthen: Tambo Shire Council.
Argus, 30 July 1850, 4; 7 January 1861, 4–5; 3 September 1862, 4; 23 August 1870, 6; 12 August 1872, 6; 1 February 1873, 4–5; 15 February 1873; 18 March 1873, 6; 21 March 1873, 6; 25 March 1873, 6; 15 April 1873, 7; 26 April 1873, 6; 3 June 1873, 1; 19 August 1873, 6; 8 May 1879, 6; 4 July 1887, 6; 3 December 1887, 6; 9 December 1887, 8; 19 July 1888, 13; 31 March 1890, 8.
Bennett, M. 2007. 'The Economics of Fishing: Sustainable Living in Colonial New South Wales.' *Aboriginal History* 31: 85–102.
Bird, E.C.F and Lennon, J. 1989. *Making an Entrance: The Story of the Artificial Entrance to the Gippsland Lakes*. Bairnsdale: James Yeates & Sons.
Blainey, G. 2004. *Black Kettle and Full Moon: Daily Life in a Vanished Australia*. Camberwell: Penguin Books Australia.
Bolton, G.C. 1981. *Spoils and Spoilers: Australians Make Their Environment 1788–1980*. Sydney: George Allen & Unwin.
Bowen, A.M. 2007. '"A Power of Money": The Chinese Involvement in Victoria's Early Fishing Industry'. Melbourne: Ph.D. thesis, La Trobe University.
Carstairs, Agnes. Mentiplay (nee Gettens), Alice and Mentiplay (nee Carstairs), Margaret.'1891 Women's Suffrage Petition' Accessed 20 November 2012 at http://wiki.prov.vic.gov.au. Chiang, C.Y. 2008. *Shaping the Shoreline: Fisheries and Tourism on the Monterey Coast*. Seattle: University of Washington Press.
Dingle, A.E. 1984. *Settling*. Sydney: Fairfax, Syme and Weldon Associates.
Dow, C. 2004. 'Tatungalung Country: An Environmental History of the Gippsland Lakes'. Melboune: Ph.D. thesis, Monash University.
Dow, C. 2008. '"A Sportsman's Paradise": The Effects of Hunting on the Avifauna of the Gippsland Lakes', *Environment and History* 14 (2): 145–164.
Duncan, B. 2006. 'The Maritime Archaeology and Maritime Cultural Landscapes of Queenscliffe: A Nineteenth-Century Australian Coastal Community'. Townsville: Ph.D. thesis, James Cook University.
Dunlap, T.R. 1997. 'Remaking the Land: The Acclimatization Movement and Anglo Ideas of Nature' *Journal of World History* 8: 303 –319.
Evans, S.J. 2003. *Fins, Scales, and Sails: The History of Fishing At Port Fairy 1845 to 1945*. Daylesford: Jim Crow Press.
Garner, A. 2005. *A Shifting Shore: Locals, Outsiders, and the Transformation of a French Fishing Town, 1823–2000*. Ithaca: Cornell University Press.
Gippsland Mercury, 20 May 1879, 3.
Gippsland Times, 22 November 1865, 3; 22 November 1865, 3; 26 April 1878, 4; 1 May 1878, 3; 30 August 1878, 3; 21 May 1879, 3–4; 25 August 1879, 2; 27 March 1882, 3; 6 July 1883, 3.
Hotchin, K.L. 1990. 'Environmental and Cultural Change in the Gippsland Lakes Region, Victoria, Australia'. Canberra: Ph.D. Thesis, Australian National University.
Illustrated Australian News, 7 June 1879, 90.
Lee, T and Ellis, J. 2002. *Casting the Net: An Oral History Project of the Lakes Entrance Family History Resource Centre*. Lakes Entrance: The Centre.
McEvoy, A.F. 1986. *The Fisherman's Problem: Ecology and Law in the California Fisheries, 1850–1980*. Cambridge: Cambridge University Press.
McKenzie, M. 2011. *Clearing the Coastline: The Nineteenth-Century Ecological & Cultural Transformations of Cape Cod*. Lebanon: University Press of New England.

Mercury, 11 September 1862, 6.
O'Brien, R. 1988. *Ballarat Fish Hatchery, a History: The First 117 Years*. Ballarat: Ballarat Fish Acclimatisation Society.
'Progress Report of Board to Prepare Amended Fisheries Bill 1873 No 48 in Victoria', *Victorian Parliamentary Papers* vol. 3 1873, 9.
'Progress Report From the Select Committee Upon the Fishing Industry of Victoria', *Victorian Parliamentary Papers* vol. 1 1892 q 3788–3871, 132–134.
PROV VPRS 1310 License Register Land Act 1862, Bairnsdale, 36.
PROV VPRS 4303 Units 1–3 Shire of Tambo Rate Book 1885–1889.
PROV VPRS 6605/P0000 Chief Commissioner of Crown Lands – Inwards Correspondence Unit 22 'Complaint Chinese fishermen Melbourne' 58/1298.
PROV VPRS 13206 Register of Applications Bairnsdale Section 47 Land Act 1869, 13.
PROV VPRS 16182/P0001 Outward Letter Book, Inspector of Fisheries and Game, Unit 1,1885–1894, Mandeville to Commissioner of Trade and Customs 21 August 1885 85/60.
PROV VPRS 16182/P0001 Outward Letter Book, Inspector of Fisheries and Game, Unit 1,1885–1894, Mandeville to Minister for Trade and Customs 9 November 1885 785/56, 94.
PROV VPRS 16182/P0001 Outward Letter Book, Inspector of Fisheries and Game, Unit 1,1885–1894, Mandeville to Minister for Trade and Customs 1885 785/56, 82.
PROV VPRS 16182/P0001 Outward Letter Book, Inspector of Fisheries and Game, Unit 1,1885–1894, Mandeville to Minister for Trade and Customs 11 February 1886 786/3, 118.
PROV VPRS 16182/P0001 Outward Letter Book, Inspector of Fisheries and Game, Unit 1,1885–1894, Report to the Consul for Sweden and Norway 29 June 1886, 150–153.
PROV VPRS 16182/P0001 Outward Letter Book, Inspector of Fisheries and Game, Unit 1,1885–1894, Annual Report of the Inspector of Fisheries on the Fisheries of Victoria 1889, 278.
'Report From the Select Committee Upon the Fisheries' Acts Together With the Proceedings of the Committee and Minutes of Evidence,' *Victorian Parliamentary Papers* vol. 2 1861–1862.
'Report of Board to Prepare Amended Fisheries Bill 1873 No 48 in Victoria,' *Victorian Parliamentary Papers* vol. 3 1873.
'Report of the Royal Commission to inquire into and report upon the actual state and prospect of the Fisheries of this Colony; together with Minutes of Evidence, and Appendix' *New South Wales Parliamentary Papers* vol. 3 1879–80.
Sydney Morning Herald, 19 December 1879, 2.
Robinson, R. 1996. *Trawling: The Rise and Fall of the British Trawl Industry*. Exeter: University of Exeter Press.
Roughley, T.C. 1951. *Fish and Fisheries of Australia*. Completely rev. and enl. ed. Sydney: Angus and Robertson.
Smith, T.D. 1994. *Scaling Fisheries: The Science of Measuring the Effects of Fishing, 1855–1955*. Cambridge: Cambridge University Press.
Synan, P. 1989. *Highways of Water: How Shipping on the Lakes Shaped Gippsland*. Drouin: Landmark Press.
Taylor, J.E. 1999. *Making Salmon: An Environmental History of the Northwest Fisheries Crisis*. Seattle: University of Washington Press.
The Acclimatisation Society of Victoria, *Third Annual Report of the Acclimatisation*

Society of Victoria (1864), 59.
The Australasian Sketcher, 3 August 1878; 31 January 1880.
The Yeoman and the Australian Acclimatiser, 1862.
Township of Metung, Parish of Bumberah, County of Tambo, Accessed 11 November 2010 at http://handle.slv.vic.gov.au/10381/137649 .
Tyrrell, I. 2005. 'Peripheral Visions: Californian-Australian Environmental Contacts, 1850s–1910.' *Journal of World History* 8 (2): 275–302.
Victorian Government Gazette 1 March 1872, 460 Accessed 18 August 2012 at http://www.austlii.edu.au/au/other/vic_gazette/.
Victorian Government Gazette 8 January 1878, 72 Accessed 18 August 2012 at http://www.austlii.edu.au/au/other/vic_gazette/.
Victorian Parliamentary Debates vol. XXXI Session 1879–80, 1060–1061.
Victorian Parliamentary Debates second session 1883, vol. XLIV, 1696.
Victorian Parliamentary Debates 1886, vol. LIII, 2385.
Wheelwright, H.W. 1979. *Bush Wanderings of a Naturalist.* Melbourne: Oxford University Press.

Chapter 8

The Currie family approach to land settlement

Kerry Nixon

In 1874, John Currie ventured to Gippsland to select some land. His initial destination was Brandy Creek, but a slow response at the Victorian Land Department scuttled that attempt to secure land in the region. John had selected land prior to this date; as per the newly legislated amendments to the Land Act, his entitlement was subsequently reduced by his previous land selections. Given that this would have left him with an unviable plot of only 110 acres (instead of the maximum 320 acres allowed under the Act), John utilised his brother, James – an aging, partially blind and ostensibly penurious farm servant – to select another 320 acres next door, making a total plot of 430 acres south of Drouin, near Lardners Track.[1] John thus circumvented the restrictions of the Act to maximise his opportunity to farm successfully. John Currie was not the first, and was by no means the last, to use kin in this manner; it was a common means of expanding one's landholdings.[2]

These facts are on the public record because John Currie commenced writing a diary in March 1873. He tired of it quickly, and from October of that year, until only three months before her death in 1908, Sarah Ann Catherine Currie, known as Kate by her husband, maintained the diary, recording all manner of minor details relating to farm, family and community

---

[1] Some discrepancies in the survey process relating to the actual size of the plots allocated were, in this case, accounted for at a later date, and the total land holdings in John Currie's name were amended to 429 acres. However, it should be noted that the probate file of Albert Bryce Currie (the first adult child of John and Kate to die, in 1933) describes the size of the landholding left by John Currie as being 428 acres: PROV VPRS 28/P3/2529.

[2] J.M. Powell, 1970. *The Public Lands of Australia Felix: Settlement and Land Appraisal in Victoria 1834–91, With Special Reference to the Western Plains.* Melbourne: Oxford University Press.

life.³ The diary is in private hands, but a copy resides in the Victorian State Library. This chapter outlines how, by fulfilling their obligations to the Victorian Government, the Currie family affected the landscape. In doing so, I highlight the benefits and deficiencies of the Land Acts (mainly the 1869 and the 1874 amendments) in service of the legislators' original ideals, and also elucidate how the Currie family utilised the Acts to serve their own purposes. Information in the diary has been supplemented and verified by land selection files and probate files held at the Public Records Office Victoria.

The Curries were not alone, either in their vision of the future or their choice of destination. By the time they moved to the hamlet of Lardner in the county of Buln Buln in December 1875 the area already accommodated a number of settler families. Some of these farmers had been neighbours of the Curries at their previous abode in Ballan, a small town between Ballarat and Bacchus Marsh. Since John Currie married Sarah Ann Catherine Wells in 1865 the couple had made their life on the land, initially at Parwan, south of Bacchus Marsh, but mainly near the town of Ballan on land John selected. This was supplemented with some leased and purchased land; they were well versed in the intricacies of what was required of them as farmers. The sale of their previous property had netted the Curries nearly £600 with which to commence their new farming venture. This is significant; the Land Acts were seemingly designed to assist anyone – even those without funds – to select land. The repayments of the land purchase price were spread over many years, and often beyond the seven years anticipated by the legislators. However, the absence of capital was a major barrier to success, particularly in regions where the land had to be 'improved' before it could be cultivated; that is, cleared of indigenous vegetation. The Selection Acts were clearly not framed with the clearance of forests in mind. That the Curries arrived with the proceeds of their previous farm's sale in their pockets meant they could afford to purchase labour and seed and, most importantly, time. Given the enormity of the clearing project – the landscape was filled with trees as tall as 60 metres and dense undergrowth – time was a precious commodity.

---

[3] Other authors have not used the same logic in referring to Kate. For instance, Ailsa McLeary refers to her as Catherine, and Tony Dingle calls her Ann. Ailsa McLeary, 1998. *Catherine: On Catherine Currie's Diary, 1873–1908*. Carlton: Melbourne University Press; Tony Dingle, 1984. *Settling*. Vol. 2 of *The Victorians*. McMahons Point: Fairfax, Syme and Weldon Associates. See The Currie Diary 1873–1914. MS 10886. State Library Victoria Australian Manuscript Collection. Photocopy. 7 vols: Vol 1, 6 Jan 1875; Vol 1, 3–4 February 1876; Vol 1, 29 July 1877; Vol 2, 16 February 1881; Vol 2, 21 February 1881 (hereafter The Currie Diary).

The land laws in Victoria required the settler to fulfil a number of requirements once the licence was granted but before the lease was issued. Anyone over the age of 18 could select land, except married women and previous selectors, of more than 320 acres. The selector held his block under licence from the Crown for three years, during which time he was required to make improvements to the value of £1 per acre, erect a permanent dwelling and live on that property, clear and cultivate 10 per cent of the land, fence the property, and pay two shillings per acre per annum in rent. If these conditions were met a lease was issued and the rental payment would continue for another seven years, at the end of which the land became the freehold property of the settler. Alternatively the settler could buy the land at the end of the licence period for 14 shillings per acre. These regulations were designed to encourage farms rather than, or instead of, more pastoral runs. These would be peopled by independent yeoman farmers who, through their intensive cultivation practices, would create a hive of rural activity and whose self-interests would serve to break the stranglehold on land and politics that the squatters enjoyed.[4]

This was a particularly Victorian vision – the NSW Robertson Land Act was not nearly so encouraging to the development of agriculture and yeoman farmers.[5] This vision of independent farmers providing for themselves and sending their surpluses to market has been viewed as a response to industrialisation: a harking back to a vision splendid of a simple life attuned to the seasons, and nary a factory nor machine in sight; a glorification of a time that had not ever really existed in England, when the 'small man' was in control of his own destiny.[6] It was protectionist in essence, and an antidote to

---

[4] Dingle, *Settling*, 58.

[5] D.W.A. Baker, 1958. 'The Origins of Robertson's Land Acts'. *Historical Studies*. 8 (30), 166–182; Bill Gammage, 1990. 'Historical Reconsiderations VIII: Who Gained, and Who was Meant to Gain, from Land Selection in New South Wales?' *Australian Historical Studies*. 24 (94), 104–122. Baker notes that the main purpose that Robertson had in mind when framing the land acts was to address the inequalities of opportunity faced by small settlers and landowners in general, as opposed to the squatters. The Robertson land acts are far more accommodating of *laissez faire* economic principles. 'Robertson's speeches are barren of plans for the settlement of a yeomanry'. Baker, 'The Origins', 181.

[6] This issue deserves far more attention than it has been given here and is the subject of further investigation by the author. Overton discusses the various descriptions of yeoman in use, including peasantry, self-sufficient family farmers, small landholding farmers, and the like. He concludes that yeoman were ranked below gentlemen and were more likely to perform some of the manual labour themselves; they did not necessarily own the land they worked, but their holdings were of the middling size, a size he chooses not to quantify. Dingle quantifies the landholdings as being between 30 acres and 300 acres in size, and therefore large enough to require the assistance

market forces which were subject to wild and unpredictable fluctuations. '[A]grarian ideology... embodied a vision of a society restored to its natural state – one of non-competitive, independent agriculture'.[7] It was a rapacious idea, following the British settlers to all corners of the New World, perhaps as a result of the seemingly endless vistas of land stretching out in all directions, empty as the day was long. In the United States, the yeoman ideal became 'part of the national ideology', albeit suffused with individualism in a manner peculiar to the Americans.[8] As Hesseltine observes:

> the yeoman tradition rested firmly on the belief that the tiller of the soil was the primary source of American wealth, that all other ways of life were dependent upon him, and that he who was surrounded by his own farm and enjoyed the fruits of his own industry owed no man anything but love and goodwill'.[9]

A liberalisation of the land laws followed, which promoted the quest for expansion across the west. And yet 'the values and attitudes underpinning the belief in the yeoman myth were already anachronistic by 1862' when President Lincoln signed the Homestead Act which allowed for an increase in land claims.[10] Land speculation and business-like attitudes to farming had already overrun the idealism of yeomanry, as new and cheaper forms of transport transformed the ability to move goods to market and people to the latest frontier. An 'adequate sufficiency' quickly proved to be inadequate in the face of rising living standards attributable to industrialisation. Nonetheless, the yeoman myth survived, drawing succour from the influx of European migrants, desperate to get access to their own piece of paradise.

---

of hired labour. Thus Dingle's yeoman is sufficiently large to be more capitalist than self-sufficient. In either case, the yeoman hardly represented the 'small man' per se; rather, more of a middling to large man, given that 22.5 per cent of farmers identified in the 1851 English census worked landholdings of less than 20 acres, and only 37.8 per cent of farmers worked more than 100 acres of land. Mark Overton, 1996. *Agricultural Revolution in England: The Transformation of the Agrarian Economy, 1500–1850*. Cambridge Studies in Historical Geography. Cambridge: Cambridge University Press, 40; Dingle, *Settling*, 58; Leigh Shaw–Taylor, 2005. 'Family Farms and Capitalist Farms in Mid Nineteenth–Century England'. *Agricultural History Review*. 53 (2), 165.

[7] David Goodman, 1994. *Gold Seeking: Victoria and California in the 1850s*. Stanford: Stanford University Press, 123.

[8] J.M. Powell, 1978. *Mirrors of the New World: Images and Image Makers in the Settlement Process*. Canberra: Australian National University Press, 65.

[9] William Best Hesseltine, 1961. 'Four American Traditions'. *The Journal of Southern History* 27 (1), 3–32.

[10] Powell, *Mirrors of the New World*, 67.

At this time the Victorian legislators also employed the concept of yeomanry to 'sell' an American model of land selection to the large post-mining population, who were agitating for the land to be unlocked.[11] As early as 1852, newspapers such as the *Argus* had been promoting the benefits of putting the small man on the land. Diggers and urban reformers joined together to form the Victorian Land League in 1856, developing a list of demands which bore remarkable similarity to the Chartist reforms of 1840s England, albeit adjusted for local conditions.[12] The delegates to the reformist Land Convention of 1857 were eager to promote 'the Chartist ideal of the yeoman farmer' because it best suited the political needs of the Convention.[13] The final list of demands included free selection before survey, payment by instalment, abolition of squatting tenure, free commonage on all unalienated Crown land and protection of native industries.[14] Land reforms along these lines were implemented from 1862.

Nonetheless, the vision of the yeoman farmer was not a sophisticated piece of propaganda, be it to sell the land or convince potential urban dwellers of the virtues of a life of hard work and dubious reward on the land. Indeed, David Goodman argued that agrarianism was a response to the dislocation and disruption of society that gold-seeking provided; that gold was akin to industrialisation in this manner. Goodman notes that both conservatives and radicals arrived at a similar agrarian ideal as a solution to the discontents that they noted: the conservatives were concerned about the 'spectre of a society of masterless men, an anarchy of self-employment in which the "great social bonds" no longer held'; the radicals wanted to create a 'new land which could be made free from the political corruption and aristocratic domination of the old'. Even so, consensus was difficult to find, and the fundamental issue was how land should be utilised. The conservatives, most of them with squatting interests, dominated the Legislative Council (the Upper House) and were divided on the issue of free trade in land versus protection of their landed interests. The radicals, predominantly middle-class urbanites, were dominant in the Legislative Assembly (the Lower House).

---

[11] David Denholm draws attention to an alternative use of the yeoman ideal, as the means to maximise the sale of Crown land. David Denholm, 1979. *The Colonial Australians*. Ringwood: Penguin.

[12] Powell, *The Public Lands of Australia Felix*.

[13] John McQuilton, 1979. *The Kelly Outbreak 1878–1880: The Geographical Dimension of Social Banditry* Carlton: Melbourne University Press, 25.

[14] Margaret Kiddle, 1961. *Men of Yesterday: A Social History of the Western District of Victoria 1834–1890*. Carlton: Melbourne University Press.

Notwithstanding their lack of knowledge on the specifics of farming or the particular challenges presented by the climate and poor soils of the Victorian landscape, the members of the Lower House were eager to accommodate the miners, labourers and tradesmen demanding access to land.[15]

In their attempts to create an agrarian ideal for yeoman farmers the Victorian legislators had scant understanding of the onerous burden the regulations created; even when the Royal Commission inquiring into the Progress of Settlement under the Land Act 1869 (1878–79) determined as much, the amendments to the legislation were only marginal improvements. In some instances the changes created more problems than they solved. The most significant changes to arise from the Royal Commission were the extension of the licence period, and the removal of restrictions on mortgages over the settlement blocks.[16] The main criticism – that the amount of land that could be selected was too small to provide a sufficient income – was entirely ignored. But equally, the amount of work required to bring the land to productive capability was utterly underestimated, particularly in Gippsland. The government was also criticised for the lack of transport infrastructure that was so critical in assisting farmers to reach the main markets. Gippsland was serviced by a series of bullock tracks in 1875, when the Curries first moved to Lardner. In 1879 the train began servicing Drouin, but the roads were a constant source of agitation. Warwick Frost, building on a theme initially raised by Stephen Legg, contends that this reluctance to invest in the 'wet frontier' was not confined to Victoria.[17] It was an affliction common to all governments along the east coast of Australia because the output of these areas did not justify the expense involved in development, until the arrival of international dairy opportunities in the late 1880s.[18]

The land clearing provisions and the fencing requirements were of particular difficulty in Gippsland. This was a region of primeval forest bounded by the swamps of Koo-wee-rup to the east, the tee-tree scrub near the south coastal regions and as far as the eye could see in the north and east. Frost concluded that the best description of the 'wet frontier' was provided by the first-hand accounts found in *The Land of the Lyre Bird*:

---

[15] Goodman, *Gold Seeking*, 115-129.

[16] Powell, *The Public Lands of Australia Felix*.

[17] Stephen Legg, 1984. 'Acadia or Abandonment? The Evolution of the Rural Landscape in South Gippsland, 1870–1947'. Melbourne: Masters thesis, Monash University.

[18] Warwick Frost, 1997. 'Farmers, Government, and the Environment: The Settlement of Australia's "Wet Frontier", 1870–1920'. *Australian Economic History Review*. 37 (1), 19–39.

the scrub itself was, generally speaking, a dense growth of many kinds of trees – hazel, musk, blackwood, wattle, gum, saplings, etc. etc.– growing so thickly together as to present an appearance of a forest of bare poles, with the foliage at the top and a ruck of undergrowth and rubbish in the bottom; while all through it grew a forest of very large eucalyptus trees... Coming now to the scrub itself, that tremendous jungle forty to sixty feet in height that filled in the spaces between the great trees, a wonderful variety of flora was to be found in it.[19]

The Curries chose land set back from Lardners Track, which joined Brandy Creek to the Old Sale Road in the north.[20] At the bottom of the selection was a large swamp, into which ran the runoff from the gentle hill on which the balance of the selection was positioned. The effort required to clear the land was considerable, expensive and dangerous. For instance, swordgrass was an ever-present danger, growing in height from three to eight feet, and springing up plentifully after the scrub clearing initially took place. 'A cut from one of the leaves was severe', stated Mr T.J. Coverdale, who wrote of his own experiences in 1920.[21] However, the most significant danger lay in the methods and tools used to clear the forest. A sharp axe was essential, and unrelenting effort was required to make headway; accidents and strains were a regular occurrence. Wiregrass, which wound itself around the trunks of trees and saplings, could also provide a painful reminder of the challenges of clearing, with its raspy texture biting into the skin as the axeman tried to extricate his axe from its tendrils, and it provided a further opportunity for the settler to gash himself if he made contact with it unwittingly.[22] Getting lost was another risk, as Kate noted in her diary on 14 July 1877: 'I don't feel very well. think [sic] I hurt myself when I was lost yesterday'.

John and James Currie had made a number of forays to Lardner prior to their permanent settlement there, in order to commence the land clearing process. When Kate initially arrived, with her three children (Katie aged eight, Tom aged six and Bertie just five months of age) John showed her the

---

[19] Ibid.; T.J. Coverdale, 1972. 'The Scrub'. In *The Land of the Lyre Bird: A Story of Early Settlement in the Great Forest of South Gippsland*. Korumburra: The Shire of Korumburra for the South Gippsland Development League, 20–21.

[20] The contemporary rendering was commonly 'Lardner's Track', but this is now standardised to 'Lardners Track'.

[21] Coverdale, 'The Scrub', 25.

[22] W.H.C. Holmes, 1972. 'Scrub Cutting'. In *The Land of the Lyre Bird: A Story of Early Settlement in the Great Forest of South Gippsland*. Korumburra: The Shire of Korumburra for the South Gippsland Development League.

site he had chosen for the house, and she wrote, 'I like it fine'. Enough land had been cleared by John and James for the family to set up camp until the house was built (they moved to the new house – which had dirt floors and no chimney – in early March 1876). A small plot had already been cleared for a garden, and within days of arriving 'John planted one row of Kidney potatoes, the new ones with the biggest of the early rose at the top of the row'.[23]

Transforming the forest into farmland required destruction of the very fabric of its being, and fire was one of the most effective tools the settlers had at their disposal. On 13 January 1876 Kate wrote, 'Very hot. John set fire to the place where the House is to be – it is tearring [sic] away'. But on 14 January the news was not so good: 'Came on rain at night so John thinks the fire did not do so much good.' Burning was used by all the Gippsland settlers to clear the land, and work occurred all year round to prepare for the next burn. First, trees would be ringbarked and, much later, the next year or possibly later still, some would be cut down. This would also bring down some of the undergrowth. However, most of the preparatory work was in cutting the scrub, which would then allow the undergrowth to dry off. As a patch of forest was cut, and light penetrated into areas that hadn't seen much previously, new plants sprang up to take advantage of their chances, particularly swordgrass. Thus the burns would often be hampered by green undergrowth, which resisted attempts to ignite. Burns took place when the weather was right: on a fine, relatively calm day in mid January in the early years of the settlement at Lardner, and late summer once the district had some land in productive usage. In 1876 burns took place all round the district; necessity dictated that land clearing be done as quickly as possible. For the Curries, the early burn in January was for a particular purpose: to clear a site for a house. This purpose could hardly be stalled for another couple of months as all that protected the family on this wet frontier was a canvas tarpaulin. On 3 February Kate noted, 'Lots of fires to day – McKay burned his – he says it did not do well.' The next day she commented on John's afternoon absence: 'Away with G Grants lot to burn – 9 of them here altogether say they had a good burn.'[24] From January to March of 1876, Kate mentioned that either John or one of their neighbours were burning on 12 separate occasions. Burns were a communal affair, as neighbours worked together to maximise their chances of having a 'good burn'.

---

[23] Currie Diary, Vol. 1, 6 January 1875.
[24] Currie Diary, Vol. 1, 3–4 February 1876.

These burns did not always work out as planned. Sometimes they failed through a change of weather, other times they got out of hand when the wind sprang up unexpectedly. James Currie's shack, a one-roomed separate home built on 'his' selection to fulfil the permanent settlement requirement, was burnt to the ground two years running, in 1879 and 1880. In 1881, John set fire to an old tree that had previously fallen onto the potato paddock. On 15 February it was a very hot day, and despite John's preventative actions, the fire quickly spread to some trees close by: 'We passed a very anxious night up almost all the time. Two or three large trees have been showering sparks in all directions. John said he would chop them all down in the morning'.[25] However, the axe had been left near the fire the previous day by Tom, and the handle had burnt out. 'James [sic] axe would not cut, and we had not another handle'. Kate went to borrow the McEvoys' axe before daylight, such was her worry, but their axe was at Greenshields to whose house she duly went.[26] The distance Kate covered before dawn was considerable, approximately three miles through the bush, and there is no indication in the diary that she caught, saddled one of the horses and rode; she most probably walked. A few days later, Kate wrote of her exhaustion: 'I suppose I was so frightened and I was very hot'.[27] In March another controlled burn caused high anxiety, but major fire was averted by a timely change in the weather.

By the 1880s Kate was constantly on edge all summer through fear of fire. She recorded in 1888 a particular row that she had with John over when to burn. This was very distressing for her for a number of reasons. She was not querying the need to burn, but the timing. Kate had advised her husband to hold off burning until after 1 March, partly because later burns tended to be more effective as the bush had more time to dry out. However, timing was well against her. After begrudgingly delaying the annual burn until March, the weather promptly turned bad, and rain set in. The cost of such a delay in putting newly cleared land into production weighed heavily on Kate's heart. 'This, of course, entailed serious loss, as there was the loss of the area cut for the whole season, and the carrying forward for another year also meant additional labour, as undergrowth, such as dogwood, wiregrass, swordgrass, musk and firewood made a prolific growth in the following Spring'.[28] To

---

[25] Currie Diary, Vol. 2, 16 February 1881.

[26] The McEvoys were the neighbours on the northern side of the Currie property. Whilst not the nearest neighbours, they were the nearest neighbours who would have been able assist in this sort of crisis. The Greenshields lived considerably further away.

[27] Currie Diary, Vol. 2, 21 February 1881.

[28] Holmes, 'Scrub Cutting', 67.

make matters worse, Tom Currie, aged 19, had temporarily left the family to work on the railways, leaving the family short on labour. Katie, aged 21 years, was taking his place at her father's side. The lack of a good burn exacerbated the slow progress John was making in land clearing in the summer of 1888.

The work of clearing did not finish with the burn. This was when the heavy axe work commenced in order to bring down the trees that had been burnt but not felled. As many of the trees that were burnt were reasonably green prior to the burn (unless they had been ringbarked many years earlier) they were relatively easy to cut. However, they needed to be cut into small pieces, as they were heavier than dry wood. The whole family would then be employed in 'picking up'; that is, collecting the pieces of wood left behind and the ones just cut. These were heaped together to await another burn: 'pileing'. as Kate Currie calls it. These smaller fires were more easily controlled, and as the wood had already been burnt, it was drier. The heat of summer was not essential for these burns and they generally took place in the autumn, around April, sometimes smouldering for days.[29] Stumps would have fires set inside them in order to eliminate them from the landscape. Sometimes they would be left, as permanent reminders of how the landscape had been 'tamed', but only just. Once the land was cleared of wood, the ground could be prepared: ploughed and harrowed then planted with potatoes or turnips to break up the ground more deeply, or sown with various grass seeds, such as rye grass, cocksfoot seed and clover.[30] It is timely to remember exactly how this was done at the time, and the skill involved in doing it: the era of mechanisation had come to the harvest, but not yet to the sowing. John drove a horse – which was harnessed to a heavy piece of machinery – back and forth across the paddock being prepared, first in one direction, and then across the grain. Having harrowed, the plough was then employed in a similar fashion. As Kate noted on 22 April 1876: 'Fine day. John Ploughing – can't do much as he is short of feed for the horses and he does not like to work them.' It was hard work for both man and horse.

The Currie family struggled to fence their property due to the thick bush covering the land. The need to fence was regularly noted, through the unwanted arrival of other people's stock and the misplacement of their own. For instance:

---

[29] Holmes, 'Scrub Cutting'; W.H.C. Holmes, 1972. 'Picking Up'. In *The Land of the Lyre Bird: A Story of Early Settlement in the Great Forest of South Gippsland*. Korumburra: The Shire of Korumburra for the South Gippsland Development League; Frost, 'Farmers, Government, and the Environment'.

[30] This is how John Currie processed the land, and his seeds of choice.

3 May 1876: Showery John clearing. Murdies cattle back twice to day – all over the ploughed ground.

19 June 1877: the horses knocked down the rails and are at the furtherest end of the place. John ploughing.

Fencing was critical in the maintenance of animal husbandry, a fact of which farmers were well aware, so it is curious that the Victorian legislators saw need to specify the necessity of fencing. Whilst John Pickard has paid considerable attention to the type of fences constructed, the suitability of such, and for what purposes fences were constructed, he has not considered other meanings that could be conveyed by the erection of fences.[31] Paul G Boucier discusses at length the symbolism of fences to the gentleman farmer of nineteenth-century America, noting that fences were 'symbols representing various facets of his conceptions of land, nature and society'. A fence conveyed 'the virtue of the farmer who constructed it' with 'virtue' connoting an amalgam of all the positive attributes of the yeoman farmer as God's favoured sons of the earth: patient, frugal, honest, industrious, independent, self-reliant, vigilant, benevolent and honourable.[32] The fence also indicated order and control, a separation of the farmer's sphere of managerial influence from the wild, the area over which he could be justifiably proud, the portion of earth which he alone owned.[33] Thus the fence:

> was linked inextricably to contemporary ideas about property. Lockean philosophy formed the basis for the gentleman farmer's notion of private property, the right of a person to exclude others from the use or benefit of a resource. The fence was a physical manifestation of the exploitation of that right.[34]

---

[31] John Pickard, 1999. 'The First Fences: Fencing the Colony of New South Wales, 1788–1823'. *Agricultural History*. 73 (1), 46–69; John Pickard, 2005. 'Post and Rail Fences: Derivation, Development and Demise of Rural Technology in Colonial Australia'. *Agricultural History*. 79 (1), 27–49; John Pickard, 2007. 'The Transition from Shepherding to Fencing in Colonial Australia'. *Rural History*. 18 (2), 143–162; John Pickard, 2010. 'Wire Fences in Colonial Australia: Technology, Transfer and Adaptation, 1842–1900'. *Rural History*. 21 (1), 27–58.

[32] Paul G. Bourcier, 1984. '"In Excellent Order": The Gentleman Farmer Views his Fences, 1790–1860'. *Agricultural History*. 58 (4), 547, 552.

[33] The notion of a family farm was utilised more so in relation to the labour of those who worked it, than with regard to legal ownership.

[34] Bourcier, '"In Excellent Order"', 557.

Further, the fence was a sign of science and civilisation being imposed on the 'worthless wilderness'. Fences ran along survey lines, which took little notice of the topography or the natural features of the land itself. The surveyor simply applied his straight lines in an effort to make as equitable as possible the various selection plots.[35]

Fencing was a time-consuming and expensive pursuit. Philip McMichael estimated the expense of fencing to be £50 per mile, without giving specifics about when, where or what sort of fencing.[36] Sources do not indicate that it cost the Curries so much, particularly the fences John made himself, but research into this issue is as yet incomplete. John erected much of his own fencing, but he also paid Mr Greenshields, one of his neighbours who had also made the move from Ballan, to make fences for him. Mr Greenshields in turn employed a team of men to do the job, the identities of whom are not known, but probably included some of the other settlers in the area.[37] In May of 1877 Mr Greenshields was paid £18 for 45 chains of chock and log fencing; that is about 1 kilometre or 5/8 of a mile, or approximately £30 per mile. The fencing contractors continued to work all June and July, and as of 29 July, when 'we all went round the new fence' the fence was 'not quite finished'. By 1878 the farm, known as 'Brandie Braes' (a name harking back to John's Scottish ancestry) had 156 chains (around 3.5 kilometres) of various fences on John's selection, and another 238 chain (5.3 kilometres) of fencing on James Currie's selection, at a total cost of £197.[38] The fencing included slab fencing, dog-leg logs, log and brush, brush, as well as chock and log. Brush fences were the cheapest to erect, but the most flimsy and least durable. Chock and log fences were the most impermeable to livestock, but required considerable manpower to erect. John Currie made a number of chock and log fences himself, but generally his manpower was better directed towards other activities, such as preparing the paddocks for crops or planting. Wire fences did not feature on the Currie farm until the 1890s, even though the technology had been introduced from the USA in the 1850s.[39]

---

[35] Denholm, *The Colonial Australians*.

[36] Philip McMichael, 1984. *Settlers and the Agrarian Question: Capitalism in Colonial Australia*. Cambridge: Cambridge University Press.

[37] There are many instances where the local settlers worked for wages or as in-kind labour on each other's properties, particularly at harvest time, and the additional income became a critical addition to most settlers' farm incomes. This contracting opportunity may well have included some of the neighbours, but Kate does not provide any further details.

[38] PROV, Land Selection Files VPRS 626, 299/ 19.20 and VPRS 626, 300/19.20.

[39] Pickard, 'Wire Fences in Colonial Australia'.

John Currie was not a gentleman farmer in the same sense that Boucier considered, but the diary suggests that his approach to his fences was equally as laden with symbolism. The fence was a powerful signifier of his mastery over the bush. Every log was a tree felled, the brush came directly from the undergrowth; each chain of fence thus indicated John's encroachment on the wilderness, and his taming of it for the purposes of making a living. His pride in his efforts was regularly demonstrated by his surveying the land, with Kate by his side, when 'we all went round the new fence'.[40] There is no discernible sense in the diary that the Curries were conscious of displacing anyone or anything by their selection. And yet we know that this country was inhabited, by local fauna and the Kulin and GunaiKurnai peoples. There were indigenous fauna that were particularly bothersome to the Curries, such as wallabies, possums, wild dogs, snakes and the occasional hawk, and a variety of methods of ridding themselves of the unwanted 'intruders'. 'Bears' were also a feature, and whilst frightening to the children – Bertie in particular – were not considered vermin; at different times the term was used for koalas as well as wombats. The only inkling of an Aboriginal presence in the Currie Diary is Kate referring to a 'Dreadful coroberee (sic) among the wild dogs' on 7 July 1876. This strange comment is an uneasy reminder of the previous custodians of the land.

These ideas are hardly likely to have figured in the thinking of the legislators when they mandated that all properties be fenced within the initial licence period. Fencing had another important rationale at state level. In all likelihood, the fencing requirements were designed to dissuade the squatters from selecting more land than they could manage. Labour shortages prevalent throughout the gold rush ensured that most squatters already acknowledged the necessity of fencing their runs in order to maintain their herds, and had made the necessary capital improvements. Perhaps the government considered fencing essential to the peacefulness of neighbourhoods. A lack of fences could impose upon the goodwill of neighbours if wandering stock or disputes about boundaries were regular occurrences. By insisting on delineation of ownership of land via fencing, the legislators were possibly signifying their belief that productive farming would occur in Victoria, just as the enclosure movement in Britain had ushered in a massive increase in the productive capacity of land holdings.[41] Either way, there is research to be

---

[40] Currie Diary, Vol. 1, 29 July 1877.

[41] Of course there are many reasons for the rapid increase in agricultural productivity seen in England from the late eighteenth century to the mid nineteenth century, but enclosure

done on better understanding the intentions of the land legislators, and their various motives for action on this front.

The Currie family provide an excellent example of how individual farmers transformed the nature of farming into the business of farming: they adapted to the legislative requirements, amended the natural landscape, and persevered through the vagaries of climate. As Charles Fahey has demonstrated, farmers needed to employ capitalist farming techniques if they were to succeed; relatively small plots of land did not allow for the economies of scale required for a farmer to profit from the poor soils of Australia, and restrictions on the usage of the land only compounded the problems.[42] The Currie family bought the yeoman ideal, with its spirit of independence and industriousness, autonomy and self-sufficiency. Like all those who succeeded in farming a smallish plot of land, the Curries accepted the anti-capitalist rhetoric that lay behind the legislators' conceptualisation of an honest living on the land, subverting yeomanry to become mini-capitalists, running small organisations, using kinship and inheritance as organisational tools to maximise the use of labour for the benefit of the family project rather than the various individuals. Family members were exploited, but they expected to inherit the property one day as recompense. Inheritance of the family farm was fundamental to the yeoman ideal. They worked for less than they would have been paid were they working for anyone else, in order to benefit from their 'collective' ownership, which would be realised at some point in the future, with the death of the patriarch. Families without a ready-made workforce had a more difficult path to success because they had to rely on outside labour and neighbours would not work for such a pittance. JM Powell proposes that this was all within the ambit of the original legislative vision, when he notes:

(encompassing the removal of commons, the transformation of 'wastelands', the shift from customary tenants to leaseholders, and the consolidation of farm layouts) was a catalyst, without which the shift towards new farming techniques employing new technologies, crops and animal husbandry on larger farms could not have occurred. G.E. Mingay, 1989. 'Conclusion: The Progress of Agriculture 1750–1850'. In *The Agrarian History of England and Wales 1750–1850*, Vol. 6 Part 2. Edited by G.E. Mingay 8 vols. General Editor Joan Thirsk, Cambridge: Cambridge University Press; Overton, *Agricultural Revolution in England*. As Overton notes, '[t]he main incentive for landlords to enclose was that enclosed land was worth more than open commonfield land'. Overton, *Agricultural Revolution in England*, 162.

[42] Charles Fahey, 1984. 'The Wealth of Farmers: A Victorian Regional Study, 1879–1901'. *Historical Studies*. 21 (82), 29–51; Charles Fahey, 2011a. 'The Free Selector's Landscape: Moulding the Victorian Farming Districts, 1870–1915'. *Studies in the History of Gardens & Designed Landscapes*. 31 (2), 97–108; Charles Fahey, 2011b. '"A Splendid Place for a Home": A Long History of the Australian Family Farm, 1830–2000'. In *Outside Country*, edited by Alan Mayne and Stephen Atkinson. Kent Town: Wakefield Press.

[c]ontemporary parliamentary debates, newspaper campaigns, official emigrants' guides and the like all suggest that the idealism which assisted in spawning the land legislation assumed that the independent Australian farm, like the homestead in the United States, would be a home for the pioneer's lifetime, to be carefully tended and eventually passed on to his children'.[43]

Malcolm Voyce, in a provocative article, argues that the yeoman ideal was utilised by the state as a mechanism for the orderly governance of rural areas. 'In this way, property in the family farm context inevitably becomes a socio-technical arrangement which combines notions of kinship, residence and technicalities of property ownership'.[44] Thus the yeoman ideal was utilised to populate the countryside, and anchor families into the country economy.

The transformation of Gippsland from primeval forest to a patchwork of manicured paddocks occurred through the toil of settler families such as the Curries, and the generation who followed them. The clearing of the farm was still incomplete in 1901, when John Currie passed away. According to his probate records, some 100 acres of uncleared bush remained intact on the property, and another 185 acres of trees that had been ring-barked but not removed also existed. Less than half the total land selected was cleared and productive before John Currie died.[45] Nonetheless, the cleared portion of the farm provided income enough to keep five adults (John, Kate, Katie, Tom and Bertie) and two children (Fern and Rose). The Curries' impact on the landscape was irreversible, changing forever the ecosystem and possibly the rainfall patterns. The radical transformation was seen in terms of progress, from savagery and wilderness to civilisation. Whilst we may not condone such brutal desecration of the natural environment today, we are all the richer for the food bowl that was thus created.

Not all pioneer settler families were able to reap the benefits of their efforts in material ways, given how long it took to transform the countryside. The Currie family did see success in their lifetime, not extreme wealth, but the independence and comfortable sufficiency to which John and Kate had initially subscribed was most certainly their reward. Moreover, upon his death John left his farm to be shared amongst his five children after the

---

[43] Powell, *Mirrors of the New World*, 76.
[44] Malcolm Voyce, 2007. 'Property and the Governance of the Family Farm in Rural Australia'. *Journal of Sociology* 43 (2), 136.
[45] Public Records Office Victoria, John Currie, VPRS 28/P2/ Unit 587 and VPRS 28/ P0/1021.

death of his wife, Kate, who was bequeathed an income from the farm proceeds for the duration of her life. This was in accordance with the yeoman ideal. Katie, Tom and Bertie each received a share of the estate, whilst the youngest children, Rose and Fern jointly received a share. All the Currie children continued to farm the property together, bequeathing to the surviving siblings their share upon death.[46] Bertie was the only sibling to marry. In 1968 Fern Currie died, ending 94 years of family connection to the farm; she did not acknowledge Bertie's children in her will. Rumour has it that she never forgave her brother for extending the family.

References

Baker, D.W.A., 1958. 'The Origins of Robertson's Land Acts'. *Historical Studies*. 8 (30): 166–182.
Bourcier, Paul G., 1984. '"In Excellent Order": The Gentleman Farmer Views his Fences, 1790–1860'. *Agricultural History*. 58 (4): 546–564.
Coverdale, T.J., 1972. 'The Scrub'. In *The Land of the Lyre Bird: A Story of Early Settlement in the Great Forest of South Gippsland*. Korumburra: The Shire of Korumburra for the South Gippsland Development League.
The Currie Diary 1873–1914. MS 10886. State Library Victoria Australian Manuscript Collection. Photocopy. 7 vols: Vol 1, 6 Jan 1875; Vol 1, 3–4 February 1876; Vol 1, 29 July 1877; Vol 2, 16 February 1881; Vol 2, 21 February 1881.
Denholm, David, 1979. *The Colonial Australians*. Ringwood: Penguin.
Dingle, Tony, 1984. *Settling*. Vol. 2 of *The Victorians*. McMahons Point: Fairfax, Syme and Weldon Associates.
Fahey, Charles, 1984. 'The Wealth of Farmers: A Victorian Regional Study, 1879–1901'. *Historical Studies*. 21 (82): 29–51.
Fahey, Charles, 2011a. 'The Free Selector's Landscape: Moulding the Victorian Farming Districts, 1870–1915'. *Studies in the History of Gardens & Designed Landscapes*. 31 (2): 97–108.
Fahey, Charles, 2011b. '"A Splendid Place for a Home": A Long History of the Australian Family Farm, 1830–2000'. In *Outside Country*, edited by Mayne, Alan and Atkinson, Stephen. Kent Town: Wakefield Press.
Frost, Warwick, 1997. 'Farmers, Government, and the Environment: The Settlement of Australia's 'Wet Frontier', 1870–1920'. *Australian Economic History Review*. 37 (1): 19–39.
Gammage, Bill, 1990. 'Historical Reconsiderations VIII: Who Gained, and Who was Meant to Gain, from Land Selection in New South Wales?'. *Australian Historical Studies*. 24 (94): 104–122.
Goodman, David, 1994. *Gold Seeking: Victoria and California in the 1850s*. Stanford: Stanford University Press.

---

[46] Victorian Public Records Office Probate files of Albert Bryce Currie, VPRS 28/P3/Unit 2529; Thomas Bryce Currie, VPRS 28/P3/Unit 4082; Rose Currie, VPRS 28/P4/Unit 507; Catherine Currie, VPRS 28/P3/Unit 4962; Caroline Calfernia Currie, VPRS 28/P5/Unit 297.

Hesseltine, William Best. 1961. 'Four American Traditions'. *The Journal of Southern History*. 27 (1): 3–32.
Holmes, W.H.C., 1972. 'Picking Up'. In *The Land of the Lyre Bird: A Story of Early Settlement in the Great Forest of South Gippsland*. Korumburra: The Shire of Korumburra for the South Gippsland Development League.
Holmes, W.H.C., 1972. 'Scrub Cutting'. In *The Land of the Lyre Bird: A Story of Early Settlement in the Great Forest of South Gippsland*. Korumburra: The Shire of Korumburra for the South Gippsland Development League.
Kiddle, Margaret, 1961. *Men of Yesterday: A Social History of the Western District of Victoria 1834–1890*, Carlton: Melbourne University Press.
Legg, Stephen, 1984. 'Acadia or Abandonment? The Evolution of the Rural Landscape in South Gippsland, 1870–1947'. Melbourne: Masters thesis, Monash University.
McLeary, Ailsa, 1998. *Catherine: On Catherine Currie's Diary, 1873–1908*. Carlton: Melbourne University Press.
McMichael, Philip, 1984. *Settlers and the Agrarian Question: Capitalism in Colonial Australia*. Cambridge: Cambridge University Press.
McQuilton, John, 1979. *The Kelly Outbreak 1878–1880: The Geographical Dimension of Social Banditry*, Carlton: Melbourne University Press.
Mingay, G.E., 1989. 'Conclusion: The Progress of Agriculture 1750–1850'. In *The Agrarian History of England and Wales 1750–1850* Vol 6 Part 2. Edited by Mingay G.E. 8 vols. General Editor Thirsk, Joan. Cambridge: Cambridge University Press.
Overton, Mark, 1996. *Agricultural Revolution in England: The Transformation of the Agrarian Economy, 1500–1850*. Cambridge Studies in Historical Geography. Cambridge: Cambridge University Press.
Pickard, John, 1999. 'The First Fences: Fencing the Colony of New South Wales, 1788–1823'. *Agricultural History*. 73 (1): 46–69.
Pickard, John, 2005. 'Post and Rail Fences: Derivation, Development and Demise of Rural Technology in Colonial Australia'. *Agricultural History*. 79 (1): 27–49.
Pickard, John, 2007. 'The Transition from Shepherding to Fencing in Colonial Australia'. *Rural History*. 18 (2): 143–162.
Pickard, John, 2010. 'Wire Fences in Colonial Australia: Technology, Transfer and Adaptation, 1842–1900'. *Rural History*. 21 (1): 27–58.
Powell, J.M., 1970. *The Public Lands of Australia Felix: Settlement and Land Appraisal in Victoria 1834–91, With Special Reference to the Western Plains*. Melbourne: Oxford University Press.
Powell, J.M., 1978. *Mirrors of the New World: Images and Image Makers in the Settlement Process*. Canberra: Australian National University Press.
PROV, John Currie, VPRS 28/P2/ Unit 587 and VPRS 28/ P0/1021
PROV, Land Selection Files VPRS 626, 299/ 19.20 and VPRS 626, 300/19.20
Shaw–Taylor, Leigh, 2005. 'Family Farms and Capitalist Farms in Mid Nineteenth–Century England'. *Agricultural History Review*. 53 (2): 158–191.
Victorian Public Records Office Probate files of Albert Bryce Currie, VPRS 28/P3/Unit 2529; Thomas Bryce Currie, VPRS 28/P3/Unit 4082; Rose Currie, VPRS 28/P4/Unit 507; Catherine Currie, VPRS 28/P3/Unit 4962; Caroline Calfernia Currie, VPRS 28/P5/Unit 297.
Voyce, Malcolm, 2007. 'Property and the Governance of the Family Farm in Rural Australia'. *Journal of Sociology*. 43 (2): 131–150.

Chapter 9

'If the worst comes to the worst, you can always milk cows'
Creating the 'dairy industry problem' in Gippsland

Charles Fahey

Introduction

In the introduction to his 1943 sociological survey of the Victorian dairy industry, Maurice Rothberg wrote:

> The problems of the dairy farmer in Victoria are very much those of the farmer in Australia. The isolated farm and house, the distance to large centres, the poverty of schooling facilities, uncertain prices for farm produce, seasonal hazards – those are but a few of the factors in common. The routine of milking and feeding operations carried on day in and day out and often year in and year out constitute one of the more unique characteristics which sets dairy farming apart from other branches of agriculture. In return for this arduous labour, returns are disappointingly low, it being commonly felt that the incomes of dairy farmers rank among the lowest of all farming people. All these factors, and others, are perhaps implicit in the statement of one discontented dairy farmer: 'well after all, if the worst comes to the worst, you can always milk cows'.[1]

Less than 20 years after Rothberg studied Victorian dairy farming, and after one of the more prosperous decades in Australian agricultural history, many agricultural economists considered that dairying was a major problem

---

[1] Maurice Rothberg, 1948. 'Victorian Dairy Farming: A Social Survey.' Chapel Hill: PhD thesis, University of North Carolina, 1.

for the rural economy. These economists were concerned that there were too many small dairy farms in Australia where farmers did not milk sufficient cows to make an adequate income. This unprofitable industry relied on government support and was financed by the local consumer who paid a higher price for milk than the export price. This was the 'dairy industry problem'.[2]

This chapter will look at the historical origins of the 'dairy industry problem' by examining settlement in the county of Buln Buln in Gippsland (with particular reference to the shires of Buln Buln and South Gippsland) from the mid 1870s through to the 1930s. The chapter will argue that the 'dairy industry problem' had its roots in colonial and later state government policies of land settlement. In the 1870s, the colonial government offered land to 'free' selectors, on credit payment spread over ten years. The Act allowing free selection also obliged the settler to place improvements to the value of £1 per acre within three years. This was intended to place onerous financial burdens on former pastoral tenants and thus prevent them from acquiring large freehold holdings. This made sense on the open woodlands and grass lands of the Northern and Wimmera plains where large pastoral holdings pre-dated the act. It made little sense in large parts of Gippsland where there were few extensive pastoral runs and where there were dense forests. The Act also opened up land without any regard to the economics of Australian farming. Questions of products and markets were simply not considered. The driving ambition was the vague notion of creating a class of self-reliant yeomen farmers.[3] This was a resilient ideology that persisted well into the twentieth century. In the first two decades of the twentieth century closer settlement legislation perpetuated the folly of small, uneconomic farms. It was in the forests of Gippsland that this misguided ideology had its most tragic consequences.[4]

To examine the problems of the original settlers, the chapter will draw on the diaries of Sarah Ann Catherine (Kate) Currie, who settled with her family in the parish of Longwarry. Located in the county of Buln Buln, near the town of Drouin, the parish of Longwarry was dense forest when settlers arrived in the mid 1870s. Historians have been drawn to the Currie

---

[2] N.T. Drane, and H.R. Edwards, 1961. *The Australian Dairy Industry, An Economic Study*. Melbourne: Cheshire, 5.

[3] Marilyn Lake, 1987. *The Limits of Hope: Soldier Settlement in Victoria, 1915–1938*. Melbourne: Oxford University Press, 12–24.

[4] Patrick Morgan, 1997. *The Settling of Gippsland: A Regional History*. Leongatha: Gippsland Municipalities Association.

diary for the vivid portrait it paints of the private life of a selector family.[5] The tragedy of the drowning of an infant child and the subsequent mental breakdown of Kate has also captured the attention of historians. However, the diary is primarily a record of the work carried out on the Currie farm, as the previous chapter in this collection by Kerrie Nixon shows, and it is one of the best accounts we have of the project to settle and farm in the coastal forests of Australia in the nineteenth century, a region described by Warwick Frost as the 'wet frontier'.[6] To place the Currie family in context, all extant land selection files from their parish of Longwarry have been examined. Correspondence in these files ends when the land was purchased from the Crown, and municipal rate books have been examined to trace the land holdings of settlers beyond the date they were issued with a Crown grant. Rate books were examined for 1891 and 1911 to determine the area of land (in acres) held by all residents in the parish. Nineteenth century agricultural statistics tell us little about Victorian farming other than the area sown to crops and the aggregate numbers of livestock in each shire. Official agricultural statistics throw little light on the management of individual farms. However, from the early twentieth century detailed agricultural statistics are available for small areas. These have been examined to analyse farm activity from 1911 to the 1930s. Details have been collected for the Curries' parish of Longwarry and the adjoining parish of Drouin West. Comparative data from the neighbouring parishes of Mirboo South and Wonga Wonga has also been compiled. When the closer settlement scheme fell into financial disarray, closer settlement officials collected considerable information on all settlers' farming activities to refinance debts. The chapter will examine the fortunes of twentieth century 'closer settlers' through their revaluation files.

The Currie family's selections

Land Acts were introduced in 1860, 1862, 1865 and 1869 to break up large pastoral estates. As historians from Stephen Roberts to JM Powell have shown, the first three acts were a failure in breaking up the large pastoral estates and much Crown land passed from control by pastoralists as Crown

---

[5] Ailsa McLeary, 1998. *Catherine: On Catherine Currie's Diary, 1873–1908*. Melbourne: Melbourne University Press.

[6] Warwick Frost, 1997. 'Farmers, Government, and the Environment: The Settlement of Australia's "Wet Frontier", 1870–1920.' *Australian Economic History Review*, 37 (2) (March): 19–38.

licensees to freehold ownership by the same squatters.[7] There is not space here to outline why the Acts failed but I would argue that their main problems were that they offered land at too a high a cost for poor settlers, and most of the land open for selection was isolated from markets. Recently, Domytro Ostopenko has also argued that there was little demand for much of the land opened up under the first two Acts.[8] Yet there were, as I have argued elsewhere, a large number of settlers who did establish small farms under the first three Acts.[9] These farms became a staging point for settlers under the 1869 Land Act, and migrants moved in three major streams: to the Wimmera Plains, to the Northern Plains and to Gippsland.[10] Settlers on the plains confronted a mixture of open woodlands and grasslands and relatively quickly produced marketable crops. In the forests of south Gippsland settlers faced the much more daunting task of clearing dense forests and finding marketable products. John and Kate Currie were among the intrepid settlers who pioneered Gippsland.

We first meet the Currie family when John Currie, a farmer in Ballan, commenced a diary in 1873. He soon abandoned this chore and the recording of farm activities was left to his wife Sarah Ann Catherine, or Kate as she was known by her family.[11] John Currie was born in 1834 in West Calder, near Edinburgh, the son of a farmer. He migrated to Victoria in 1854 and took part in the great alluvial gold rushes. We know little of his early colonial life. He appears in official records for the first time in 1864 when, at the age of 30, he married Sarah Ann Catherine Wells. Kate, then aged 19, was born in Suffolk and migrated to Australia as an infant. Her father Leper Harry Wells was a brick maker operating in the Ballarat and Ballan districts. John described himself as a farmer on his marriage certificate and he may have been a tenant farmer. Kate at marriage had no stated occupation. It was not

---

[7] J.M. Powell, 1970. *The Public Lands of Australia Felix: Settlement and Land Appraisal in Victoria 1834–91 with Special Reference to the Western Plains*. Melbourne: Oxford University Press.

[8] Domytro Ostopenko, 2012. 'Facing Economic Maturity: Farmers of the Colony of Victoria in 1840–60.' Melbourne: PhD thesis, La Trobe University.

[9] Charles Fahey, 2008. 'Moving North: Technological Change, Land Holding and the Development of Agriculture in Northern Victoria, 1870–1914.' In *Beyond the Black Stump, Histories of Outback Australia*, edited by Alan Mayne, Adelaide: Wakefield Press: 183–190.

[10] Charles Fahey, 2011. 'The Free Selector's Landscape: Moulding the Victorian Farming Districts, 1870–1915.' *Studies in the History of Gardens and Designed Landscapes*. 31 (2): 97–108.

[11] Sarah Ann Catherine Currie, Farm Diary 1873–1916, State Library of Victoria, Australian Manuscript Collection, MS 10886 (hereafter Currie, Diary).

until the year after his marriage, in June 1865, that John took up 203 acres in the parishes of Gorong and Parwan near Bacchus Marsh under the 1865 Land Act. Finding that part of the land was more unsuitable for grazing than cropping, he transferred his interest in 90 acres to a neighbour. John continued to farm the balance of his selection, and by 1871 he also rented a further 120 acres.[12] Ballan, with access to the Ballarat market, was in a good position to establish a successful farm and for almost 10 years the Currie family sold grain, livestock and dairy products to the populous gold mining market.[13]

By the mid 1870s notes appear in the diary about preparations to move and select in Gippsland.[14] John was clearly encouraged to move by a neighbour, and his first trip to Longwarry parish was with Lachlan Grant of Ballan. The diary offers no explicit account of why the Curries moved; in a later letter to the land bureaucrats John did explain that he needed more land. By October 1874 John had pegged 110 acres; as he had selected previously, he could only take up the balance of his entitlement to 320 acres. The fact that he had transferred an earlier selection aroused the suspicions of authorities as to his *bone fides* as a selector. This delayed his moving onto the land. However, John appealed for support from his local member. This proved successful and he was granted a licence to occupy 109 acres in September 1875.[15] To extend the family holding, James, his brother, selected an adjoining block of 320 acres.

In absolute area the new farm was over twice the size of the original 1865 Ballan farm; it would, however, take many years before the area available for agriculture was equal to the old farm. The former holding in Ballan was a key to re-settlement in Gippsland; the sale of the first property was used for subsistence and to finance clearing on the new farm. In October 1875 Kate recorded in her diary:

> John got a note from Mr Musgrave to say he had paid Mr Grant's cheque in to the bank for the three hundred pounds to our credit and his bill for £283 for collection due 10 April 1877.[16]

---

[12] The genealogical information is based on death and marriage records for John and Kate Currie. John's selection history is detailed in his file Victorian Public Records Office Series (VPRS) 626/P0/1557/299. For the rented land see Ballan Rate Books 1871, VPRS/5557/P0/5.

[13] Currie, diary. The diary for 1873–1875 describes the farming routine in Ballan.

[14] John's first trip to Gippsland was on 4 March 1874.

[15] There were some discrepancies in the size of the selection due to rounding errors and surveying inaccuracies. The Curries' final property was 429 acres.

[16] Currie, diary 21 October 1875. The diary was very rarely punctuated and the quotations in this chapter maintain the original punctuation, or more often lack of punctuation.

Table 9.1: The Currie work routine, 1877

January	Clearing	Gardening	Harvesting oats and barley	Burning off
February	Finished reaping	Burning off	Piling up for fire	Carting palings
March	Piling	Burning off	Started Sowing	
April	Piling	Burning	Fencing	Cutting logs and digging potatoes
May	Building	Digging potatoes	Chopping and splitting logs	Piling logs
June	Fencing	Ploughing	Chopping logs	Planting Fruit trees
July	Ploughing	Digging around fruit trees	Fencing and log pulling	Digging carrots sowing grass
August	Clearing and rolling logs ready for burn	Planting cocksfoot	Fencing & gathering sticks with the children	Sowing
September	Clearing and chopping logs	Burning logs and raking fire	Erecting chock and log fence	Planting flower seeds
October	Collecting bark	Building dairy	Pulling logs, making fires, chopping and clearing	Planting potatoes and maize
November	Planting potatoes	Pulling logs, burning off	Shearing and fencing	Cutting logs for stockyard
December	Clearing	Getting chocks ready	Cutting hay	Burning off

(Currie diary, 1877.)

This sum was over five times the yearly wage of an urban labourer. In applying for his lease in 1878 John Currie testified that he placed improvements on James' property to the value of £578 and on his own property to the extent of £384. Although family labour was employed in clearing, fencing and planting, John also paid contractors to help him improve his selection.[17] He in turn extended his own cash reserves by working as a contractor. At the end of three years much had been accomplished on the combined area of 429 acres. The family cut the scrub and ringbarked the trees on 200 acres of land. They very quickly built a family house, a hut for James, and a barn. Particularly burdensome was the erection of fences: 235 chains (approximately 4727 metres) of chock and log fences, 26 chains (523 metres) of stub and picket fences, 54 chains (1086 metres) of log fences, 35 chains (704 metres) of brush

---

[17] Currie, diary 3 January 1876 for a contract to employ labour.

and log fences, and 44 chains (885 metres) of brush fences.[18] Clearing a farm from the forest was both expensive and backbreaking labour, which quickly ate into savings. After three years John valued his total improvements at £962. This was almost £400 more than he received from the sale of his Ballan farm. In addition to ringbarking, chopping, piling and burning scrub, normal farming activities such as ploughing, shearing, harvesting and gardening had to be undertaken. Table 9.1 lists the main activities undertaken by the family during each month in 1877 and highlights the unremitting labour required to win a farm from the forest.

The search for an income

Due to the rigours of clearing the Curries faced almost two years without substantial farm income, and it was not until late in 1877 that Kate's journal regularly reported dairying activity. On 2 October 1877 Kate recorded that 'John home left a cow at Dandenong and returned with five cows and bull'. Less than a fortnight later she made her first of many entries in her diary about milk production. She wrote, 'Rintel (a local storekeeper) came for butter – gave him 6 lb I have to send him 2 quarts of milk every day at 4d per quart'. Four days later she observed 'commenced to get back room ready for dairy'. On 10 November she received her first payment of £2/10/10 for four weeks, and purchased a pair of ordinary boots for herself and Sunday boots for her oldest child, Katie. However, her optimism was soon tested. On 12 December she wrote that Rintel was unwilling to take any more, 'He says I must not send any more yet as he does not know what to do with it'.[19]

With the arrival of the railway to Gippsland the Curries could diversify their markets. In April 1878 Kate Currie gleefully recorded, 'John went to Melbourne… his first trip by train'. With access to the train and the possibility of placing butter in the Melbourne market, she sought out merchants by mail. However, even with the extra markets in Melbourne, dairying remained a risky and uncertain enterprise. One of the major problems was the fluctuating price for butter. In March 1885, when she received a cheque for £25 for her butter, she wrote, 'I wish he [Mr Tait] would put a ticket in the box to let me know the price every week'. Feeling that she was not getting the best price, she wrote a letter of complaint to Tait. He in return wrote an indignant letter. He rebuked Kate Currie that those who informed her that

---

[18] John and James Currie Land Selection Files VPRS 626/P0/1557/299 and VPRS 626/P0/1557/300.

[19] Currie, diary, 2 October, 17 October, 4 November, 10 November and 12 December 1877.

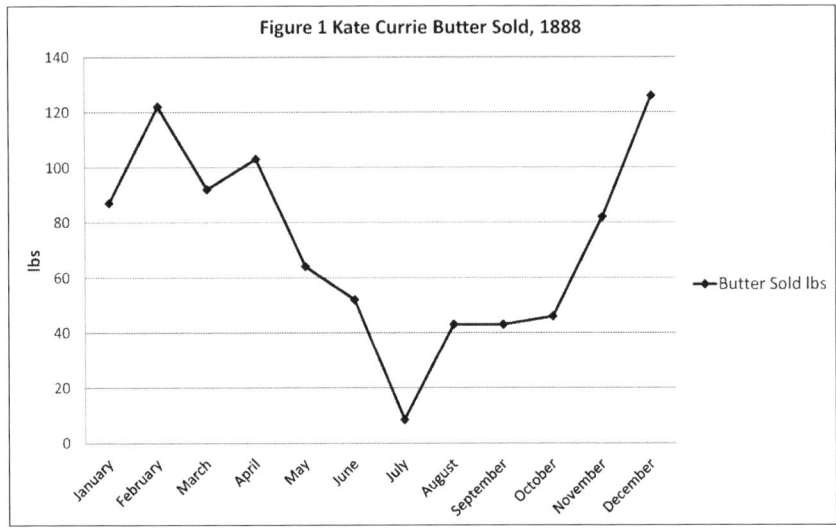

Figure 9.1: Kate Currie butter sold, 1888

he was not giving the best price were liars. She was, however, convinced that her letter had 'helped the price as it rose 7d per lb in one week, it never rose so much as that at once before'. Three years later she was still selling to Tait and price was still an issue. On 9 July 1880 Kate wrote:

> Got a cheque from Mr Tait for May's butter £8/5/5 think it very good he does not say anything about the cheque for £5 he sent me I suppose it is a present a very genteel way of giving me a little more for my butter it was very acceptable many thanks to him.[20]

The Curries slowly built up their herd and by 1888 were milking 30 cows at the season's peak. However, as pasture died off in winter cows were dried off. The Curries tried to deal with this by hand feeding their cattle. On 24 May 1888 Kate wrote, 'some smart showers. Agnes Hardie's wedding Day Daa ploughing his young folk clearing for him it takes such a long time to give the cows the maize'. A few days later she observed, 'Fine day cloudy but no rain Daa ploughing Katie went out with him this morning I milked all myself we are just milking 8 I now hope to have more soon'.[21] However, the Curries were not able to break out of this seasonal cycle and their butter returns for 1888 show the characteristic pattern of a spring–summer flush and the drying off of herds in winter (Figure 9.1).

---

[20] Currie, diary 4 April 1878, 9 July 1880, 19 March 1885, 4 April 1885 and 9 July 1888.
[21] Currie, diary 24 and 26 May 1888.

Small farms

Although the Curries came to their selection with good capital reserves, the costs of farm making in Gippsland soon exhausted their savings. As well as attempting to earn an income from dairying, the Curries also sought employment outside the farm. The eldest son, Tom, briefly joined the railways (1888) and Kate, with the help of her children, established a postal round (1878). To increase their income John also became the manager of the property of Mr George Stoving, a Melbourne broker who purchased 480 acres near the Curries. In the 1891 shire rate book Kate, her eldest daughter, Katie, and Tom are rated on the original selections, and John and George Stoving were joint occupiers of the latter's farm.[22] By the late 1890s Bert, John and Kate's second son, had entered the farm workforce and he also provided off-farm income from timber cutting. In 1901, when John died, the Curries had carved a farm of 145 acres from their forest selections of 429 acres, with other land not yet fully cleared. Yet after a quarter a century of tireless labour 100 acres was still native vegetation. The Curries were resourceful farmers and, in addition to the mainstay of dairying, they earned income from oats, potatoes and a small flock of sheep (Table 9.2).

With their detailed diary, the Curries provide us with a rare insight into the management of a Gippsland selection as it was transformed from bush to a productive farm. The success or failure of other farms must be inferred from less eloquent sources. One obvious source is the files kept on each selection block by the Lands Department, and one possible indicator of success is the proportion of selectors who went from licensee to leaseholder and, when all rents were paid, to freeholder. In the most detailed study of Gippsland land settlement, Stephen Legg sampled 259 selections under the 1869 Land Act.[23] Of the original selectors, only 134 licence holders, or members of their family, stayed on their blocks long enough to receive a Crown grant. In the Currie's parish of Longwarry there was also a significant turnover. Of the 121 files located for this parish, only 61 per cent of original licence holders, or their family members, received a Crown grant. In Longwarry, high turnover did not cease when a grant was issued. Shire rate books reveal that by 1891 the number of original settlers had declined to only 45 per cent. However, as the original settlers moved out new settlers moved in;

---

[22] Buln Buln Shire Rate Books 1891, VPRS 9571/P1/10.

[23] S.M. Legg, 1984. 'Arcadia or Abandonment: The Evolution of the Rural Landscape in South Gippsland – 1870 to 1947.' MA thesis, Melbourne: Monash University, 219–220.

the rate books of 1891 list 101 farmers in Longwarry parish, or 80 per cent of the number who took up selections. In this churning of residents, some consolidation and amalgamation of farms took place. By 1891, 10 per cent of farmers in Longwarry were taxed on properties of 474 or more acres. This elite, like John Currie, clearly had the sufficient acreage to milk large herds and diversify into cropping and other sources of income. However, the costs of clearing and the uncertainties of markets had also created a large number of small farms; in 1891 a significant proportion of Longwarry farms, 40 per cent, occupied farms of 120 acres or less. Small farms reduced the number of cows that could be carried and constrained the opportunities for additional income from cropping. In Drouin West, a parish adjoining the home of the Curries, Alexander Howden on 158 acres in 1898 ran 11 head of cattle and cropped 1.5 acres of oats (Table 9.2).

Table 9.2 The Probate Inventory of John Currie and Alexander Howden

	John Currie, Longwarry (died 1901)	Alexander Howden, Drouin West (died 1898)
Land	429 acres 28 acres sown with grass 145 acres cleared 185 acres rung 100 acres native state	158 acres
Crops	Last year's crop 150 bags of oats (about 19 acres; before 1910 there were about 4.3 bushels per bag) 16 tons of potatoes	1.5 acres of oats (would have produced about 12 bags)
Horses	1 aged draught horse 2 draught horses 2 young horses	1 mare
Cattle	35 dairy cattle 15 young cattle	11 cows (type unstated)
Sheep	25 Lincoln Sheep	
Pigs		
Implements	Reaper and Binder Double furrow plough Single furrow plough Scarifier Disc harrows Winnower Chaff Cutter Cream Separator	Plough Chaffcutter Harrow
Gross Estate	£2489	£517
Net Estate	£2284	£517

(Probate Inventories VPRS 28/P2/79/993 and 28/P2/69/43.)

The folly of closer settlement

In the 1870s and 1880s the south Gippsland dairy farmer was trapped by the twin forces of endless clearing and the production of a labour-intensive product for which there were uncertain markets. In the late nineteenth century technological developments offered some way out of this dilemma. In 1888 the Victorian colonial government initiated the first of a long series of measures designed to provide state support to the dairy industry. The initiatives included a system of bonus payments made for exported produce and for the establishment of manufacturing works; the purchase of the Melbourne Refrigerating Works at Newport to cheaply store produce destined for export; a decrease in freight rates for dairy produce; and the erection of storage sheds at local railway stations. Finally, the railways provided rapid transit to Melbourne.[24] Technology also came to the aid of the dairymen with the invention of the cream separator. First introduced in large numbers in the 1890s, the separator permitted farmers to prepare cream from milk on the farm. Cream delivered to butter factories had 30 per cent butterfat compared to the more bulky milk, which had only 3 per cent. This drastically reduced transport costs. Skim milk from the separator could be used to feed pigs. The invention of the Babcock tester allowed factories to accurately measure butter content and reward farmers for the quality of their cream.[25]

These technological advances were undermined by government schemes of closer settlement. This policy compounded the problem of small farms, creating more small and heavily indebted properties. In the early twentieth century closer settlement legislation authorised the purchase of large estates and broke these up into smaller farms, which were sold to settlers at market prices. The cost per acre for closer settlers was well above the £1 per acre paid by nineteenth century selectors and burdened many with heavy debts as soon as they occupied their blocks. Estates were of two kinds: irrigated and dry. The former were generally small (less than 100 acres) while the latter varied in size, but were generally smaller than existing farms. Although small farms had limited carrying capacity for livestock and little room for diversified farming, bureaucrats boosted their potential. In 1909, for example, the *Journal of Agriculture* reported on the farm of Mr Swingler on the Moe Swamp. Mr W.A. Herkes, the Department's Dairy Inspector, wrote that Swinglers had farmed their block of 37 acres freehold and 35

---

[24] Greg S.J. Brimsmead, 1990. '1888 – Turning Point in the Victorian Dairy Industry.' *Australia 1888*, (5) (September): 67–79.
[25] Rothberg, 'Victorian Dairy Farming', 23–24.

acres leasehold for nine years. During 1908 Swingler had 25 acres under cultivation, including seven acres maize, six acres of oats (for hay), three acres of oats (for green fodder), seven acres of potatoes and two acres of Japanese millet. The farm grazed 28 dairy cattle and produced an average income of £12/14/6 per cow per year in cream, or £356/5/10. Calves to the value of £25 were sold or kept for herd use. They also received £93 from potatoes and £10 for millet seed.[26] Mr Herkes claimed that the total farm income in 1908 was £484, over three times the wage of a skilled urban tradesman (about £130–£150). Parish Statistical Returns, as we shall see below, reveal that few small farms carried this number of livestock or were as intensively cultivated.

The allure of such promotion was strong and closer settlement quickly found willing applicants. In the Currie's parish of Longwarry a few of the larger farms were acquired by the state and sold to closer settlers. The number of farmers rose between 1891 and 1911, and the average size of farms fell (Table 9.3). Despite the introduction of the separator, the dairy farm remained labour intensive. In his study of dairying in the 1940s, Rothberg estimated that a good milker could handle 10 cows per hour.[27] The daily milking of 30 or more cows and the associated work of separating and cleaning was beyond the labour of one couple, and the labour of children was critical to the dairy industry. By 1901 the Currie family had several grown children who could work in the dairy and they had sufficient land cleared to run 35 dairy cows (Table 9.2). Over the period 1890 to 1911, Bruce Davidson estimated that 30 cows produced a gross income of £127 from butterfat and the rearing of 21 pigs raised a further £42.[28] Taking no account of expenses, this sum was little more than the wage of a skilled urban worker. His estimate of the gross income from 20 cows was only £113, not much more than the wage of an unskilled labourer. However, data from parish statistics indicates that in the early twentieth century average herd sizes in south Gippsland were well short of 30 cows (Table 9.4). In the Currie district the average was closer to 15 cows. While the Department of Agriculture promoted dairying as a profitable industry, in reality the average farmer in south Gippsland was probably worse off than an unskilled urban labourer. His income would have fallen well short of a skilled manual worker. In addition to low income, settlers frequently carried major debt burdens.

---

[26] *Victorian Journal of Agriculture*, VII, 1909.
[27] Rothberg, 'Victorian Dairy Farming', 195.
[28] Bruce R. Davidson, 1981. *European Farming in Australia: An Economic History of Australian Farming*. Amsterdam: Elsevier Scientific Publishing, 210.

Table 9.3 Farms Sizes in the Parish of Longwarry (acres)

A farm was defined as 10 or more acres. A percentile is the value of a variable below which a certain per cent of observations fall. Thus 40 per cent of Longwarry farms were 142 acres or less.

	Original Selections	1891	1911
Mean	193	228	177
Median	160	160	143
40 percentile	142	120	114
20 percentile	320	320	304
10 percentile	320	474	320
Number of farms	124	101	142

Land Selection Files Longwarry VPRS 626 and
Buln Buln Shire Rate Books VPRS9571/P1/10 and 30.

Table 9.4 Parish Statistics Selected South Gippsland Parishes

	Wonga Wonga and Mirboo South		Longwarry and Drouin West	
	1911–12	1931–32	1911–12	1931–32
Average (acres)	197.7	178.0	145.4	111.7
Wheat and oats (acres)	1.1	3.6	9.8	4.6
Green fodder (acres)	0.2	2.3	1.9	2.3
Sown grass (acres)	136.1	92.2	32.1	18.9
Native pasture (acres)	51.8	35.0	65.0	45.3
Unproductive (acres)	6.8	42.0	76.5	35.0
Milk cows per dairy farm	23	21	13	14
Pigs	12	10	7	4
Number of farms	167	101	350	442

Parish Statistics Australian Archives MP570 Boxes 6, 7, 34 and 35.

New settlers in Gippsland in the early twentieth century commenced farming with heavy debts to the Closer Settlement Board. A sample of closer settlement estates from Buln Buln county in Gippsland shows that settlers, constrained by a maximum credit limit of £2500, selected small estates. Closer settlement dairy farms in Gippsland were often little more than 100 acres. Settlers had to borrow to purchase their farms and quickly found that they were not able to carry sufficient stock to meet the repayment of their loans to the Closer Settlement Board. Debts rapidly mounted, reducing settlers' initial low equity in their properties. By the mid 1930s the equity of most closer settlers in their farms was less than one-third, and average debts stood at £2700. Faced with spiralling farm debts, the Closer Settlement Board drew up detailed lists of settlers' assets and liabilities. Settlers were assessed on their efficiency, a concept that could be as vague and judgemental as the neatness of the farm buildings. Most critically, the process introduced the concept of a 'living area' or a measure of the farm's ability – based on size and productivity – to provide the farmer and his family with a decent income. In this process debts were written off and additional land was made available to farmers judged to be efficient. Among the sample of Gippsland closer settlers, farm sizes were increased from an average of 116 acres (median 103) to 139 acres (median 127) and average debts were written down to just over £1000.[29]

The problems of the dairy farmer in Gippsland were compounded by developments in other parts of the state, and closer settlement encouraged the growth of dairy farms in the Western District and on irrigation farmers in the north.[30] During the Great War Britain purchased dairy produce and export prices were high. This only encouraged more settlers to try their hand at dairying. In the years after the war export prices collapsed at the very time supply of dairy produce was growing rapidly from closer settlers. In other states foolish policies of promoting closer settlement were also adopted, which further undermined export prices. In the end the system was only supported by schemes that essentially established two prices for milk: a subsidised high local price and a lower export price. The urban consumer thus bore the burden of closer settlement.[31]

---

[29] This analysis is based on a random sample of 84 settlers' files from Gippsland. VPRS 5714. I am indebted to Jacqui Coucill-Hope for providing me with her data. Jacqueline C. Coucill-Hope, 2004. 'Back on Track: The Closer Settlement Scheme in Victoria.' Melbourne: PhD thesis, Latrobe University 2004.

[30] Manuscript Parish Statistics show that northern irrigated dairy farms milked well under 20 cows on average. In 1921, 167 settlers on the Tongala-Wyuna estate in the north farmed an average of 109 acres and milked on average 9 cows! Australian Archives, MP570, Box 21. In Dreeite near Colac 27 dairy farmers in 1911 milked an average of 71 cows. By 1921 large pastoral properties had been broken up for closer settlement and 167 settlers now milked an average of only 23 cows. Australian Archives, MP570, Boxes 6, 7, 34, and 35.

[31] Rothberg, 'Victorian Dairy Farming', 36–47; Drane and Edwards, *The Australian Dairy Industry*, 191–209.

Conclusion

The problems of clearing dense forests in the nineteenth century and the deliberate policies of the Closer Settlement Board encouraged the growth of small and unviable farms in south Gippsland. By the early 1940s the writing off of closer settlers' debts, the removal of 'inefficient' settlers and the provision of 'living areas' enabled the dairy industry to enjoy the improvement in economic conditions brought about by war-time demand. The 1940s also saw for the first time the introduction of mechanised farming into the dairy industry (a factor that helped northern farmers from the nineteenth century) and in the beginning of the 1940s progressive farms – perhaps a fifth to a quarter – began to use milking machines. Rothberg argued that milking machines in the 1940s could double the number of cows milked per hour. The rate of uptake was more rapid in the next five years, and the number of milking machines rose from 9225 in 1941 to 38,339 in 1946 and 63,066 in 1951.[32] When he conducted his sociological study of dairy farms in the 1943, Maurice Rothberg collected data on gross farming incomes.[33] It was no surprise that his work demonstrated that incomes rose with the size of farms and the size of herds milked. And of course his work demonstrated that with rising income housing conditions improved and fewer family members were involved in the dairy. The problem was that as the dairy industry entered the 1940s it still bore the burden of past land-use decisions and small farms were a feature of the industry. As late as 1941 over a third of cow keepers in the county of Buln Buln in south Gippsland milked less than 20 cows.[34] This burden of past policies left Gippsland dairy farmers in a poor position to enjoy the 1950s, one of the more prosperous periods in the history of Australian farming.

References

Ballan Rate Books 1871, VPRS/5557/P0/5.
Brimsmead, Greg S.J., 1990. '1888 – Turning Point in the Victorian Dairy Industry'. *Australia 1888*, (5) (September): pp. 67–79.
Buln Buln Shire Rate Books 1891, VPRS 9571/P1/10 and 30.
Currie, John and James, Land Selection Files VPRS 626/P0/1557/299 and VPRS 626/P0/1557/300.

---

[32] *Victorian Year Books*, 1941–1951.
[33] Rothberg, 'Victorian Dairy Farming', 387–395.
[34] Parish Agricultural Statistics, Australian Archives, MP570, Boxes 48–9.

Currie, Sarah Ann Catherine, Farm Diary 1873–1916, State Library of Victoria, Australian Manuscript Collection, MS 10886. 1873–1875: 3 January 1876; 2 October, 17 October, 4 November, 10 November and 12 December 1877; 4 April 1878; 9 July 1880; 19 March 1885; 4 April 1885; 24 and 26 May 1888; 9 July 1888.

Davidson, Bruce R., 1981. *European Farming in Australia: An Economic History of Australian Farming*. Amsterdam: Elsevier Scientific Publishing.

Drane, N.T and Edwards, H.R., 1961. *The Australian Dairy Industry, An Economic Study*. Melbourne: Cheshire.

Coucill-Hope, Jacqueline C., 2004. 'Back on Track: The Closer Settlement Scheme in Victoria'. Melbourne: Ph.D. thesis, Latrobe University 2004.

Fahey, Charles, 2008. 'Moving North: Technological Change, Land Holding and the Development of Agriculture in Northern Victoria, 1870–1914'. In *Beyond the Black Stump, Histories of Outback Australia*, edited by Mayne, Alan. Adelaide: Wakefield Press: pp.183–190.

Fahey, Charles, 2011. 'The Free Selector's Landscape: Moulding the Victorian Farming Districts, 1870–1915'. *Studies in the History of Gardens and Designed Landscapes*. 31 (2): 97–108.

Frost, Warwick, 1997. 'Farmers, Government, and the Environment: The Settlement of Australia's "Wet Frontier", 1870–1920'. *Australian Economic History Review*, 37 (2) (March): 19–38.

[Gippsland Settlers' Files] VPRS 5714.

Lake, Marilyn, 1987. *The Limits of Hope: Soldier Settlement in Victoria, 1915–1938*. Melbourne: Oxford University Press.

Land Selection Files Longwarry VPRS 626.

Legg, S.M., 1984. 'Arcadia or Abandonment: The Evolution of the Rural Landscape in South Gippsland – 1870 to 1947.' Melbourne: MA thesis, Monash University.

McLeary, Ailsa, 1998. *Catherine: On Catherine Currie's Diary, 1873–1908*. Melbourne: Melbourne University Press.

Morgan, Patrick, 1997. *The Settling of Gippsland: A Regional History*. Leongatha: Gippsland Municipalities Association.

Ostopenko, Domytro, 2012. 'Facing Economic Maturity: Farmers of the Colony of Victoria in 1840–60.' Melbourne: Ph.D. thesis, La Trobe University.

[Parish Statistics] Australian Archives, MP570, Boxes 6, 7, 21, 34, and 35.

Powell, J.M., 1970. *The Public Lands of Australia Felix: Settlement and Land Appraisal in Victoria 1834–91 with Special Reference to the Western Plains*. Melbourne: Oxford University Press.

Parish Agricultural Statistics, Australian Archives, MP570, Boxes 48–9.

Probate Inventories VPRS 28/P2/79/993 and 28/P2/69/43

Rothberg, Maurice, 1948. 'Victorian Dairy Farming: A Social Survey.' Chapel Hill: PhD thesis, University of North Carolina.

Victorian Journal of Agriculture, VII, 1909.

Victorian Year Books, 1941–1951.

Chapter 10

Lyrical writing, lyrical campaigning
Jean Galbraith and the wildflowers of the Latrobe Valley

Meredith Fletcher

Jean Galbraith often described the place where she lived, the Latrobe Valley, as the very fabric of her being.[1] She first expressed this as a young girl, when she rejected going to Melbourne to finish her secondary schooling, claiming she would be only half a person in the city. Throughout her life she had no desire to live anywhere else. Yet isolation from research libraries, herbaria and colleagues did not prevent Jean Galbraith from becoming one of Australia's most influential writers on gardens, plants and nature. Significantly, her valley was a lifelong source of inspiration. Discovering wildflowers in the bush with her parents led to botanical study and nature writing. Collecting seed from the hills and heath led to a passion for growing native plants and her first commissioned articles. When she became an established writer and was housebound caring for relatives, she relied even more on her place and surroundings to provide material for her many writing commitments. Her garden was a living archive of plants, stories and people that she could fashion into articles and books. Her landscapes were imprinted with memories of beauty and botanical discovery that featured in nature writing and countless children's stories. Parochial in its origins, her writing became national in its significance and impact.

Jean Galbraith's story is significant for the insights it provides into Australia's botanical, gardening and environmental history. The strong

---

[1] See Jean Galbraith's draft of acceptance speech, Australian Natural History Medallion 1970, box 4106 folder 4, Jean Galbraith Papers, MS 12637, Australian Manuscripts Collection, State Library of Victoria, cited hereafter as the Jean Galbraith Papers with box and folder numbers; Jean Galbraith to John Turner, 11 January 1985, University of Melbourne Archives, John Turner Collection TURN00804; Esther Wettenhall, Interview with Jean Galbraith, 1992; Marjory Burgess (Galbraith's niece), 2008, 'Notes on Jean Galbraith'.

relationship between Galbraith and her valley also provides insights to what 'country' can mean for settler Australians; to issues of nationalism and native flora; to notions of botanical belonging; and to Australian nature writing.[2] But in response to the themes of this book – especially the interaction between people and their environments – I look at the relationship between a writer and her place to explore how the valley shaped Jean Galbraith's writing and how, in turn, she placed her writing at the disposal of the valley as she watched it being transformed into an industrial region.[3]

* * *

Galbraith claimed she was born with a love of plants. First it was the plants in her parents' garden. She could stand on the verandah, with its passion vine, and see an abundance of roses in the garden and clouds of pink and white blossom in the orchard. She could walk from 'our' garden, past her mother's perennial border and her father's roses, round a clump of shrubs and she was in her grandfather's garden, a long walk for a small child.[4] One of her earliest memories was of making a garden while she was on a holiday at the sea. She made sand gardens on the beach with flowers from the cliff tops, and remembered the satisfaction this gave her, a plump, contented two-year-old.[5] As she grew older, her love of plants spread from the garden to the bush in her valley in Gippsland, bounded by the foothills of the Great Dividing Range to the north and the Strzelecki Ranges to the south, with the Latrobe River flowing eastwards on its way to the Gippsland Lakes.

Born in Tyers in 1906, Galbraith grew up on land that had been selected and cleared by her family in the 1870s. Sheep and dairy cows grazed on the river flats and lower slopes of her valley but remnant bush was close by. Carpets of wildflowers grew along the roadsides: milkmaids, early Nancies, buttercups, bush peas, orchids, billy buttons. In spring, chocolate lilies

---

[2] See Meredith Fletcher, 2014. *Jean Galbraith: Writer in a Valley*. Melbourne: Monash University Publishing.

[3] The sources for this chapter are mostly drawn from the Jean Galbraith Papers in the State Library of Victoria, MS 12637, an extraordinary environmental history archive that includes correspondence with botanists, gardeners and writers and Galbraith's botanical observations of the Latrobe Valley for over a 70-year period, as well as Galbraith's published writing which also spanned nearly 70 years.

[4] Jean Galbraith, July 1974. 'From Day to Day in the Garden', *Australian Garden Lover*, 35.

[5] John Nicholls, 1986. 'Two Gippsland Naturalists: Jean Galbraith and Bill Cane', *Gippsland Heritage Journal*, 1 (1), 33.

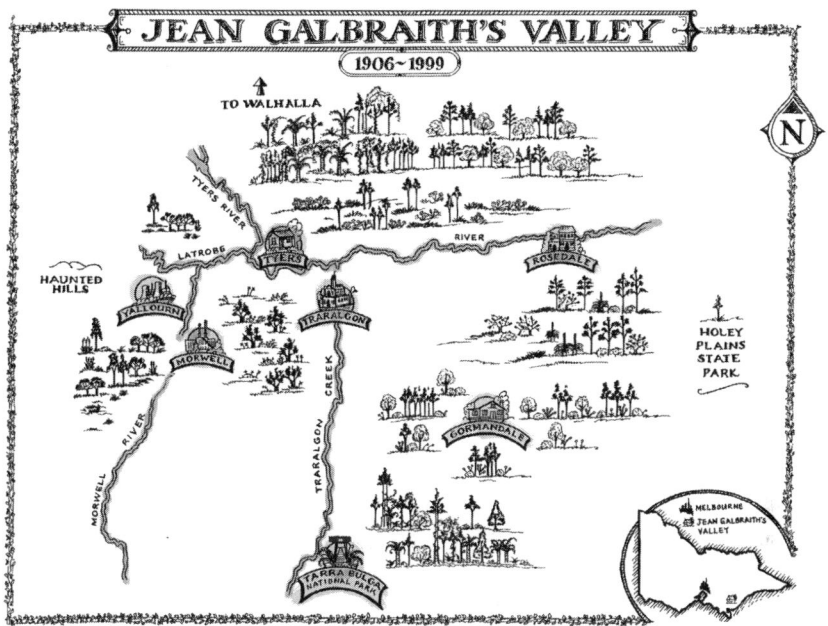

Figure 10.1: Jean Galbraith's Valley, 1906–09.
Map drawn by Sharon Harrup.

formed 'waving tides of purple bloom'. In summer, bluebells made swathes of blue along roadsides and flat hilltops and sweet bursaria filled riversides with creamy blossom and scent. In winter, red heath 'ran like fire' under the trees and white heath covered hilltops with snow.[6] Across the Latrobe River to the south, high in the Strzeleckis, was a different wonderland for Jean to explore: cool temperate rainforest and towering mountain ash in the national parks of Tarra Valley and Bulga Park.[7] She kept detailed field notes of what she saw in the bush, often resorting to invented names as there was no field guide to Victorian wildflowers to help with identification. Her notebooks show the early hallmarks of a writer, describing the texture of a styphelia, for example, with 'each flower looking like a tiny star cut out of velvet', and mistletoe whose 'red tassels hang like fringes of tapistry [sic] from their green waxy foliage'.[8] The notebooks didn't just provide a record of her early botanising (where she went and what she saw), they also record the botanical

---

[6] Taken from a selection of Jean Galbraith's writing: 'Flowers of the Wayside', *Wild Life*, September 1940; 'Australian Native Flowers', *Garden Lover*, February, 1926; 'Epacris Impressa: Who Grows it Now?' *My Garden*, June 1941.

[7] These have been combined to form the Tarra Bulga National Park.

[8] Jean Galbraith, 1919. 'Notes on Plants 1919', Jean Galbraith Papers, 3473.2.

Figure 10.2: Jean Galbraith with her father and brother out collecting plants for the garden near the Walhalla Road, c. 1920.
Ian Hyndman collection.

discovery of her place and of developing an intimate knowledge of vegetation in the valley and its hills. She knew individual plants. She wrote of a clematis she had been watching for five years and recorded when it flowered for the first time. And there were red-letter days, too, such as the day she saw her first correa. She was in the Strzeleckis, driving back to Tyers in the buggy with her father on a day of atrocious weather, and approaching Gormandale, which was famous for its roadside flowers. There was a bright flash of red – 'velvet bells woven of red fire'. She was out among the correa, the sleeting rain forgotten.[9]

Galbraith's childhood love of wildflowers extended to growing them in the garden, re-introducing wildflowers that had been cleared by her selector forebears, and surrounding herself with plants perpetuating memories of golden days in the bush.[10] Growing Australian flora was a rare occupation in the 1920s, when a very limited number of species was available in nurseries. It depended on experimentation and trial and error. Her source was the bush

---

[9] Jean Galbraith, September 1943. 'From Blackwarry to Gormandale', *Wild Life*, 7.
[10] Jean Galbraith. February 1927. 'Garden of Memories', *Australian Garden Lover*, 434.

and she went out collecting seeds and digging up small seedlings for the garden. She gained a reputation among her mentors at the Field Naturalists' Club of Victoria (FNCV), who were guiding her botanical education, as a successful grower of natives.

In 1925, Galbraith was working at the classifying table at the FNCV's wildflower show in the St Kilda Town Hall, when she was introduced to gardening magazine editor Ralph Boardman, who was keen to promote the growing of native flora in his magazine, *The Australian Garden Lover*. He commissioned Galbraith to write a series, 'Australian Native Plants', and she chose her favourite local wildflower, Correa, as a pen name. There were immediate challenges for the 19-year-old contributor when it came to writing the articles. Few of the readers were familiar with the plants she was describing and her articles were not illustrated. She had to use her skills as a writer to help readers imagine the plants. In her first article, published in February 1926, she described the summer plants that were flowering, and gave readers instructions on how to collect sweet bursaria, a practice she would later deplore. Use a sharp trowel to lift the young seedlings growing under the older plants, she wrote. Wrap them firmly in moss or grass and take them home to grow in pots. A year later, plant them in the garden. 'Put them in a corner where they have room to stretch their thorny arms and they will give you the joy of blossom, and the scent of summer riversides.'[11]

But an irrepressible urge to write about her landscapes soon took over from seasonal articles about native plants. After a visit to Tarra Valley she broke free from the template of the flowering year and helpful hints. Her articles changed from garden writing to nature writing. 'I meant to write to-day of the varied beauty of Australian Daisies,' she began her January 1927 article, 'In a Fold of Hills', 'but I have been among the hills, and the joy of their songs and silences must be told.'[12] There was an urgency to share this special place with her readers, to tell its story and to write of its beauty.

Ditching the daisies, she took her readers to the hills of mountain ash and rainforest instead, and wrote of flowers, ferns, berries, fruits, leaves and tendrils. Some of her readers may have known the mint bush that flowered in the hills – it was cultivated in home gardens – but she also drew their attention to the blossom on the mountain ash, just visible far up the hillsides 'like dim white clouds'. It was easier to see the blossom floating in the still

---

[11] Jean Galbraith, February 1926, 'Australian Native Flowers', *Garden Lover*, 581. The *Garden Lover* later changed its name to the *Australian Garden Lover*.

[12] Jean Galbraith, January 1927. 'In a Fold of Hills', *Australian Garden Lover*, 394.

water of the creeks or as fallen flowers that spotted the roadside. She wrote about clematis – its flowers finished – which now had a new beauty of feathery fruits: 'From tree to tree hang the long brown ropes of its stems and its green tendrils wreath the tree fern fronds'. She wrote of the scent of the forest. There was Austral mulberry, which people who lived in the hills called orangewood, 'its orange scented leaves breathing sweetness as one brushes past'. But even more sweetly scented was the musk-daisy bush, 'with broad leaves dark above and silver below, with white Daisy-like flowers, whose starry florets have fallen now'. She told her readers of the purple appleberry, its flowers starting to fall:

> It is a strong and graceful climber, with narrow dark green leaves, and long bells, yellow green just touched with indigo, that in the autumn will be replaced by big berries of purple blue, even lovelier than the flowers.

Many of the plants, she wrote, were easy to grow. They had been carefully moved to the Galbraith garden, or had grown from seed collected in the hills. But although she was writing for a gardening magazine, her article didn't end in the garden. She took her readers back to Tarra Valley, the 'fold in the hills':

> I have told you of some of the treasures that enrich the hills; seek them, cultivate them when you can, but to know their full beauty, visit the stream sides where they grow… where tree ferns mix with gums and maiden hair brushes the water… and the silver air is tangled in mutable loveliness of scent and song.[13]

Through the *Garden Lover* articles, Galbraith was developing a distinctive style that was lyrical, evocative and autobiographical. Her articles make a significant contribution to Australian nature writing. As nature writer Mark Tredinnick argues, Australia, compared with America, has a very limited tradition of nature writing, a genre he prefers to call a literature of place. In her 'Australian Native Flowers' series, Jean Galbraith was writing a literature of place infused with lyricism and wonder. She gave voice to place, told its stories and, as Tredinnick claims of nature writers, brought the landscape to a second life in the imagination of readers.[14] She was inspired by her valley

---

[13] Ibid., 395.
[14] Mark Tredinnick, 2003. *Place on Earth, an Anthology of Nature Writing from Australia and North America*. Sydney: University of New South Wales Press, 31–39.

and her deep Christian faith. Through her writing, she wanted people to experience the beauty that God had created; she wanted 'to tell the beauty of all those things which have their being in Him'.[15]

Her life and daily routine provided the material for her articles. Following 'The Fold in the Hills', she wrote regularly about the rainforest of Tarra Valley and Bulga Park. Articles on coastal flowers at Inverloch followed family camping holidays. The 'rainbow shores' of Tooradin found their way into an article after an FNCV excursion to the mangroves of Westernport Bay. Visits to her grandmother in the granite country of Beechworth in north-eastern Victoria provided material for articles, while the flowers and trees from the Grampians featured in the *Garden Lover* after Jean returned from her post as resident botanist at the Tourist Bureau's nature study camps. But most of the articles were inspired by familiar haunts close to home: along the Walhalla Road or the Tyers River; the heath lands in the hills near her house; the view from a log looking towards the Latrobe River; a peppermint tree growing by the west fence. The series continued until 1935, when she worried her writing was becoming stale. She began to write for the *Garden Lover* readers about her own garden with its mix of native and exotic plants, articles that would eventually become an Australian gardening classic, *Garden in a Valley*, first published in 1939.[16]

Still in her early twenties, Galbraith began writing for other publications, continuing to draw on the landscapes around her for content. The originality of her writing elicited enthusiastic responses. She launched into children's writing after being commissioned to write a bird article for the *Victorian School Paper*, an Education Department publication distributed to primary schools throughout Victoria. She based the article on her childhood experiences of exploring the tea-tree thickets along the Latrobe River, and revealed to young readers the secret world of birds making their home there. The article received fulsome praise from the editor of the *Victorian School Paper*, who emphasised how original her contribution was. 'You have filled my day with refreshment,' he wrote after receiving her article and told her of reading it aloud to a 'rapt' audience. 'I have not seen for years a bird article that moved me so... It will bring joy to thousands.'[17] This article began a long association with the *Victorian School Paper*, and she also contributed

---

[15] Jean Galbraith to John Lothian, 24 May 1931; 25 June 1928, Jean Galbraith Papers, 3462.3

[16] Jean Galbraith, 1985. *Garden in a Valley*. 2nd edition, Melbourne: Five Mile Press.

[17] Gilbert Wallace to Jean Galbraith, 14 September 1927, Jean Galbraith Papers, 3461.1.

to the *New South Wales School Magazine*, writing child-centred adventure stories in the bush that introduced children to the wonders of nature. The stories were mostly based on observations in the bush around her. Galbraith also began writing regular nature articles for the *Leader*, and contributing to *The Argus* and *The Age*.

It may have taken some courage for a young Jean Galbraith to send unsolicited articles on Victorian wildflowers to the *Gardeners' Chronicle of America* in New York in 1929, but she was encouraged by her writing mentor, John Lothian, who considered her nature writing a rare gift. The editor sent a very prompt reply to Mr Galbraith of Tyers, inviting him to send regular contributions.[18] Unlike the *Garden Lover* series, these articles had the advantage of being illustrated. Soon, readers in America were writing to Mr Jean Galbraith of Tyers via Traralgon with requests for seeds.

This was all achieved before Galbraith was 25. At her twenty-fifth birthday in 1931 she spoke of not wanting to be elsewhere than 'this perfect valley which becomes each year more beloved'.[19] Unlike many fictional heroines who needed to leave the confines of home to realise their writing ambitions, Jean accepted the role of the daughter staying home to care for parents and relatives and achieved a career as a writer at the same time, inspired by these very surroundings, her 'perfect valley'.

* * *

In another 20 years Galbraith considered herself a freelance writer and was writing for many new outlets. She branched into broadcasting by writing nature study scripts for ABC school broadcasts and stories for 'Kindergarten of the Air'. She wrote about her garden and Australian flora for a new international readership in the boutique English gardening magazine *My Garden*, and she wrote for new Australian gardening magazines. As a botanist, she was commissioned to write field guides to Australian wildflowers and was a regular contributor to the Australian nature magazines *Walkabout* and *Wild Life*. She also continued her connection with the *Garden Lover*, now writing a long-running series 'From Day to Day in the Garden' which, among other things, told stories of happenings in her garden and celebrated the beauty of everyday life. Her writing was still underpinned by her place. She continued to delight and enlighten readers with her lyrical writing. From

---

[18] Dorothy Ebels to Jean Galbraith, 3 August 1929, Jean Galbraith Papers, 3461.1.
[19] Jean Galbraith to John Lothian, 29 March 1931, Jean Galbraith Papers, 3462.3.

her home in Tyers, inspired by her landscapes, she put her vision of nature into words and helped Australians of all ages to see their own landscapes in new ways.

But Galbraith's valley was changing. When she stood on the hill behind her house at night as a fledgling writer in the 1920s she could just see the State Electricity Commission of Victoria (SECV) brown-coal mining and power-generating complex at Yallourn as a 'city of lights long and low in the west'. That changed after the war when the SECV was faced with meeting the spiralling demands for electricity in an energy-hungry Victoria. Operations at Yallourn expanded with the building of new power stations and extensions to the open-cut mine. The nearby dairying and railway centre of Moe was transformed into a dormitory town for SECV workers. The SECV began its inexorable march eastward to develop a new open-cut mine and briquette factory at Morwell in the 1950s, and build the Hazelwood Power Station in the 1960s. A massive development of a new open-cut mine and two power stations was completed 30 years later at Loy Yang, south of Traralgon. From Moe in the west to Traralgon in the east, a new region was created. The Latrobe Valley no longer meant the wide expanse of river flats between the foothills of the Great Dividing Range and the Strzelecki Ranges, but an industrial region dominated by the SECV.[20] There was little interest in protecting bushland or creating wildflower reserves. Great swathes of the Haunted Hills – widely known for their spectacular flora – were bulldozed to make way for extensions to the Yallourn open cut and for fire breaks. Local tracks notable for the carpets of wildflowers growing along them were widened to become roads. Flower-filled bush was cleared for housing estates and overburden dumps. The Latrobe River became increasingly polluted and transported industrial waste to the Gippsland Lakes. Australian Paper Manufacturers (APM) increased operations at their mill to the south of Tyers and began clearing native forest on the slopes of the valley for pine plantations. Much of the floriferous sandy country near Rosedale was cleared by APM for pine plantations too.

For 25 years, Galbraith's writing had been nurtured by her landscapes. Now she began using her writing to defend her landscapes and alert readers to the disappearing wildflowers and the importance of protecting them.

---

[20] See Meredith Fletcher, 2002. *Digging People Up For Coal: A History of Yallourn*. Melbourne: Melbourne University Press. 137–8.

From children's books to gardening articles, her writing in all its forms became a platform for conservation.

In a talk to a community group in Gippsland, Galbraith explained that her conservation stance was based on ecological and preservation reasons.[21] But she also passionately believed that people needed nature: 'unspoilt places' where people could 'watch the life of nature' and where beauty could be preserved and available to all. She argued especially that children should have access to nature to 'revel' in the unexpectedness of the bush and delight in the flowers. 'Any generation growing up without the opportunity to know the countryside in its full variety and beauty suffers irreparable loss'.[22] Her commitment to conservation was also fuelled by her deep faith: she deplored the destruction of God-given beauty. In her talks, she quoted from Genesis that God had planted a garden and put man there to dress it and keep it:

> In all reverence I can say The Lord God planted a garden eastward in Gippsland but when man entered the garden he did not dress it and keep it. He bulldozed it and chopped it and slashed it and burned it and he built all sorts of unsightly buildings which are certainly useful but rarely beautiful.[23]

Her main concern was for the disappearing wildflowers in her valley that she was observing with anguish. However, as a descendant of selectors who had cleared virgin forest to create a productive farm, she was not against development and land clearing. Instead, she argued for what she called balanced development, explaining to the premier, Henry Bolte, that any future land clearing for farms (or pine plantations, as she argued elsewhere) should always be accompanied by reserves where native bush could be preserved.[24] The reserves had benefits for farming. Remnant bush minimised erosion, created windbreaks for stock and provided shelter for native birds that kept insect pests in check.

She communicated with the SECV, APM, the Country Roads Board, the Forestry Commission, the Town and Country Planning Board and local councils to alert them to the disappearing wildflowers and to discuss the establishment of small reserves. Lyricism was vital to conveying to

---

[21] Jean Galbraith, Notes for talk to Yarram Apex, 1967, Jean Galbraith Papers, PA 93/55.
[22] Jean Galbraith to P. Atherstone, 24 November 1960, Jean Galbraith Papers, 4102.1.
[23] Notes for talk to Business and Professional Women's Club, no date, Jean Galbraith Papers, PA 92/10/1.
[24] Jean Galbraith to Henry Bolte, 22 June 1955, Jean Galbraith Papers, 4102.5.

companies and government departments the beauty and significance of areas under threat. So was her environmental memory, her intimate knowledge of the valley based on more than 30 years of detailed botanical study, and also her reputation as a botanist. These elements were evident in a letter discussing the status of a once beautiful site near the Latrobe River with a local official in 1955, and provide an insight to her campaigning.

> You probably remember a delightful 'avenue' of milk white manna gums that used to be there. Between the trees were pleasant grassy glades surrounded by wattles, eucalypts, purple-berried turnip wood trees with shiny leaves, amber fruited and waxy flowered White Elderberry bushes, Elderberry Panax with its decorative leaves and berries, fragrant Violet bush, the Twining Silkpod with its curious long fruit, white clematis, wonga vine, hemp bush. Now the wattles remain and some thickets of tea tree but little else.
>
> Once we may have said, 'These plants grow everywhere'. Now I don't know any place in the district where several of the plants I mentioned grow wild, and every patch of bush I know is disappearing. We must save something – and here is an opportunity to do so.[25]

More publicly, Galbraith's *Garden Lover* series, 'From Day to Day in the Garden', became full of laments for lost flowers and steeped in calls for wildflower protection. It might be in a passing reference to a train trip, such as in 1955, when she returned from a science conference in Melbourne and wrote of the disappearing garden of the Haunted Hills where once thousands of wildflowers had bloomed from April to November. 'If here… a strip of bush and wildflowers could have been preserved… ' she wrote.[26] Increasingly, train journeys for Galbraith became filled with trepidation. She told readers that the railway line between Traralgon and Melbourne had been a 'pageant of flowers' but now, in 1968, there were only three places where that abundance had survived. Each time she travelled on the train she was 'living in dread' that the wildflowers might be gone.[27]

It was not unusual for whole articles in the *Garden Lover* to be devoted to saving wildflowers. In an article published at the end of 1965 she told her gardening readers of an attempt with her field naturalist colleagues

---

[25] Jean Galbraith to Mr Mitchell, Draft of letter, 28 April 1949, Jean Galbraith Papers, 4102.1.

[26] Jean Galbraith November 1955. 'From Day to Day in the Garden', *Australian Garden Lover*.

[27] Jean Galbraith, November 1968, 'From Day to Day in the Garden'. *Australian Garden Lover*.

to save remnant bush from pine plantations in the Strzeleckis. The bush was especially precious to her because of her favourite correa that was still growing there. 'It is many years since red correa (known as *var. cardinalis*) was abundant in the heathlands south of our valley,' she wrote. 'We used to go out to see it every year, then during the years I could do little travelling about was an era of much clearing, and later I discovered that correa had become rare in every place where we used to see it'. Members of the Latrobe Valley Field Naturalists' Club, of which Galbraith was a member, had been alerted to an area where the correa grew at Traralgon South on a hill of wildflowers that would soon be cleared for pines, and went out to inspect the area. Jean had a special role among the field naturalists: 'I was the only one who knew those hills before they were despoiled and could tell if this was a fair sample of what they had been. It was and I shall not soon forget it'.

> Already we could see various kinds of peas and riceflowers and a shower of correa bells on the high cutting. That was only the beginning. Pink bells and orchids, correa and wattles, sundews and trigger plants and coral pea. There were more bushes of *correa var. cardinalis* than I had seen for twenty years… As we walked down the hill I found myself thinking, 'It can't be lost, not when we have found it at last'. But of course it can be, as so much else has been.

She used this opportunity to outline her ideas on balanced development, that saving this strip of wildflowers could benefit the future pine plantation as well as the local community.

> As we went towards the house above the creek flats it seemed to me that this could be a perfect example of land usage in balance. Six miles away was the town [Traralgon], encircled by farms and forest, and here were rich pastures, with their stock contented and well fed. The white house stood on a level headland jutting out from the hill, with the wildflower strip edged with manna gums behind it, while, above the wildflowers will soon be forests, well managed, and protected by the birds that will go into it from the strip of bush. Very few native birds will live in a pine forest, but they will make an excursion into it for insects if they have a strip of native bush to which they can retreat… [The company] can reserve the strip along the fence for the good of the community as well as their forests.[28]

---

[28] Galbraith, December 1965, 'From Day to Day in the Garden'.

Figure 10.3: Jean Galbraith in 1969, when she was writing the *Field Guide to Wild Flowers of South-East Australia*, commissioned by Collins.
Ian Hyndman collection.

Her botanical work was also dominated by the plight of the wildflowers. When Galbraith was commissioned to write field guides to Australian wildflowers in the postwar years, she undertook the commissions with the firm conviction that spreading knowledge of native flora could help with its protection: the more people who knew about native plants, the more that native plants could be conserved. She wrote with her readers in mind – writing accessible field guides filled with 'vivid mental pictures' that would help a layperson to identify a plant.[29]

In her preface to the first edition of *Wildflowers of Victoria*, published in 1950, she described wildflowers as 'the robe of the countryside' – there for all Victorians to enjoy – and she hoped that her book would 'help to preserve this fast vanishing robe'.[30] When she was revising the field guide for a third

---

[29] Jean Galbraith to Commonwealth Literary Fund, 11 May 1965, 28 May 1965; A.L. Moore to Jean Galbraith, 24 May 1965, Jean Galbraith Papers, 3466.2.

[30] Jean Galbraith, *Wildflowers of Victoria*, Melbourne: Horticultural Press, 15.

edition in the 1960s, her conservation motive was so strong that she offered to take a lower royalty. She told her publishers the book had value in 'helping people to protect plants through being able to recognise them,' and a reduced royalty would keep the cost of the book down.[31]

Conservation themes also permeated her children's writing, which reached thousands of school children. Take her Grandma Honeypot stories, for example. The nature adventures of Grandma Honeypot with her grandchildren were first heard on the ABC's 'Kindergarten of the Air' and then appeared as stories for young readers in the *New South Wales School Magazine*. They were collected and published as a book, and a second edition of *Grandma Honeypot* was issued as a reader for grade two pupils in New South Wales. A Braille edition was also published. The stories were all inspired by events around Galbraith's garden and locality – blue wrens, growing wildflowers in the garden, wildflowers in the bush, gliders, possums, bower birds – all with underlying messages of respecting and caring for the natural world. Other stories in the *Victorian School Paper* and the *New South Wales School Magazine* were fashioned around combatting erosion, the important role of trees as windbreaks and in stabilising river banks, revegetation projects, and improving degraded landscapes. There were stories about seeds and the importance of leaving flowers to reproduce. A strong stance on wildflower protection was also evident in the primary school botany texts that she wrote.

In the 1960s, the editor of the *Educational Magazine*, a journal for Victorian teachers, commissioned Galbraith to write a series he called 'Beauty in Distress' that would discuss the urgency of preserving native plants, with the hope that a heightened awareness among teachers could be passed on to children. The articles Galbraith wrote were direct and sober as she described the loss of heathlands, disappearance of orchids or vulnerability of ferns. She didn't absolve herself from the part she had played as a child in irresponsible picking, remembering especially the tree ferns she had gathered to decorate the hall for school Christmas concerts, 'their fronds nailed pitifully to the walls'.[32] She urged teachers to instil in pupils an interest and enthusiasm for wildflowers: to encourage them to admire the flowers' beauty without taking them.

Again, she turned to her landscapes and memories for this series. There was a particularly arresting article about the Tyers school ground, a cross between a story for teachers and an environmental history. Her article

---

[31] Jean Galbraith to L. Godfrey, Longmans, 23 May 1964, Jean Galbraith Papers, PA 9738.
[32] Jean Galbraith, May 1965. 'Ferns'. *Educational Magazine*, 182.

stressed the importance of reserving bush for children and highlighted the experiences the bush could offer. Based on her father's stories of his schooldays in the 1870s, her own school experiences before 1920 and the impact of an increasing postwar population in Tyers, it was a history of changing landscapes and of how children interacted with nature over time. She began with the earliest days of the school, when it was surrounded by dense tea-tree and children played in its mysterious underworld, free to roam every afternoon while their teacher slept at his desk recovering from a lunchtime drink. As settlement spread, the tea-tree was cleared for a horse paddock used by children riding in from the hills. The paddock was soon covered with wildflowers as plants shot up in the clearings, and children explored among them. When bicycles and bus travel replaced horses, trees started growing in the horse paddock and it became a park with groves of blackwood, scattered peppermints and red box. Children climbed the trees and made cubbies when they wanted something more adventurous and creative than the school playground could offer. The paddock became too shady for the former abundance of wildflowers, but they still grew in sunny spaces. Eventually, though, as more families began settling in Tyers and enrolments increased, a progressive school committee bulldozed all the trees and levelled the horse paddock to form an oval with a cricket pitch. Organised games replaced child-centred adventures and discovery in the bush; blackwood, eucalypts and wildflowers were gone.[33]

* * *

Writing from her dining room in Tyers for nearly 70 years, Jean Galbraith introduced Australian and international readers to the beauty of Australian flora through her gardening and botanical publications, her children's writing and nature writing. In the postwar years her writing alerted readers to the flora's vulnerability and loss. By the time Galbraith died in 1999, the abundance of wildflowers in the Latrobe Valley had disappeared. Now, her writing fulfils two important roles. It perpetuates the flowers of the Latrobe Valley by bringing them to life in the imagination of readers. And it also reveals what has been lost. Her writing makes a good companion on visits to local reserves and parks (many owing their existence to the efforts of Galbraith and her field naturalist colleagues), where fragile remnants of the

---

[33] Jean Galbraith, December 1965. 'Story of a Schoolground'. *Educational Magazine*, 521–522.

flora still grow. Sandwiched between pine plantations on the sandy ridges near Rosedale, Holey Plains State Park is now recognised as a site of national botanical significance, for the diversity of plants growing there and for some extremely rare species. Red correa is still abundant at the flora reserve at Traralgon South. In late spring, chocolate lilies still create 'waving tides of purple bloom' at the Crinigan Road Bushland Reserve. The rare epiphytic butterfly orchid survives in the Morwell National Park. And still at the Tarra Bulga National Park, in Galbraith's fold in the hills, 'the silver air is tangled in mutable loveliness of scent and song'.[34]

References

Burgess, Marjory. 2008. 'Notes on Jean Galbraith'.
Esther Wettenhall Collection, in author's possession.
Fletcher, Meredith. 2002. *Digging People Up For Coal: A History of Yallourn*. Melbourne: Melbourne University Press.
Fletcher, Meredith, 2014. *Jean Galbraith: Writer in a Valley*. Melbourne: Monash University Publishing.
Galbraith, Jean. February 1926. 'Australian Native Flowers', *Australian Garden Lover*, 581–582.
Galbraith, Jean. January 1927. 'In a Fold of Hills', *Australian Garden Lover*, 394.
Galbraith, Jean. February 1927. 'Garden of Memories', *Australian Garden Lover*, 434.
Galbraith, Jean. September 1940. 'Flowers of the Wayside', *Wild Life*.
Galbraith Jean. 1941. 'Epacris Impressa: Who Grows it Now?' *My Garden*, 555–557.
Galbraith, Jean. September 1943. 'From Blackwarry to Gormandale', *Wild Life*, 7.
Galbraith, Jean. 1950. *Wildflowers of Victoria*, Melbourne: Horticultural Press.
Galbraith, Jean. October 1955. 'From Day to Day in the Garden', *Australian Garden Lover*.
Galbraith, Jean. May 1965. 'Ferns', *Educational Magazine*, 182–183.
Galbraith, Jean. December 1965. 'Story of a Schoolground', *Educational Magazine*, 521–522.
Galbraith, Jean. 1968. 'From Day to Day in the Garden', *Australian Garden Lover*.
Galbraith, Jean, July 1974. 'From Day to Day in the Garden', *Australian Garden Lover*, 35–37.
Galbraith, Jean. 1985. *Garden in a Valley*. 2nd edition, Melbourne: Five Mile Press.
Jean Galbraith Papers, MS 12637, Australian Manuscripts Collection, State Library of Victoria.
John Turner Collection, University of Melbourne Archives.
Nicholls, John. 1986. 'Two Gippsland Naturalists: Jean Galbraith and Bill Cane', *Gippsland Heritage Journal*, 1 (1), 33–37.
Tredinnick, Mark. 2003. *Place on Earth, an Anthology of Nature Writing from Australia and North America*. Sydney: University of New South Wales Press.

---

[34] Galbraith, 'In a Fold of Hills', 395.

Chapter 11

'We have to live here'
Early days of forest activism in East Gippsland

Deb Foskey

Introduction

Conflict over the management of native forests has been a feature of Victorian politics for several decades. Forest action groups have formed, reformed and dissolved, and state and federal governments have responded with pre-election promises, legislation and regulations, committees and inquiries. Where forestry is the dominant industry, it becomes a major issue for local people who dodge timber jinkers on local roads and see close-up the impact of clearfelling and regulation breaches on waterways and landscapes. Making those concerns known requires a public campaign with wide media attention which aims to persuade governments to reconsider forest policy. This chapter describes the first campaign of this type in Gippsland.

While the broad public may see a united front and hear a single message, grassroots (local) environmental activists often have different aims to state and federal environmental organisations. National and state conservation bodies such as the Australian Conservation Foundation (ACF), the Native Forests Action Council (NFAC), the Conservation Council of Victoria (CCV) and the Victorian National Parks Association (VNPA) have the ear of governments through representation on committees and briefing and lobbying appointments. In contrast, local activists rely on these more powerful lobby groups to take their issues to the government, yet lack the influence to determine how they will present them.

This chapter explores an early instance of divergent strategies and objectives in Australian forest conservation history. The author, then Debbie McIlroy, was a local resident involved in the early East Gippsland forest campaign. This account of the first forest campaign in East Gippsland is written from that perspective and informed by later involvement in forest campaigns with national and state environmental organisations.

The Errinundra forest campaign of 1983–84 marks a turning point in the East Gippsland forest campaign, taking it from a regional focus to a state and national focus. Letters to politicians, articles in local papers and educational forest camps gave way, for a time, to blockades against logging and the involvement of state and national conservation groups and concerned scientists, elements now seen as integral to successful forest campaigns. The role of local people can be lost amid the furore, but the continuity of grassroots groups underlies the actions of the drive-in, drive-out crowd, as the Errinundra situation attests. Their role is often discounted in the literature.[1] Cohen's account of the Errinundra blockades in *Green Fire* acknowledges local forest activists but is strongly flavoured by his personal perspective.[2] This chapter attempts to redress these omissions.

From selective logging to clearfelling

East Gippsland's forests have long been seen as valuable. From the mountain ash and shining gums towering over the cool montane rainforests of the Errinundra to the coastal tea-tree pockets of lowland rainforest, East Gippsland's forests cover mountains and coastal plains. They are, however, valued differently. They were initially seen as a barrier to settlement, but a good source of timber for building and fencing.[3] The first hardwood sawmills, in the Tambo Valley in the 1860s, provided timber for Melbourne and regional towns and, as forests were cut out, roads extended and the railway line to Orbost established, production moved to the east.[4] The 1939 fires in central Victoria entrenched East Gippsland's role as an alternative

---

[1] Judith Ajani, 2007. *The Forest Wars*. Melbourne: Melbourne University Press; Rod Anderson, 2006. *Cheap as Chips: A History of Campaigns to Save Victoria's Native Forests*. Maryborough: McPherson Printing Group; Drew Hutton and Libby Connors, 1999. *A History of the Australian Environment Movement*, Melbourne: Cambridge University Press; Peter F. Morgan, 1997. 'Contested Native Forests: A Theoretical and Empirical Study', Faculty of the Constructed Environment, Royal Melbourne Institute of Technology, Accessed 20 November 2011 at http://www.mcmullan.net/pmorgan/docs/phd/TCH7.htm.

[2] Ian Cohen, 1997. *Green Fire*. Sydney: Angus & Robertson. Accessed 22 November 2012 at http://iancohennsw.blogspot.com/2011/01/green-fire-ebook.html.

[3] Meredith Fletcher, 2005. 'East Gippsland Shire: Heritage Gaps Study', Vol. 3, Environmental History. Accessed 22 November 2013 at http://www.egipps.vic.gov.au/Page/page.asp?Page_Id=1059&h=1.

[4] Ibid.

source of mountain ash.[5] By the mid 1960s, cutting of the Errinundra and Goongerah forests north of Orbost was in full swing.[6] While there has always been conflict over forest management it was not until the 1980s that conflict occurred in the forests.[7] Several factors escalated dissent about forestry methods: the replacement of selective logging with clearfelling; the influx of new residents; and an increase in the influence of well-resourced state and federal environmental interest groups responding to increased public concern about environmental issues.

Selective logging (the cutting of sound individual trees) was the main timber-harvesting method until the 1970s. By then, axes, two-man saws and bullock drays had given way to chainsaws, powerful bulldozers and articulated timber jinkers. These technologies, along with a new market for lower quality logs in the form of woodchips, entrenched clearfelling as the preferred method of timber extraction. Clearfelling is the cutting and bulldozing of all trees and other vegetation on a coupe. Selective logging maintains a diverse age and size range of trees while clearfelling produces regrowth of similar-sized and aged trees. Early coupes were large and continuous, chosen without regard to ecological values. The evaluation of early trials in the 1960s ignored environmental impacts, although biological research informed later studies of clearfelling.[8] Over time, public outrage led to changes: a decrease in the size of clearfelled areas from 2000 hectares to less than 100; retention of 'seed' and 'habitat' trees; a 'patchwork' approach rather than continuous tracts of clearfelled land; regulations protecting watercourses, gullies and rainforest; and legislation protecting endangered species. Routley and Routley wrote a searing critique of native forest clearfelling which was widely read by concerned forest lovers and hostilely received by timber interests.[9] Peter Rawlinson alerted conservationists to

---

[5] Tom Griffiths, 2001. *Forests of Ash: An Environmental History*. Cambridge: Cambridge University Press.

[6] Meredith Fletcher and Linda Kennett, 2001. *Thematic Environmental History*, Churchill: Centre for Gippsland Studies and East Gippsland Shire Council, 39–40; Fletcher, 'Heritage Gaps Study', 26.

[7] Hutton and Connors, *A History*.

[8] Native Forests Action Council, 1981. 'Clearfelling'. An unpublished briefing paper prepared for Evan Walker; Vicki Pattemore, 1982. 'Forestry and Wildlife: Problems and Perspectives'. *Wildlife in Australia* (September): 83–85; Harry Recher and Wyn Rohan-Jones, 1981. 'Forests and Wildlife Management: Conflict or Challenge?' *Living Earth* (June), 14.

[9] Richard Routley and Val Routley, 1973. *The Fight for the Forests, the Takeover of Australian Forests for Pines, Woodchips and Intensive Forestry*. Canberra: Research School of Social Sciences ANU.

the threat of woodchipping in the first publication on the topic.[10] The terms 'woodchipping' and 'clearfelling' became such derogatory terms that they now rarely appear in government and industry literature.

Scientists were discovering that the Errinundra was a biological hotspot and that its rare species – like the long-footed potoroo and the powerful owl – were further endangered by clearfelling.[11] Botanist David Cameron was employed by the government department with responsibility for forest management, the Forests Commission Victoria (FCV), in the late 1970s to study the vegetation of the Errinundra Plateau, already heavily impacted by logging. He found that some of the oldest extant vegetation survived there and that its plant communities were unique.[12]

The demand for pulpwood for paper-based products opened up new markets and an export woodchip mill was established at Twofold Bay, south of Eden, to draw on the cheap 'waste' timber left after the removal of sawlogs in the south-east NSW and East Gippsland forests. The NSW Askin Government approved tenders from the Harris Daishowa company, then 51 per cent owned by Australian Harris Holdings and 49 per cent owned by the Japanese Daishowa Paper Manufacturers.[13] Woodchips were first exported from NSW in 1969 and from East Gippsland in 1981 when a 100,000 tonne trial began near Cann River. Currently, 85 per cent of timber taken from East Gippsland ends up as woodchips and the export woodchip mill is now owned by Nippon and called South East Fibre Exports.[14]

The environmental effects of woodchipping were not assessed until the industry was well established; it was just one chapter of WD Scott's impact study.[15] Although the report supported the industry expanding into East Gippsland, the authors of the environmental chapter called for more research and monitoring of woodchipping's impact on terrestrial and aquatic ecosystems, reduction in the size of clearfelled areas, lengthening of the cutting cycle, avoidance of the use of fire before and after clearfelling,

---

[10] Peter Rawlinson, 1977. *Woodchipping in Victoria*. Melbourne: Patchwork Press and Native Forests Action Council.

[11] David Cameron, Robin Adair and John Blyth, 1984. Letter, unpublished. 24 January.

[12] Ibid.

[13] Colong Wilderness, 1999. NSW RED Index. Accessed 9 November 2011 at http://www.colongwilderness.org.au/RedIndex/NSW/clgubr99.htm.

[14] Dan Cass, 2011. 'Woodchips no Longer a Renewable Energy Fuel', *The Drum*. 11 July. Accessed 22 November at http://www.abc.net.au/unleashed/2790222.html.

[15] W.D. Scott, 1975. *A Study of the Environmental, Economic and Sociological Consequences of the Woodchip Operations in Eden, New South Wales*. Sydney: Harris Daishowa.

and reservation of forest areas of significance.[16] In 1980, FCV prepared an Environmental Effects Statement (EES) prior to extending pulpwood harvesting to forests in the Shire of Orbost. Two weeks before the publication of the EES the Victorian Liberal Government announced the commencement of a 'trial' woodchip scheme despite earlier assurances that woodchipping would not occur in East Gippsland until the environment assessment procedures had been completed.

Residents become concerned

Since many East Gippsland communities have, at some time in their history, relied on timber cutting and processing for employment they are unlikely to be critical of logging. But other local opinions emerged when, in the 1970s, the population of East Gippsland was augmented by a new wave of settlers who were attracted by cheap, relatively uncleared land and inspired by ideas of self-sufficiency and the formation of intentional communities. These newcomers valued forests for their huge old trees and habitat values rather than as an economic resource.[17]

When the proposal to log south-east NSW and East Gippsland forests for woodchips was announced, some residents saw the need for a local group to lobby on behalf of the forests. Thus, at Tubbut in 1981, Concerned Residents of East Gippsland (CROEG) was formed at a meeting attended by about 30 people. Appreciating the forest environment and disconcerted by its destruction, most members of early CROEG were 'new settlers', all living in East Gippsland. With rumours rife and little reliable information, CROEG 'saw and sees its task as being primarily one of research and communication of its findings within the local area, and at State and Federal levels where appropriate'.[18] Some members of this group were already working to influence Australian Labor Party (ALP) forest policy through a newly formed Bonang-Bendoc branch. Unusually, Gippsland then had an ALP Member, Barry Murphy MLC, whose concern about woodchipping had been aroused by a Timber Worker Union organiser who feared that woodchipping would come to dominate the hardwood industry

---

[16] Harry Recher, Stephen S. Clark and David Milledge, 1975. *An Assessment of the Potential Impact of the Woodchip Industry on Ecosystems and Wildlife in South Eastern Australia*. The Australian Museum, Sydney, 137–138.

[17] Ian Watson, 1990. *Fighting over the Forests*. Sydney: Allen and Unwin.

[18] CROEG, 1984. Letter to the Victorian Premier, unpublished. 22 January.

and exacerbate job losses. As Dargavel shows, subsequent developments proved these fears well founded.[19]

CROEG's first letter, written directly after the group's formation, was sent to the Victorian Minister for Forests and the Forests Commissions of NSW and Victoria, informing them that 'in the Orbost Shire… Concerned Residents of East Gippsland' had formed '[i]n response to the increased interest in State Forests in East Gippsland, and the lack of coordinated, factual information about present and future operations in these forests'.[20] The letter sought essential information (which was never received): figures indicating the upper limits of forest resources; assessment of regrowth rates in clearfelled, chipped areas; details about whether costs of regeneration were covered in the revenue of Orbost–Cann River Forest Districts; and results of a CSIRO study on nutrient and microclimatic changes. CROEG rejected the EES and called for the replacement of clearfelling with selective logging managed on a sustained yield basis.[21]

Other groups were also concerned. Orbost District Environment Group's (ODEG) submission on the EES questioned the economic viability of pulpwood extraction and recommended a decrease in coupe size.[22] In May 1981 it hosted a conference of 'thirty people… representing ACF, NFAC, CCV, Canberra, Bonang and Orbost… to establish a joint and coherent thrust against the proposed export woodchip industry and to lay out the logistics of the coming campaign'.[23] Harris-Daishowa indicated that it would establish a pulp mill at a later stage, mentioning a possible site on the Snowy River at Orbost. This ensured stiff opposition from farmers and Orbost residents, causing the Victorian Government to shelve the project.[24]

In its early years, CROEG members were as concerned about the economic future of East Gippsland as they were about the environmental

---

[19] John Dargavel, 1995. *Fashioning Australia's Forests*. Melbourne: Oxford University Press.

[20] CROEG, undated. Letter to NSW and Victorian Forests Commissions and Victorian Minister for Forests. unpublished.

[21] W.D. Scott & Co., Forests Commission Victoria and Victorian Ministry for Conservation, 1981. *Environment Effects Statement on Pulpwood Harvesting from State Forest in East Gippsland*, Melbourne: W.D. Scott & Co.; Robert McIlroy, 1981. 'A Guide to a Second Reading of the Environment Effects Statement on Pulpwood Harvesting from State Forest in East Gippsland'. CROEG unpublished, 26.

[22] ODEG, undated. 'Terms of Reference', unpublished.

[23] ODEG, 1981. 'Conference against Woodchips in Orbost District Environment Group, Reference', Newsletter, 25 June.

[24] *Snowy River Mail*, 28 June 1989, 1.

impacts of woodchipping. CROEG's approach to employment was influenced by a report prepared for the Tasmanian Forestry Commission which showed that harvesting a small volume of seasoned and dressed timber could generate a higher number of skilled jobs than the woodchip industry.[25] CROEG advocated value adding through small-scale local industries such as furniture making and charcoal production and envisaged more jobs in the bush in repair of clearfelled and damaged areas, riparian zone management and pest control. A 1986 study by Peter Christoff and Margaret Blakers, commissioned by CCV, investigated employment prospects in East Gippsland and reinforced that view; it anticipated imminent changes in the Australian timber market and proposed a seven-year transition strategy to restructure the timber industry 'to re-establish it on a sustainable basis'.[26] It recommended the establishment of value-added projects and programs, including seasoning and veneer plants, a model mill and a woodcraft centre. In the bush, they claimed, better utilisation of felled trees, timber salvage operations, plantation and intensive silvicultural work, environmental rehabilitation, establishment and maintenance of national parks and employment in fire protection services would bring 250 jobs. Woodchipping, by contrast, would not bring many local jobs and 'provid[ed] few advantages for other regional industries and potentially [acted] as a disincentive to regional tourism development'.[27] A later study for south-east NSW presented similar findings.[28]

Like CROEG, Christoff and Blakers were prepared to accept some logging in East Gippsland forests. CROEG members respected local timber and many used it in their owner-built houses. The view that a small amount of selectively logged timber could feed small local industries still has adherents in East Gippsland.[29] CROEG's successor, Environment East Gippsland (EEG), says of its work: 'We campaign to make the logging industry truly sustainable which to a large degree involves a shift into existing plantations'.[30]

---

[25] Ronald Sinclair, 1979. *Tasmanian Wood Skills Resource Investigation*. Hobart: Tasmanian Forestry Commission.

[26] Peter Christoff and Margaret Blakers, 1986. *Jobs in East Gippsland: a Transitional Economic Strategy*. Melbourne: Conservation Council of Victoria, 3.

[27] Ibid., 5.

[28] Keith Tarlo, 1986. *Forests and Jobs in Eden*. Sydney: Total Environment Centre.

[29] Personal comments to author, 2009.

[30] Environment East Gippsland, 2014. 'What we do'. Accessed 22 November 2012 at http://www.eastgippsland.net.au/?q=about_us.

Although East Gippsland's forests were mentioned in the state's major newspapers in 1981, CROEG members were more interested in having their localised view of woodchipping and its impact on East Gippsland broadcast to people in the Orbost Shire.[31] Thus, typewriters, pens, scissors and glue were employed in producing a four-page supplement inserted in the *Snowy River Mail* on 15 April 1981.[32] The broadsheet reflects its kitchen table origins. The main piece of information is presented in bold type in a hand-ruled frame.

> The woodchip industry is coming to town! We have all heard what a boost it will be to the economy of our area: more services, more jobs, GROWTH! How nice, if it were true.

> The big decision will soon be made on whether to open the doors of East Gippsland's forests to woodchipping for export. An Environment Effects Statement is due to be released in about May this year [1981]. The public will be given a brief chance to react to this Statement and the FINAL DECISIONS will be made.

Following several bite-sized articles, quotations from government reports, a map of pulpwood concession zones, cartoons and tables, CROEG's view was summarised:

> It is unsound management to:
>
> - sell off our forests at their lowest value to a wasteful and greedy packaging industry
> - allow employment numbers to fall dramatically
> - place native forests at severe risk.

Finally, the broadsheet asked, 'East Gippsland, why play with fire? The stakes are higher than you think'.[33]

While the broadsheet was aimed at a local audience, the main responses came from outside the area. Two weeks later, industry responded with a Forest Industries Resource Management Group leaflet, 'Woodchip Facts', distributed in the *Snowy River Mail*.[34] A month later, Minister for Forests

---

[31] *The Sun*, 16 October 1981, 36–37; *The Age*, 21 October 1981.
[32] *Snowy River Mail*, 15 April 1981.
[33] Ibid.
[34] *Snowy River Mail*, 29 April 1981.

Jock Granter called the CROEG broadsheet 'a very biased view'.[35] The head of the Victorian Sawmillers' Association, TR Brabin, described it as 'regurgitation of misstatements, unsubstantiable [sic] assertions and inaccurate conclusions'.[36] The knives were out.

In the larger world, forests were on the agenda too. The Sassafras Declaration, organised by NFAC, was signed by 48,000 people calling for the protection of East Gippsland's forests and presented to the Minister for Conservation on World Environment Day in 1982.[37] CROEG was now working strategically with the NFAC, the CCV and the ACF in their campaign activities, but continued to initiate events and prepare its own submissions, write letters and lobby politicians as local residents who cared about the forests.[38] It began the longstanding practice of Forests Forever camps aimed at increasing understanding and support for East Gippsland forests with locals and scientists to inform visitors about the ecosystems and species of the montane forests. The first, held in 1981, was an organised two-day walk along the Yalmy Road from Bonang to Goongerah, with a camp in an area slated for clearfelling. Sculptor William Ricketts, poet Judith Wright and respected bureaucrat and conservationist Nugget Coombs attended. About 100 people completed the walk, which ended with films, speeches and a social gathering in the Goongerah Hall. The second Forests Forever, in 1982, was held on the Errinundra Plateau to draw attention to its threatened ecosystems.[39]

From Forests Commission to Department of Conservation, Forests and Land

In March 1982 the ALP, led by John Cain, was elected to government in Victoria after 27 years of Liberal Party government. This was good news to forest protection advocates since:

> [i]n February 1982, the ALP platform was rewritten to include: opposition to woodchip schemes, the establishment of a public inquiry

---

[35] *Snowy River Mail*, 13 May 1981.
[36] Ibid.
[37] Anderson, *Cheap as Chips*.
[38] Deborah McIlroy 1983. 'A Local Opinion', in 'East Gippsland, Forests or Woodchips', Supplement to *Chain Reaction* 13, Friends of the Earth.
[39] Deborah McIlroy, 1983. 'The Errinundra Plateau: Forest Magic'. *Habitat*, 11 March, Australian Conservation Foundation, 25.

into all aspects of forestry in Victoria, and the creation of a new national park of real significance in East Gippsland.[40]

The Bonang-Bendoc ALP branch may have played a role in these changes but the real work was done in Melbourne at policy working group meetings where conservation-minded ALP members argued the case with timber union representatives. NFAC, ACF and CCV were lobbying politicians and the media with national parks proposals. Their thrust for national park declarations was based on the belief that all major habitats and ecosystems would thereby be protected and lead to an increase in tourism, thus providing new jobs in this industry.[41] The groups also wanted a national park to protect the Rodger River forests at the northern end of the Orbost Forest District; this demand gained urgency after the Labor Government released logging plans for the area in June 1983. CROEG contributed to the campaign by writing submissions and by holding its 1983 Forests Forever on Waratah Flat in the heart of the Rodger River forests.

The ALP Government made several decisions which changed forest management in East Gippsland. First, the Minister for Forests, Rod MacKenzie, established an interdepartmental inquiry to report on the environmental and economic effects of pulpwood harvesting and cancelled the controversial trial pulpwood harvesting operation in East Gippsland before it began. Second, in August 1982, MacKenzie instituted a new pre-logging planning procedure in logging zones and forest management blocks where less than 50 per cent of the timber had been removed. The findings of flora and fauna surveys conducted by botanists and wildlife biologists from the Arthur Rylah Institute were to be transferred to forest maps. Survey reports, maps and a harvesting plan were to be released for public comment before a final harvesting plan was adopted.[42] These surveys 'provided public information which was used in developing pro-environmental public discourses as was the case in the soon-to-be-fought battles over the Rodger River and Errinundra Plateau forests'.[43]

---

[40] Morgan, 'Contested Native Forests'.

[41] Peter Durkin, 1983. 'Public Inquiry and National Park Proposals'. In *East Gippsland: A Natural for National Parks*, background notes for a meeting with the Hon. Rod McKenzie, Minister for Conservation, Forests and Lands 27 October 1983, unpublished.

[42] Morgan, 'Contested Native Forests'

[43] Ibid.

Third, in January 1983, several land management bodies were merged into the Department of Conservation, Forests and Lands under Minister Mackenzie; for the first time, conservation was explicitly recognised as a function of government departments. Finally, in December 1983, Minister MacKenzie ordered a 'comprehensive inquiry into the timber industry' and a two-year moratorium on logging in the Rodger River until the inquiry was completed.[44]

With their attention focused on the Rodger, environmentalists missed the release of logging plans for Compartment 3 in the West Errinundra block. A later government report noted:

> These plans, following a direction from the Minister of Forests, were made available for public perusal and consideration. At no stage did any member of the public or any conservation group avail themselves of the opportunity to review in advance logging for last season. If they had, the controversy over logging in Compartment 3 may well have been avoided as the significance of the site from their point of view would have been brought to the Government's attention in advance and this would have enabled alternative arrangements to be made as was done in the case of the Rodger Management block if necessary.[45]

Botanist David Cameron, however, in his comprehensive biological survey of the Errinundra Plateau for FCV, became alarmed about the threat logging posed to its unique plant communities. He wanted to share his knowledge with other scientists and in the winter of 1983 organised a two-week 'happening' at Goonmirk Rocks on the Plateau which was attended by a number of scientists who later became important in the Errinundra campaign. A group with a different mode of activism also became inspired by Cameron's knowledge of the flora of the Errinundra.

Confrontation: the Errinundra blockade

The Nightcap Action Group (NAG) took its name from the rainforest in northern NSW where it began as a local group opposed to logging at Terania Creek. It went on to prevent the logging of Mount Nardi in an area later protected in national park. Flushed with their success and

---

[44] Ibid.
[45] Victorian Government, no date. Errinundra Plateau: Resolution of Conflict, Victorian Government, 18.

Figure 11.1: The greenies of the Errinundra Plateau, demonstrating against the destruction by logging of East Gippsland rainforest Victoria, 1984.
Bruce Postle for *The Age*, 12 January 1984, from the collection of the photographer.

evangelical in their methods, a delegation including Ian Cohen travelled south to attend the 1983 Forests Forever camp at Waratah Flat. There they met David Cameron who invited them to the Plateau where he explained its significance.

This determined the site of the next blockade and, during the summer of 1983–84, about 50 protesters journeyed from the Down to Earth festival at Wangaratta to the Errinundra. The 'young naïve revellers' in their scant hippy gear were unprepared for conditions on the Errinundra.[46] They camped on a logging access road now known as Blockade Track. They were not ready for the physical conditions of the rainforest and the opposition of loggers:

> Small groups of workers came to see our camp. Invariably, arguments developed and little communication was achieved. We were culturally poles apart. Unlike many areas of forest confrontation where interaction had achieved some level of understanding and a begrudging mutual respect, the loggers were tough and intransigent. No less so were the

[46] Cohen, *Green Fire*.

'fundamentalist' greenies, who pushed a one-dimensional extremist line which aggravated loggers and damaged our cause. Yet the very fact that we had the audacity to be there was enough. Lacking clear direction for a constructive campaign, we had only a fervent desire to empower ourselves and save the forest.[47]

CROEG members learned about the blockade after the campers arrived and sought a meeting with the re-named Nomadic Action Group (NAG). They offered logistic support (bathrooms, food, firewood and warm clothing) but did not join the blockade. As Debbie McIlroy explained to the blockaders, 'We have to live here'.

CROEG and NAG organised a public meeting on 12 January so the two groups could explain their respective aims to local people and the media. CROEG was concerned that locals would not be differentiated from the new arrivals, an obfuscation which would suit logging interests very well. The meeting took place at Bonang, with the editor of the *Wyndham Observer* chairing. As it turned out, there was no opportunity to make any points of differentiation at all.

Cohen provides a vivid account of the meeting:

> The Bonang Highway for hundreds of metres near the Bonang Hall was lined with timber jinkers and cars. Aggressive men wearing specially-designed T-shirts and swigging beer greeted anyone who looked like a greenie with jeers and insults. Standing on the podium to speak required great courage and the volume of shouted obscenities ensured that no-one was heard. NAG representatives tried first and the CROEG representative, Debbie McIlroy, followed; being local did not ensure her a hearing and the audible comments were deeply insulting. It is difficult to say whether the presence of television cameras and journalists checked or incited violence. There were no police present but union representatives tried to quell aggression.[48]

Despite the media's interest, the campers did not present a coherent or achievable message at this meeting or in later encounters.[49] Divisions between would-be leaders hampered the group's ability to develop a unified voice.

---

[47] Ibid.
[48] Ibid.
[49] David Cameron, 2013. Personal conversation, 27 January; Cohen, *Green Fire*.

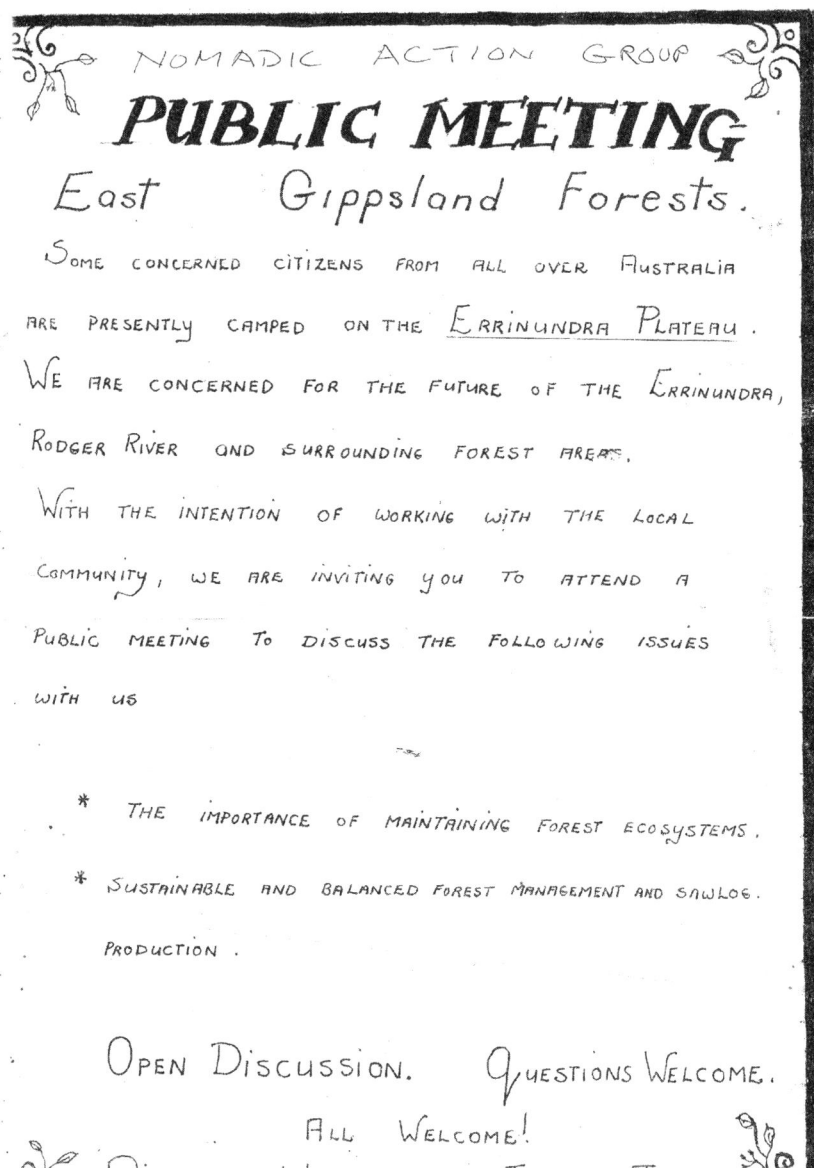

Figure 11.2: Invitation to a meeting, prepared with the best available technology.
(Collection of the author)

Our blind stumblings stirred a media storm in Victoria. TV and press crews descended on us in the forest. However, the consensus of the group, led by Jules and Doug, was to refuse to communicate with the media but rather to stand apart and act as wild tribal people. The group, manipulated by them, would not permit me to speak to the media. So instead of discussion and explanation, they were treated to an extravaganza of naked, painted bodies, ritual dancing, totem pole construction and primitive revelling. After a few days of this the media decided to quit. An historical opportunity was lost to communicate our concerns for the forest.[50]

The Forest People, as they now called themselves, brought the media to the Errinundra Plateau and placed themselves – if not the issues – on the front pages. Helicopters gyrated above the camp almost daily. CROEG hoped that the blockade would alert the broader public to threats to East Gippsland's forests: 'once people from all over Victoria know something of the quality of its forests, they will request preservation.'[51] CROEG's attempts to distance itself from the protesters proved fruitless and members were 'dismayed at local reaction which has obscured and manipulated all the fair, sensible things we have been saying since 1979'.[52] Furthermore, CROEG members could see that their agenda of stopping woodchipping in favour of forest conservation and management to enhance local jobs and environmentally sustainable management would sink under the louder and more organised claims of ACF and CCV for new national parks.

The Forest People's decision to camp and protest was made in ignorance of the Victorian political context and existing forest campaigns; they had not contacted any of the groups working on East Gippsland forest issues. The Forest People's 'baseline belief is that logging of Native Forest on the continent of Australia should be stopped until a whole new understanding on resources, ecological balance and sustainability of resources and employment, is reached'.[53] This, however, was never likely to be achieved and more specific and pragmatic demands were required. Fortunately, David Cameron's science-based approach was available to fill the vacuum and the Forest People adopted his proposals:

---

[50] Ibid.
[51] CROEG, Letter to Victorian Premier.
[52] Ibid., 3.
[53] Morgan, 'Contested Native Forests'.

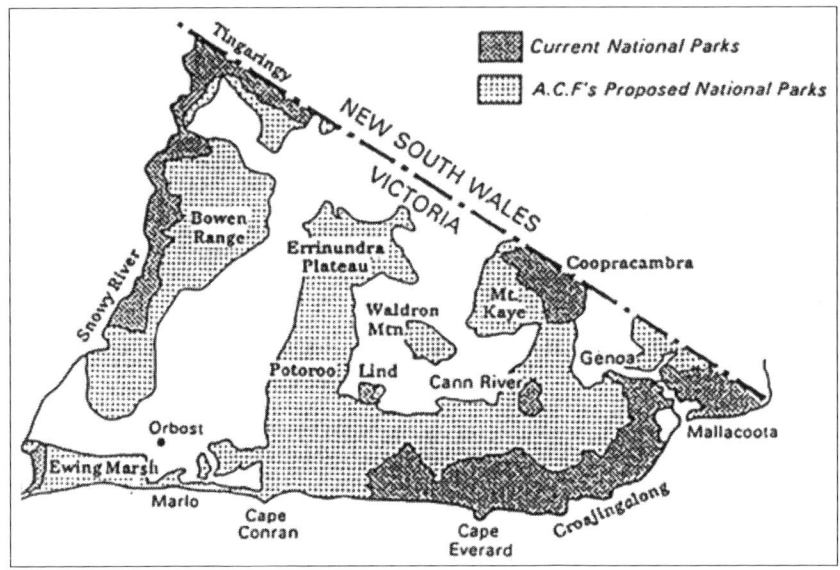

Figure 11.3: Thinking big: the Australian Conservation Foundation's plans for national parks in 1982.
Morgan 'Contested Native Forests', Chapter 7.

1. An immediate moratorium on logging of all forests within those areas proposed for National Parks by the Native Forest Action Council, in the following areas (a) Errinundra Plateau, (b) Rodger River-Yalmy-Bowen Range-Gelantipy Plateau, (c) Mt. Kaye.

2. A thorough investigation of all virgin forests stands in Victoria, carried out by the Victorian Government, in conjunction with concerned conservation bodies, the aim being to protect in a like manner further areas of outstanding significance.[54]

The minister wanted to remove the protesters and asked Cameron to assist by arranging a secret meeting between McKenzie, his staff and representatives of NAG on the Plateau. However, word leaked to the logging community who prepared to turn up in intimidating numbers, which resulted in the conservation movement deciding to muster as many numbers as they could.[55] The media was also there. The minister arrived on the Errinundra by helicopter on 13 January. Representatives of NFAC,

---

[54] Ibid.
[55] Cameron, Personal conversation.

CCV, ACF, VNPA and CROEG were there, resembling the minister's staff in their attempt to distance themselves from the Forest People. They viewed the blockade as premature and counter-productive since the ALP Government was in the process of implementing much of their agenda; they saw their role as controlling the damage to a campaign taken off the tracks. However, the national and state lobby groups were pleased that their proposed national parks were now included in the Forest People's log of claims. In contrast, CROEG members were disappointed that the destruction caused by clearfelling and woodchipping was lost in the focus on national parks.

The meeting did not change the government's plans for the Errinundra. Under pressure from the timber industry, Mackenzie announced four days after the meeting that logging would continue as scheduled for the next three years. Thus, the blockade continued, embarrassing the ALP Government. The Forest People had delayed logging for three weeks but there was no legal framework for removing them. Government and bureaucrats blamed lack of:

> leadership or executive and no person with whom meaningful negotiations could be held... [making it] virtually impossible, despite numerous attempts, to get the group to voluntarily shift from the Plateau or discuss any proposals to resolve the dispute.[56]

Legislative change was needed. A special meeting of Cabinet was called to amend the Forests Act to prohibit 'people from entering or remaining in a timber harvesting area for the purpose of or with the effect of obstructing, hindering or impeding any timber harvesting operations'.[57] Those who failed to leave when directed by an 'authorized officer' would face hefty fines. Although environmentalists and civil libertarians criticised these regulations as draconian, the government insisted that, without them, its position was untenable. Thus, the first forest blockade in East Gippsland was brought to a close on 2 February when 90 police arrested 20 protesters and caused the camp to be disbanded.[58] In March, the timber industry celebrated with a public meeting in Orbost attended by 2000 to 3000 supporters. Timber Towns Association, a grassroots Victoria-wide lobby group, was born.

---

[56] Victorian Government, Errinundra Plateau, 20.
[57] Ibid., 21.
[58] Morgan, 'Contested Native Forests'.

Aftermath

The blockade had achieved its aim of attracting public attention but it did not stop logging. Moreover it led to the introduction of legislation to prevent people from entering logging coupes, which remains to this day. Concerned scientists sought other ways of continuing the momentum to protect sensitive forests. Cameron and two colleagues wrote to hundreds of scientists and conservationists (including CROEG) urging them to write to the minister demanding the exclusion of Cameron's 'red line areas' from logging and calling for a review of land use options on the Errinundra.[59] Following this, on 25 February 1984 'The Errinundra Declaration' was published in Melbourne's morning newspapers with more than 400 signatures, effectively echoing David Cameron's recommendations:

> We, the undersigned biological and environmental scientists, call on the Victorian Government to redirect forestry operations away from areas on the Errinundra Plateau designated as environmentally sensitive and biologically significant, pending the outcome of the Timber Industry Inquiry and a full scale review of land use in East Gippsland by the Lands Conservation Council of Victoria.[60]

The Victorian Government could see that a few arrests had not solved the controversy of the Errinundra forests. Although the Forest People had gone, scientific and conservationist voices in Victoria had not been silenced. Soon after the declaration was published, government officials began a series of meetings with NFAC, CCV and ACF focusing on how to identify and protect sensitive areas. As a result, the Cutting Areas Review Committee (CARC) was established. Initially comprised solely of departmental representatives, it was later expanded to include representatives of the Victorian Sawmilling Association and environment groups represented by the newly formed East Gippsland Coalition, which consisted of VNPA, CCV, NFAC, The Wilderness Society and CROEG. Morgan cites Mackenzie's intention that the CARC was established 'to assess all proposed cutting areas, in advance, to ensure that sensitive areas are identified and protected at the same time as ensuring that sawlog allocations are satisfied'.[61]

---

[59] Cameron *et al.*, Letter. The 'red line areas' included the areas of highest conservation value according to Cameron's assessment.
[60] *The Age*, 25 February 1984, cited in Morgan, 'Contested Native Forests'.
[61] Mackenzie 1984, cited in Morgan, 'Contested Native Forests'.

The promised Victorian Timber Industry Inquiry began in 1985 under Professor Ian Ferguson. CROEG wrote two submissions and appeared at a public hearing, believing that:

> [O]ur role as conservationists living in a timber-producing forest area compels us to participate to the best of our ability in the Inquiry, to attempt to balance the overwhelming bulk of the industry's input and put the situation into perspective.[62]

Its submissions focused on the problems with clearfelling and woodchipping, backed up by evidence of Daishowa's impacts in Bombala and Eden forestry districts.[63] CROEG had developed more sophisticated political understanding and believed participation in the inquiry was important. However, the submission expressed scepticism that 'the scope of this inquiry will not be wide or deep enough to do more than throw into relief the vast and dense mat of problems besetting the forests and present utilization of them'.[64]

After the Errinundra Blockade the Forest People left East Gippsland, moving on to the next action or returning home. If they lived interstate, it is unlikely that anybody would know that they had covered themselves with mud and banged drums on a dirt road in a rainforest in faraway East Gippsland. The city-based groups had increased their public profiles and access to ministers and bureaucrats. City greenies could remain active forest campaigners from their city base or they could lose interest in forests, change their job and take the stickers off their cars. CROEG members, by contrast, had to continue living among people who associated them with the 'blow-in' Errinundra blockaders:

> It should be noted that our cause is not a popular one in East Gippsland where the interests of timber companies retain dominance, and rumour is more rife than accurate information. Some of our members have been slandered and persecuted, particularly since the publication of the broadsheet. Nor did the blockade of logging actions on the Errinundra Plateau by the Forest People further the flow of information, as it suited influential timber industry representatives in order to gain timber

---

[62] CROEG, Letter to Victorian Premier.
[63] CROEG, 1984. Submission to Victorian Timber Inquiry from Concerned Residents of East Gippsland. unpublished document. July 1984; CROEG, 1984. Second submission to Victorian Timber Inquiry. unpublished document, November.
[64] CROEG, Submission to Victorian Timber Inquiry, 2.

workers and local support, to dump us in the same bag as the 'Forest People'.⁶⁵

Some CROEG members stepped back, in order not to be identified with the 'ferals of the forest'. Similarly, bumper stickers such as 'Give AIDS to Greenies' and 'Save the forests: pulp a greenie' encouraged campaigners to dissociate from conservation groups. Some found that forest activism reduced employability. Others found themselves fighting battles close to home as the Orbost Shire Council sent its building surveyor to hunt for illegal dwellings, harassing new settlers who were living in sheds and caravans while they gathered resources to build a permanent 'legal' house.⁶⁶ Several people were taken to court and gaol and fines ensued; this incursion into the private space of their homes was unacceptable to some families who left the district. Others with children of high-school age, such as my family, chose to relocate to avoid schoolyard bullying and the unsustainable alternative of long daily bus rides to Bombala or boarding at Orbost. The Tubbut-Bonang-Bendoc area lost many young parents and their children, a demographic decline reflecting that of timber communities affected by mill closures. In the absence of further waves of urban-to-rural migration, the area still suffers a deficit of people between the age of 20 and 60.

Although CROEG members supported the national parks proposals, they remained concerned that export woodchipping was being allowed to dictate forest management in state forests. While the city-based groups (apart from VNPA) had strongly opposed woodchipping and clearfelling their focus on the protection of iconic areas means that, while many of these are now protected in national parks, woodchipping and clearfelling continue to lay waste to native forests of high conservation value.⁶⁷

Conclusion

The Errinundra campaign of last century, with its limited tools and resources, now seems so archaic as to be irrelevant. Nonetheless, while slow and cumbersome, word got around, as the presence of both environmentalists

---

⁶⁵ Ibid.
⁶⁶ Meredith Fletcher, 2004. 'Ferals and their Muddies'. In *History Australia*, 2 (1): 8.1–8.16. Accessed 22 November 2012 at http://arrow.monash.edu.au/hdl/1959.1/114144.
⁶⁷ ACF, undated. 'The Great Forest Sell-out: The Case Against the Woodchip Export Industry'. An Australian Conservation Foundation Viewpoint, *ACF Newsletter*, supplement.

and loggers at the minister's meeting on the Plateau attested. While environmentalists have many more tools for contemporary campaigning – including word processors, the internet and mobile phones – the requirements remain the same: grass roots activists, a cadre of professional people with access to government, people prepared to give up their time and put themselves in danger in logging coupes, and media available to expose the issues.

CROEG changed as new people replaced earlier activists who moved away. More supporters came from cities and places outside East Gippsland as Forests Forever camps and direct actions enticed them into the forests. The parochial name of 'Concerned Residents' was dropped in favour of Environment East Gippsland (EEG) and Goongerah Environment Centre (GEC). The latter developed as a direct action-oriented group working loosely with EEG.

While the 1985 Timber Industry Inquiry report did not lead to the changes that forest activists wanted in East Gippsland, the 1986 Land Conservation Council's recommendations of large new national parks incorporating a proportion of the Errinundra's shining gum and mountain ash forests did in part.[68] However, CROEG's desire to stop woodchipping becoming the major endpoint for logged forests was put aside and woodchipping now dominates the logging industry and clearfelling is the only harvesting technique. There remains, therefore, a need to defend East Gippsland's old-growth forests, and a number of area-based campaigns have been fought since 1984 for the forests of Brown Mountain (adjacent to the Errinundra Plateau in 1993–94 and 2009–13) and Goolengook (south-east of Goongerah from 1997 to 2002).

Lessons learned on Errinundra are applicable to all campaigns. First, when visitors come from elsewhere to save forests, carefully established claims may be disregarded. The Forest People's reliance on ideology developed in northern NSW illustrated the necessity of understanding the political context so that workable objectives can be developed. Second, scientific knowledge and the support of scientists add depth to slogan-like demands and provide a point of consensus for campaigners. Third, blockaders provide good photographs and film footage but incite anger in loggers and forestry officials and feed stereotypes about 'greenies' in contemporary campaigns,

---

[68] Ian Ferguson, 1985. Report of the Board of Inquiry into the Timber Industry in Victoria, Dept. of Conservation, Forests & Lands, Melbourne; Land Conservation Council 1986. *East Gippsland Area Review: Final Recommendations*, December, Land Conservation Council Victoria.

as they did in 1984.[69] However, today's campaigns more often have to resort to legal action. Governments' legislative interventions have reduced options for biological protection so that forest campaigners now have little choice but to use court processes to challenge forest management.[70] This makes the relationship between local and city-based environment groups even more crucial, in order to share expertise and human and financial resources.

Local environment groups were an important part of the campaigning fabric in 1984 and remain so 30 years later. CROEG, EEG, GECO and local groups in the Central Highlands, the Otways and elsewhere play indispensable roles in providing information and increasing local awareness while maintaining continuity between high profile campaigns. Local activists do the day-to-day work of monitoring the environmental impacts of logging and keeping authorities informed that they are being watched. Further, their presence gives legitimacy to blockades and provides on-the-ground knowledge, essential to a successful campaign, although their contribution may never be visible beyond the local area and be seen as unwelcome there.

References

The Age, 21 October 1981.
ACF, undated. 'The Great Forest Sell-out: The Case Against the Woodchip Export Industry'. An Australian Conservation Foundation Viewpoint, supplement to the *ACF Newsletter*.
Ajani, Judith, 2007. *The Forest Wars*. Melbourne: Melbourne University Press.
Anderson, Rod, 2006. *Cheap as Chips: A History of Campaigns to Save Victoria's Native Forests*. Maryborough: McPherson Printing Group.
Cameron, David, 2013. Personal conversation, 27 January.
Cameron, David; Adair, Robin and Blyth, John, 1984. Letter, unpublished. 24 January.
Cass, Dan, 2011. 'Woodchips no Longer a Renewable Energy Fuel'. *The Drum*. 11 July. Accessed 22 November 2012. Available from: http://www.abc.net.au/unleashed/2790222.html.
Christoff, Peter and Blakers, Margaret, 1986. *Jobs in East Gippsland: a transitional economic strategy*. Melbourne: Conservation Council of Victoria.
Cohen, Ian, 1997. *Green Fire*. Sydney: Angus & Robertson. Accessed 22 November 2012. Available from: http://iancohennsw.blogspot.com/2011/01/green-fire-ebook.html.
Colong Wilderness, 1999. NSW RED Index. Accessed 9 November 2011. Available from: http://www.colongwilderness.org.au/RedIndex/NSW/clgubr99.htm.
CROEG, undated. Letter to NSW and Victorian Forests Commissions and Victorian Minister for Forests, unpublished.

---

[69] Peter Vaughan, 2006. 'Old Growth Forests: Volunteers as Protesters', 1–4, In *New Community Quarterly* (Spring). Accessed 22 November 2011 at www.newcq.org/pdfs/43/ncq%20gook%2043.pdf .
[70] Jill Redwood, 2011. Personal conversation with the author, September 20.

CROEG, 1984. Letter to Victorian Premier, unpublished. 22 January.
CROEG, 1984. Submission to Victorian Timber Inquiry from Concerned Residents of East Gippsland, unpublished document. July.
CROEG, 1984. Second submission to Victorian Timber Inquiry, unpublished document November.
Dargavel, John, 1995. *Fashioning Australia's Forests*. Melbourne: Oxford University Press.
Durkin, Peter, 1983. 'Public Inquiry and National Park Proposals'. In *East Gippsland: A Natural for National Parks*, background notes for a meeting with the Hon. Rod McKenzie, Minister for Conservation, Forests and Lands, 27 October, unpublished.
Environment East Gippsland, 2014. 'What we do'. Accessed 22 November. Available from: http://www.eastgippsland.net.au/?q=about_us.
Ferguson, Ian, 1985. *Report of the Board of Inquiry into the Timber Industry in Victoria*. Dept. of Conservation, Forests & Lands, Melbourne.
Fletcher, Meredith and Kennett, Linda, 2001. *Thematic Environmental History*. Churchill: Centre for Gippsland Studies and East Gippsland Shire Council.
Fletcher, Meredith, 2004. 'Ferals and their Muddies'. In *History Australia*, 2 (1): 8.1–8.16 Accessed 22 November 2012. Available from: http://arrow.monash.edu.au/hdl/1959.1/114144.
Fletcher, Meredith, 2005. 'East Gippsland Shire: Heritage Gaps Study', Vol. 3, Environmental History, Accessed 22 November 2013. Available from: http://www.egipps.vic.gov.au/Page/page.asp?Page_Id=1059&h=1.
Griffiths, Tom, 2001, *Forests of Ash: An Environmental History*, Cambridge: Cambridge University Press.
Hutton, Drew and Connors, Libby, 1999. *A History of the Australian Environment Movement*. Melbourne: Cambridge University Press.
Land Conservation Council, 1986. *East Gippsland Area Review: Final Recommendations*, December. Land Conservation Council Victoria.
McIlroy, Deborah, 1983. 'The Errinundra Plateau: Forest Magic'. *Habitat* 11 March, Australian Conservation Foundation.
McIlroy, D, undated. 'A Local Opinion', in 'East Gippsland, Forests or Woodchips', Supplement to *Chain Reaction* 13, Friends of the Earth.
McIlroy, Robert, 1981. 'A Guide to a Second Reading of the Environment Effects Statement on Pulpwood Harvesting from State Forest in East Gippsland.' CROEG, unpublished.
Morgan, Peter F., 1997. 'Contested Native Forests: A Theoretical and Empirical Study'. Faculty of the Constructed Environment, Royal Melbourne Institute of Technology, Accessed 20 November 2011. Available from: http://www.mcmullan.net/pmorgan/docs/phd/TCH7.htm.
Native Forests Action Council, 1981. 'Clearfelling'. Unpublished briefing paper prepared for Evan Walker.
ODEG, undated. 'Terms of Reference'. unpublished.
ODEG, 1981. 'Conference against Woodchips' in Orbost District Environment Group, *Newsletter*, 25 June.
Pattemore, Vicki, 1982. 'Forestry and Wildlife: Problems and Perspectives'. *Wildlife in Australia* (September): 83–85.
Rawlinson, Peter, 1977. *Woodchipping in Victoria*. Melbourne: Patchwork Press and Native Forests Action Council.
Recher, Harry, Clark, Stephen S. and Milledge, David, 1975. *An Assessment of the Potential Impact of the Woodchip Industry on Ecosystems and Wildlife in South Eastern Australia*. The Australian Museum, Sydney.

Recher, Harry and Rohan-Jones, Wyn, 1981. 'Forests and Wildlife Management: Conflict or Challenge?' *Living Earth* (June): 11–14.
Redwood, Jill, 2011. Personal conversation with the author, September 20.
Routley, Richard and Routley, Val, 1973. *The Fight for the Forests, the Takeover of Australian Forests for Pines, Woodchips and Intensive Forestry.* Canberra: Research School of Social Sciences ANU.
Sinclair, Ronald, 1979. *Tasmanian Wood Skills Resource Investigation.* Hobart: Tasmanian Forestry Commission.
Snowy River Mail, 15 April 1981.
Snowy River Mail, 29 April 1981.
Snowy River Mail, 13 May 1981.
Snowy River Mail, 28 June 1989, 1.
The Sun, 16 October 1981. 36–37.
Tarlo, Keith, 1986. *Forests and Jobs in Eden.* Sydney: Total Environment Centre.
Vaughan, Peter, 2006. 'Old Growth Forests: Volunteers as Protesters', 1–4 in *New Community Quarterly* (Spring). Accessed 22 November 2011. Available from: www.newcq.org/pdfs/43/ncq%20gook%2043.pdf.
Victorian Government, no date. Errinundra Plateau: Resolution of Conflict, Victorian Government.
Watson, Ian, 1990. *Fighting over the Forests*, Sydney: Allen and Unwin.
W.D. Scott, 1975. *A Study of the Environmental, Economic and Sociological Consequences of the Woodchip Operations in Eden, New South Wales*, Sydney: Harris Daishowa.
W.D. Scott & Co., Forests Commission Victoria and Victorian Ministry for Conservation, 1981. *Environment Effects Statement on Pulpwood Harvesting from State Forest in East Gippsland*, Melbourne: WD Scott & Co.

Chapter 12

Reams and reams of paper
The Strzelecki Forest campaigns

Julie Constable

The Strzelecki story is complex, involving attempts at early settlement, the consolidation of a state forest, logging of forest for pines, industrial pressure, and the removal of the forest from environmental and public processes in a major privatisation. This chapter gives a brief overview of the history of the Strzelecki State Forest and campaigns for forest protection. The difficulties involved in achieving conservation outcomes in the Strzeleckis are discussed. Despite protest action and some attempts to create an orderly, transparent process, large portions of the forest have been transferred into private control, as the material below demonstrates.

The chapter pays particular attention to the period of activism that emerged around the time of the Land Conservation Council (LCC) investigation in the late 1970s and the period from the late 1990s that coincided with the Gippsland Regional Forest Agreement (RFA) process and the privatization of the bulk of the Strzelecki State Forest. In both these periods, media attention was more pronounced and provides evidence of two main concerns of Strzelecki forest campaigns: the struggle for forest reserves and parks; and the fight against the transformation of forest areas for industrial wood production. It concludes with a brief reference to more recent developments, which have only continued the themes of earlier struggles around a lack of transparency and the privatisation of the forests.

The Strzelecki State Forest: a brief overview

Named 'The Great Forest of South Gippsland', by 'Selector' in the *Australasian* in 1884, the vegetation of the Strzelecki Ranges in South Gippsland consisted of a mosaic of wet forest, damp forest and rainforest. The area was renowned for giant mountain ash, fern gullies and the superb lyrebird. After more than

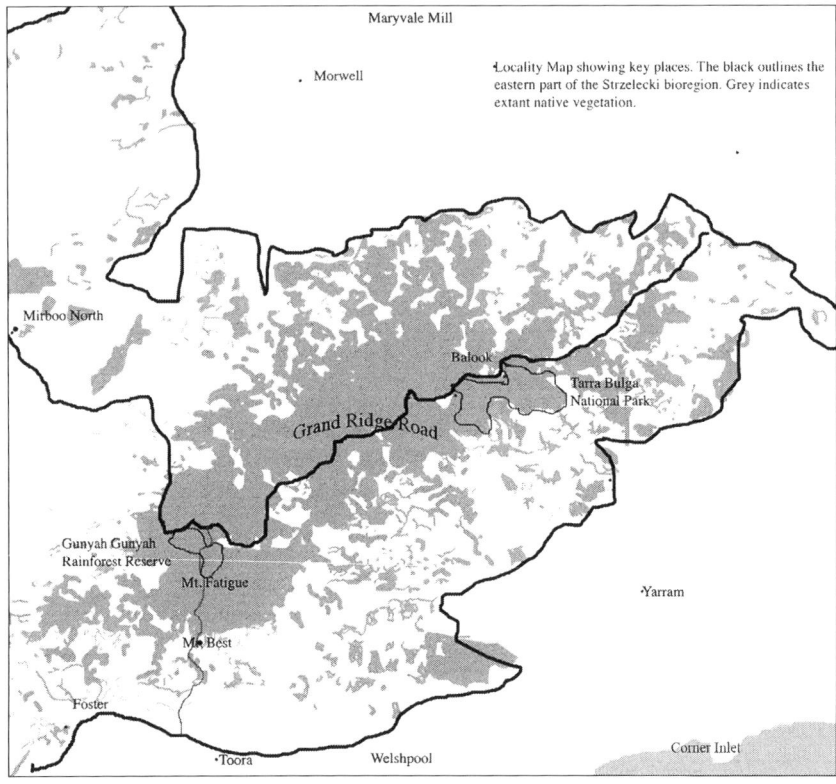

Figure 12.1: Locality map.
J. Constable, 2013.

100 years of settlement, agriculture and forestry, there is only 19 per cent of forest remaining.[1]

The Strzelecki State Forest encompasses some smaller isolated areas in the Western and Eastern Strzeleckis, but the largest continuous tract of public forest runs from north of Foster to north of Yarram in the eastern ranges. The forest contains areas of old growth forest, nationally significant rainforest and younger wet forest. Early attempts at settlement of these steeper, wetter eastern ranges largely failed in the upper catchments. Approximately 24,000 hectares remained Crown Land and a further 9000 hectares quickly reverted back to the Crown when early settlers abandoned their allotments. Not all the forest was cleared and the forest quickly reclaimed small clearings. Patrick Morgan writes 'the bush has grown

---

[1] Barry Traill and Christine Porter, 2001. Nature Conservation Review Victoria 2001. Melbourne: Victorian National Parks Association, 106.

back so quickly that evidence of earlier clearing is scarce… the trees have returned to claim their own'.[2]

From the 1930s the government bought back more land – a mixture of cleared land, regenerating forest and preserved forest – as happened in other parts of the state, placing 60,000 hectares under the jurisdiction of the Forests Commission. The Mountain Ash Reforestation Scheme, established to restore forest, was regularly misdirected. Mountain ash was regenerated, either by seed trees or seedlings; however, this usually took place after logging mountain ash or other forest types. While cleared land was often planted with *Pinus radiata*, forest was also cleared for pine production.[3] The public forest in the Strzeleckis supplied sawlogs and pulpwood from the 1940s.

Australian Paper Manufacturers (APM) established a paper mill at Maryvale in 1936 to the north of the Strzelecki Ranges. The company bought land in the Strzeleckis, mainly on the northern fall, and has cleared land and established plantations over the decades. Despite owning 25,000 hectares in the ranges, in the 1960s the company was granted leasehold over a further 8600 hectares of public land in the Strzeleckis for plantation use. While the agreements by which the state supplies the company with hardwood and softwood have affected land use and forestry practices across Victoria, the Strzelecki's proximity to the Maryvale mill make it especially susceptible to industry's influence. The LCC records that 39,000 cubic metres of hardwood was supplied to sawmillers and the pulp mill annually. This wood was sourced primarily from mountain ash trees 'felled prior to reforestation'.[4] The Department of Conservation and Environment brochure

---

[2] These hill farms were not farms in the way we think of a farm today. Patrick Morgan describes family subsistence holdings with a number of pigs, 10 cows, an orchard and a vegetable garden. See Patrick Morgan, 1997. *The Settling of Gippsland: a Regional History*. Leongatha: Gippsland Municipalities Association, 117–126.

[3] The aims of the scheme are summed up by the Land Conservation Council: 'The hardwood planting program is resulting in the restoration of a forest that will eventually have a similar structure to the original'. Land Conservation Council, Victoria. 1982. *Proposed recommendations, South Gippsland Area, District 2*. Melbourne: The Council, 33. Mansergh and Norris note that 'the policy of establishing pine plantations on *disused farmland* is a more desirable practice than clearing native vegetation. However, in some instances *disused farmland* is forested private property which would probably support a higher diversity of native species than the consequent pine plantation'. Ian M. Mansergh and K.C. Norris, 1982. *Sites of Zoological Significance in Central Gippsland*, Vols. 1 and 2. Melbourne: Ministry for Conservation, 21–22.

[4] Land Conservation Council, Victoria, 1980. *Report on the South Gippsland Study Area, District 2*. Melbourne: Land Conservation Council, 216.

Welcome to the Strzelecki Forest Drive (c. 1990) states, 'In any one year wood is harvested from about 100 ha (0.9 per cent) of the native Strzelecki forests'. It goes on to say, 'Forests are regenerated following harvesting'.[5] Reams and reams of paper have been produced from the Strzelecki mountain ash.

In 1993 the state's pine plantations were vested in the Victorian Plantations Corporation (VPC), a state-owned enterprise. In 1998 the timber rights to the pine plantations and 7000 hectares of so-called hardwood plantations were sold to Hancock Victorian Plantations (HVP). The government allowed the company to map the 'plantation' themselves by identifying mountain ash 50 years old and younger from the air. This deal has been highly controversial as the 7000 hectares claimed as plantation included regenerated logging coupes and native forest restoration. Public opposition was further aggravated by an additional 20,000 hectares of public forest being given to HVP for 'management purposes'.[6] In 2001, HVP acquired the assets of Australian Paper Plantations, a descendent of APM, giving the company management over approximately 50,000 hectares of the 60,000 hectares of native forest that remains in the Strzelecki Ranges.

In April 2008 a suppressed 1993 LCC report, *Review of Victorian Plantations Corporation (VPC) Vested Lands*, appeared.[7] The review had informed the Victorian Government that vested lands included 'extensive areas of mountain ash reforestation in the Strzelecki Ranges' and stated that 'the objective of the hardwood planting was to restore the forest so that it will eventually have a similar structure to the original forest. A range of uses was to be provided and no differentiation was made (by the LCC) between reforested areas and areas retaining the original forest cover'. It referred to the vesting process as 'a major change of use and is unclear whether major changes in silvicultural practice and the provision for non-timber uses is envisaged.' This report was only brought to light under pressure from the Ombudsman while investigating a complaint made by the author and another activist, K. Devenish, regarding the vesting and leasing of native forest in the Strzeleckis.

---

[5] The *Foster Mirror* reported that 5000 cubic metres of timber was logged annually for 40–50 years. The article does not mention the amount of pulpwood. 'CF&L to Consider Reforestation Scheme', *Foster Mirror*, 15 October 1986.

[6] 'Management purposes' was the bureaucratic term used over the phone and in letters when campaigners asked why the HVP should have public native forest.

[7] Land Conservation Council, 1993. *Review of Victorian Plantations Corporation (VPC) Vested Lands*. as requested by the Minister for Planning, August 1993.

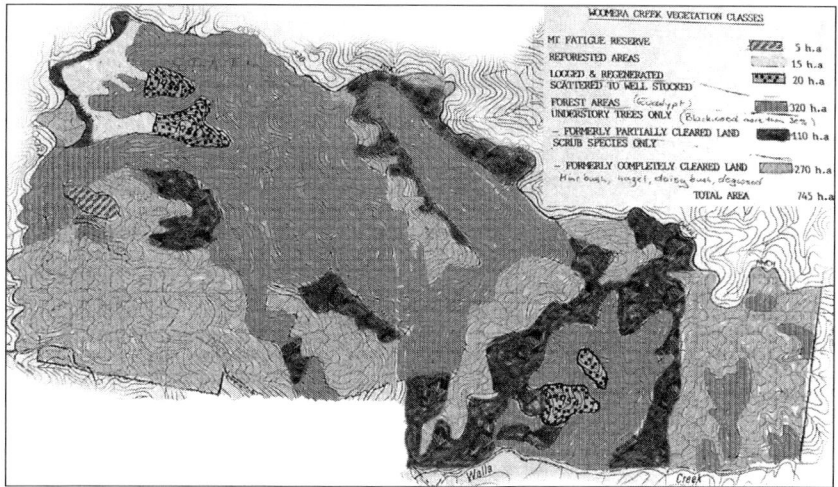

Figure 12.2: Forestry Commission map of the Woomera Creek area, given out at a public meeting at Mount Best in 1990.

This map has been modified to allow the added areas to be seen in black and white reproduction but it remains an accurate rendering of the original boundaries of the source maps.

Reproduced with permission of Arthur Webley, Mount Best Concerned Residents.

The Woomera Creek catchment serves as an interesting microcosm demonstrating the relabelling of native forest as 'plantation' across the Strzelecki State Forest.[8] The first map (Figure 12.2) was issued by the Forests Commission at a public meeting in 1990; the second (Figure 12.3) overlays in black the areas claimed as hardwood plantation by HVP.[9]

The Forests Commission map distinguishes between regenerated native forest logging coupes, reforestation and regrowth forest. The areas coloured brown (reproduced in black) indicate formerly partially cleared land and yellow (reproduced in medium grey) formerly completed cleared land. Both are described as regenerating bush with understorey species. The green areas

---

[8] There is an avenue for further research here beyond the reaches of this chapter. Given time and resources, it would be useful to check other forestry maps (if they can be obtained) against the HVP maps. However, as records are lost or hidden, and forest areas are cleared and replaced with non-indigenous species, and seed trees and habitat trees removed from these areas, it will become more difficult to see the traces of past silvicultural practices and the original forest cover. The map came to my notice when investigating the Mount Best Concerned Residents archive.

[9] At the public meeting, members of the Mount Best Concerned Residents Association hand coloured the areas from a master map. The overlay for figure 2 has been derived from the areas designated 'plantation' on the map. Hancock Victorian Plantations, 1999. 'Woorarra Plantation Block 10'. 8 February. The map and its context are discussed again in the campaigns section.

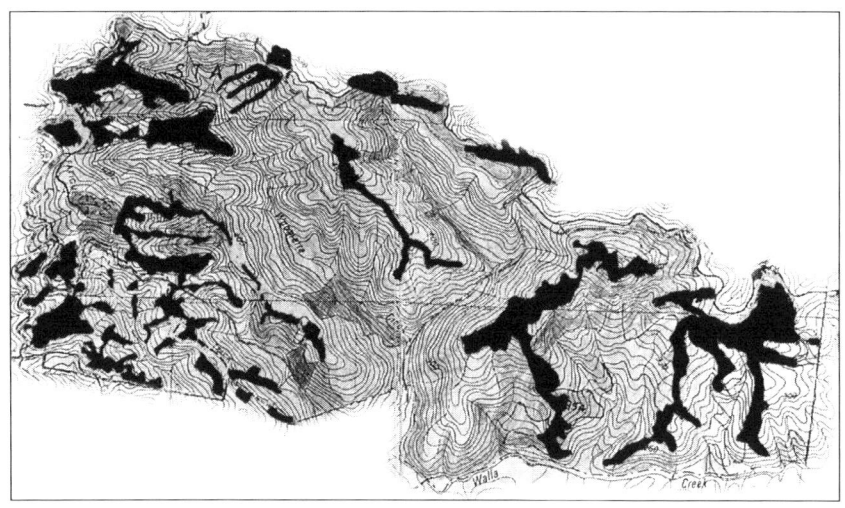

Figure 12.3: Woomera Creek catchment with overlay of 'plantations' claimed by Hancock Victorian Plantations.
Hancock Victorian Plantations Pty Ltd, 1999. 'Woomera Block Plantation Areas.' From a map prepared by HVP LaTrobe Zone, March 1999.

(reproduced in dark grey) are forest. There is no plantation marked in this catchment or any areas identified as being in a cleared state. The red area (reproduced as diagonal lines) is the Mount Fatigue look out. Two 1986 logging coupes are coloured orange (shown as dotted grey) and the aqua blue area (light grey) shows reforestation. HVP now classify these areas as 'eucalypt plantation'. HVP have also claimed as 'eucalypt plantation' logging coupes that occurred in 1990, 1991 and 1998 in the areas of forest and regrowth forest. As it has been against state policy to replace public forest with plantation since 1986, the contention here is that these areas cannot legitimately be classed as plantation.[10]

In a depleted forest area the Strzelecki State Forest is vital for the conservation of the region's flora and fauna, yet it has been appropriated for intensive industrial wood production by the lease arrangements. Mountain ash regeneration is being logged and replaced with non-indigenous plantation species and logging rotations have increased from the 100 years or never-again state forestry policies to as little as 10 years under HVP logging operations. The public forest under HVP management was deemed unavailable for reservation during the Gippsland RFA and the forest was exempt from the Sustainable Yield Review and other environmental legislation.

---

[10] Victorian Government, 1986. *Timber Industry Strategy Victoria*. Melbourne: Government Printer, 95.

Campaigns

Conservation reserves in the Strzeleckis amount to approximately 5000 hectares. Nationally agreed criteria recommend the protection of 15 per cent of each bioregion for the protection of biodiversity; however, less than two per cent of the Strzelecki bioregion is formally reserved. The main reserves are Morwell National Park, Tarra-Bulga National Park, Gunyah Rainforest Reserve, Mount Worth State Park, and Turtons Creek Reserve. These small and scattered reserves have largely been the result of local campaigns. Small additions have been made to some parks over time, but no major viable reserve system exists in the Strzelecki Ranges.

In the late nineteenth century there were calls to open up forest areas for settlement, while dissenting voices proposed the protection of valuable timber and rain catchments in the steep hill country. In an editorial, John Rossiter, owner of the *Yarram Standard*, expressed concern about the effects of clearing on rainfall, the destruction of the Agnes and Franklin headwaters, and opening up the area for selection.[11] In this era a few timber reserves were established, the largest of 25,289 acres (or 10,235 hectares) at Gunyah. Near Balook an area was set aside for the building of an agricultural college. The college didn't materialise, but the area known as College Creek, an area of nationally significant rainforest, was included in the 1960 leases to Australian Paper Manufacturers.

The parks created in the early twentieth century were often implemented with the support of shire councils. The Shire of Alberton was instrumental in the creation of the Bulga and Tarra parks and the Shire of Warragul assisted the local Naturalists Club in the establishment of the Mount Worth State Park in 1948. After many years of lobbying by the Latrobe Valley Field Naturalists, the Shire of Morwell facilitated finance for the Morwell National Park in 1967. This support and advocacy by local councils featured in later grassroots campaigns in the Strzeleckis.

The period from the late 1970s and through the 1980s was characterised by public criticism of forest management and appeals for forest conservation. A number of factors converged, perhaps allowing simmering concerns to boil over: the logging of controversial areas of forest, increasing replacement of forest with pine plantations, and the announcement that a major land use survey – the LCC's investigation into the South Gippsland area – was imminent.

---

[11] Barry Collett, 2009. *Wednesdays Closest to the Full Moon: A History of South Gippsland*, 2nd edition. Melbourne: Melbourne University Press, 129.

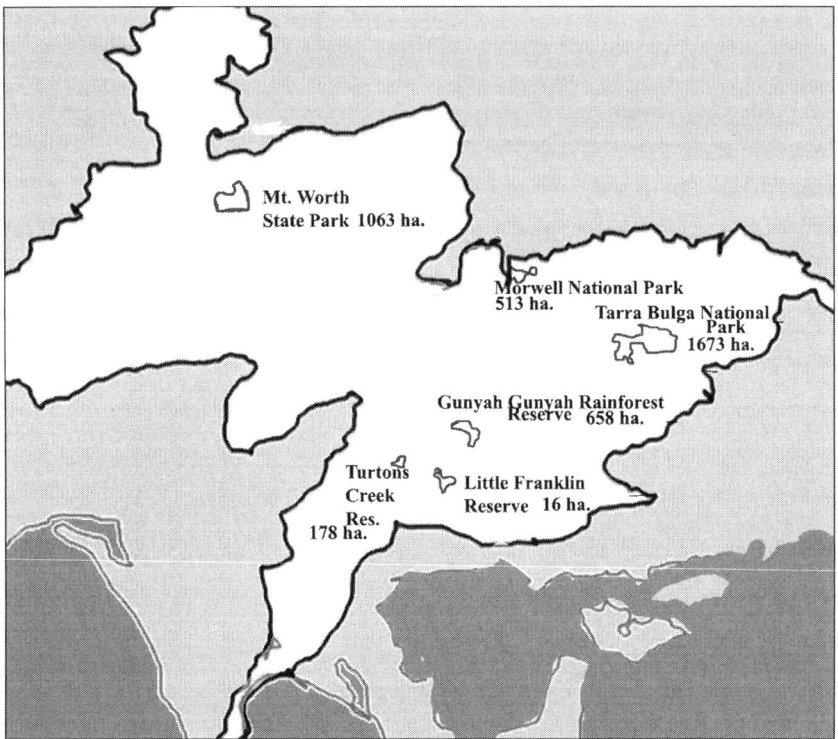

Figure 12.4: Conservation reserves in the Strzelecki bioregion.
K Devenish.

When the Forests Commission began logging some areas of the Gunyah forest in 1979, the *Foster Mirror* featured a front-page article 'Replanting of Gunyah Forest'. The accompanying editorial slammed the logging:

> Another area in our shire, which for many years has been compared with Bulga and Tarra Valley Parks for its natural beauty, is the Gunyah Forestry Reserve. However, a large area near Mt. Fatigue has recently been denuded by the Forests Commission, without prior knowledge of the residents of this shire, and without a protest from self-termed conservationists.[12]

Two of the many protest groups included the South Gippsland Conservation Society and the Yarram and District Conservation Group. The South Gippsland Conservation Society was founded in 1976 and had branches throughout South Gippsland, including Foster, Leongatha and Strzelecki

---

[12] *Foster Mirror*, 11 April 1979.

Ranges. Ellen Lyndon began the campaign to save the tree orchids (*Sarochilus australis*) in Fosters Gully, which eventually became the Morwell National Park. Lyndon received an Order of Australia in 1990 for community and conservation work. She described the onsite crowd at Gunyah as 'a typical mum, dad and the kids sort of crowd'.[13] The Yarram and District Conservation Group, formed in 1976, gradually faded out in the early 1990s.

These existing groups were spurred to protest by the Forests Commission's actions in the Gunyah Forest. Members from the South Gippsland Conservation Society and the Yarram and District Conservation Group met with the Forests Commission on site. Lyndon criticised the Forests Commission for logging the area prior to the LCC's assessment of the area. She noted: '[n]othing in the way of parks or reserves has so far been set aside on the southern fall of the range', and suggested a forest park stretching from the Midland Highway to the Foster-Gunyah Road and the Grand Ridge Road.[14] The only concession the Forests Commission made was to regenerate with mountain ash and not establish a pine plantation.

In the following years, letters to the editors of local papers questioned the actions of the Forests Commission for removing bush (including old growth forest), planting pines, the use of 1080 baiting (usually baited carrots to reduce wallaby numbers), and for logging.

Balook, a small settlement high in the Eastern Strzeleckis, was by the 1970s nestled between the Tarra and Bulga national parks. The Balook and District Residents' Association (BADRA) was concerned about the small size of the Tarra and Bulga parks and general forest management. A letter to the premier Mr Hamer complained that natural forest areas were being diminished and compromised 'partly by APM Ltd, but to a far greater extent by the Forests Commission itself'.[15]

The Balook and the Yarram group shared concerns that proposed logging in the Macks Creek Valley would pre-empt the LCC's investigation. Both groups, along with the Shire of Alberton, requested the government to end logging in the Macks Creek Valley and to make this part of an extended and joined Tarra Bulga National Park. Tensions escalated when logging commenced in November 1978 and telegrams and letters of protest were sent to the Minister of Forests, shire councils, environment groups and the

---

[13] 'On-Site Meeting about Gunyah Forest Logging'. *Foster Mirror*, 9 May 1979.
[14] Ibid.
[15] Letter from Balook & District Residents Association to Mr Hamer, 2 October 1978, signed Ms Margaret Long, President.

media. Another on-site meeting was held, with local and Melbourne media attention. Logging was temporarily halted.[16]

Across the Strzeleckis various groups held concerns about particular patches of forest, but were united in their opposition to the widespread logging and industrialisation of the landscape. They utilised the LCC process to submit proposals for enlarged and new reserves in the Strzelecki Ranges. There was agreement between the Yarram group, the South Gippsland society, the Conservation Council of Victoria (CCV), and the Bird Observers Club to extend the Morwell National Park, join and extend the Tarra-Valley National Park, and to establish a reserve in the Gunyah-Mount Fatigue area.[17] The Yarram group advocated a 4000-hectare Gunyah-Gunyah State Park and a wildlife link between it and the Tarra-Bulga National Park. The Conservation Council of Victoria supported these local suggestions.

Between the release of CCV's proposed recommendations and the final deliberations, the Forests Commission logged part of the Macks Creek Valley and established a pine plantation. This provocative action triggered a public meeting in Yarram. The meeting called for an end to pine planting and a public inquiry into forestry practices in the Yarram Forest District. Many groups made further submissions complaining about the lack of conservation objectives in the management of the forest, the belated release of flora and fauna studies, and increasing their demands for reserves.

The introduction to the LCC's *Final Recommendations* included this apologia: 'It is desirable that as much of the public land as possible is placed under forms of use that do not have a major impact on the natural ecosystem… but, because of the study area's importance to Victorian industry, this has been difficult to achieve.'[18] CCV recommended that the Tarra and Bulga parks be joined and enlarged to 1300 hectares, incorporating some of the Macks Creek valley.[19] A Mirboo Regional Park was proposed, but has never been enacted. The ambiguous quality of the recommendation – that 'the attractive wet gully plant communities (including mountain ash, myrtle beech, blackwoods, and tree ferns) in the headwaters of the West Morwell,

---

[16] 'Stand-off in Bush Meeting', *The Age*, 19 November 1978.

[17] The Land Conservation Council of Victoria was formed in 1969, becoming Environment Victoria in 1994. It is Victoria's chief non-government environmental body.

[18] Land Conservation Council. *South Gippsland Area*, 5.

[19] In 1986, legislation enlarged the park to only 1230 hectares (Esther Anderson, 2000. *Victoria's National Parks, A Centenary History*. Melbourne: State Library of Victoria. 225). The park was proclaimed in 1991 and a further small extension was made during the Gippsland Regional Forest Agreement process.

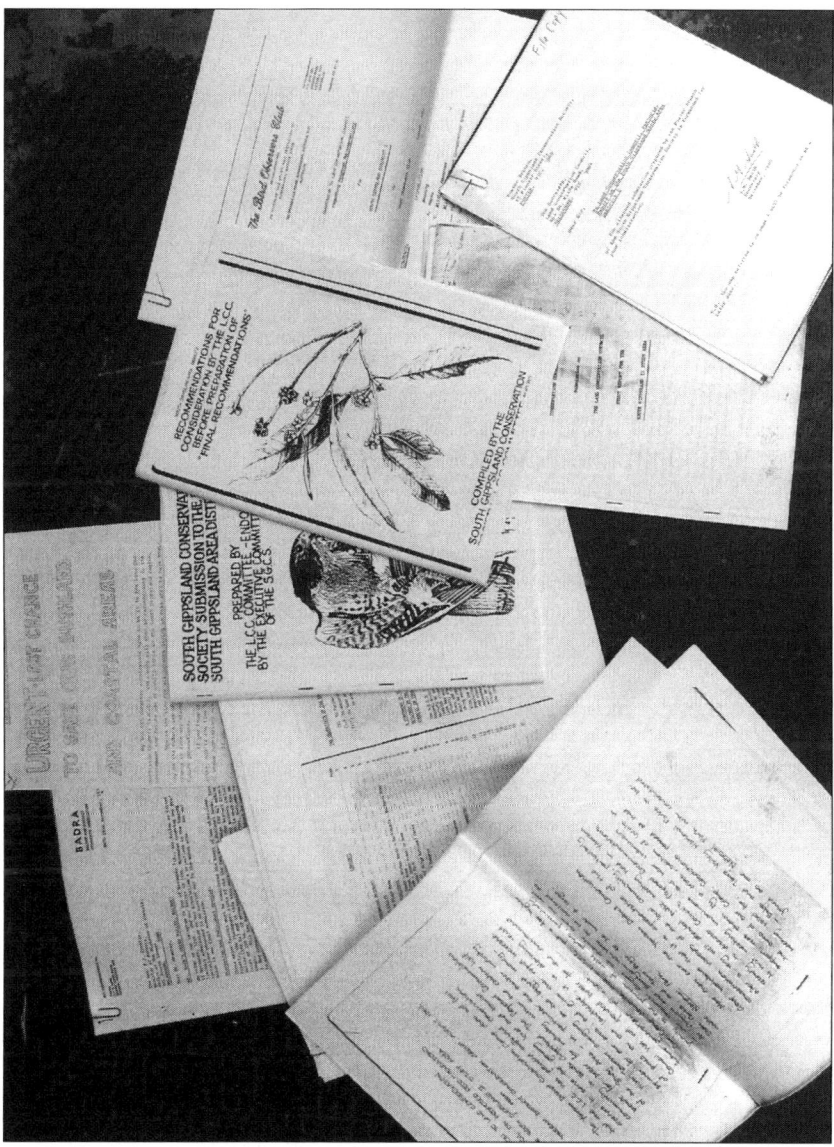

Figure 12.5: The paper chase: a selection of submissions to the Land Conservation Council investigation from the South Gippsland Conservation Society, Yarram and District Conservation Group, South Gippsland Conservation Society, Foster Branch, Conservation Council of Victoria, Balook and District Residents' Association.

Photo: J. Constable, 2013.

Dingo, Franklin, and Agnes Rivers' be 'protected by section 50 reserve or prescription' – led to further conflict between the public and the Forests Commission, and remains an unresolved issue. LCC advised that public land in the Strzeleckis 'remain or become reserved forest under the provisions of the Forests Act 1958 and be managed by the Forests Commission'.[20]

In 1983, after the Forests Commission replaced more forest with pine plantations and amid growing concerns about the focus on woodchipping, the Yarram and District Conservation Group called for a 'full and open public inquiry into the timber industry in Victoria'.[21] Support was received from some politicians, the Sawmillers Association and environmental groups. The government initiated an inquiry in 1984. One welcome outcome for conservationists statewide was the government's acceptance of the recommendation that from 1986 the conversion of public forest into plantation would cease in Victoria.

The South Gippsland Conservation Society in its second submission to Council had warned that the lack of a park system in the Eastern Strzeleckis and leaving protection of the forest to management prescriptions 'would clear the way for the last remnants of "mature" and "over-mature forests"… to be sacrificed to the timber/pulpwood industry'.[22] Government inaction on implementing a reserve in the headwaters recommended for protection was heading towards further conflict. Considerable logging occurred in the Dingo Creek catchment in 1982 and in the Agnes and Dingo Creek headwaters in 1985 and 1986.

In 1986 a residents group formed at Mount Best in South Gippsland, not far from the Gunyah and Dingo Creek areas. The Mount Best Concerned Residents Association began to lobby politicians and write to councillors opposing the logging and reforestation operations. At the July 1986 meeting of the Leongatha Branch of the South Gippsland Conservation Society, Mr Barry Traill, zoologist at Monash University, discussed the Cool Temperate Rainforest community in the Gunyah forest. After a field trip to the area, the society focused on a submission for a flora and fauna reserve in the Agnes and Franklin headwaters. The Mount Best Group investigated the flora of the Woomera Creek area close to the Mount Fatigue lookout. In 1987, the

---

[20] Land Conservation Council, *South Gippsland Area*, 35.
[21] 'Yarram Group Calls for Timber Inquiry', *Weekly Times*, 27 July 1983.
[22] South Gippsland Conservation Society Sub-committee, 1982. 'Recommendations for consideration by the Land Conservation Council before preparation of Final Recommendations', 11.

Yarram and District Conservation Group collected 350 signatures at the Yarram Show opposing further logging in the Dingo Creek catchment.

In 1987, then premier Joan Kirner announced the 680-hectare Gunyah Rainforest Reserve in the headwaters of the Agnes and Franklin rivers. However, the reserve was much smaller than the original Gunyah Timber Reserve of the 1880s and did not include all the 3000 hectares classified as a site of botanical significance in the Gunyah-Mount Fatigue and Rytons, Dingo Creek area in 1984.[23] It fell a long way short of the park proposals submitted during the LCC process and did not include all the headwaters of the Agnes and Franklin rivers or any of the headwaters of Dingo Creek, which had been recommended for protection by Council. This was, however, a significant conservation win, protecting nationally significant areas of cool temperate rainforest and some old-growth mountain ash.

Throughout 1986 the Mount Best group met with the Department to discuss the clearing of forest, and aerial baiting for foxes and wild dogs. In October 1986 the Department pronounced that the Woomera Creek area, containing 1914 regrowth of mountain ash, blackwood and tree ferns, would be 'spared from further development for the hardwood timber industry'.[24] Hopeful, the Mount Best residents proposed a 750-hectare flora and fauna reserve south of Devils Pinch Road at Mount Fatigue.[25] The proposal recommended the area of natural 1914 mountain ash regrowth – with its potential for succession, rainforest gullies and diverse understorey – be reserved, and was critical of the removal of mature trees and understorey during the 'reforestation' activities.

Negotiations became protracted and in 1989 a public meeting was held at Mount Best where forestry officials, despite their earlier stance, outlined the 'Mt Fatigue Reforestation Plan', which proposed to clear and replant approximately 200 hectares. This reforestation was described as being part of the 'overall native forest reforestation plan for the Yarram region'.[26] At this meeting the map of the Woomera Creek area was produced (see Figure 12.2) and hand coloured by members of the Mount Best group. The Department

---

[23] See P.K. Gullan, G.E. Earl, S.J. Forbes, R.H. Barley and N.G. Walsh, 1984. *Sites of Botanical Significance in Central Victoria*. Melbourne: Department of Conservation, Forests and Lands.

[24] *Foster Mirror*, 15 October 1986.

[25] Dorrington, Susan and Percival, Paddy, 1986. 'Submission on the Future of the Regrowth Forest on the Southern Face of the Mount Fatigue-Woomerra Creek Ridge'.

[26] Ross Pridgin, 1989. 'Mt. Fatigue Reforestation Plan'. Yarram: Department of Conservation, Forest and Lands, 1.

insisted that they would not regenerate with plantation species and that the plan was to restore the forest to its original condition.[27] The Mount Best group objected to the Department's plan and remained adamant that the entire area be left undisturbed 'to be managed as a community resource, rather than as an adjunct of the timber industry'.[28] The reserve was not granted and the Department continued logging operations in the Woomera Creek and Dingo Creek areas in 1989–91 and again in 1998. These coupes are currently marked as plantation on the HVP maps, although the public had been reassured specifically and emphatically that after logging the areas were to be regenerated as forest.

There was a lull in media coverage of Strzelecki forest issues until the late 1990s when large-scale logging of the Gunyah forest and plans to privatise the public forest in the Strzeleckis served as a catalyst for a period of intense and public campaigning. These events coincided with another federal–state governmental process: the Regional Forest Agreements (RFAs). In 1997 the Gunyah forest was logged from Stronachs Road, continuing south about 6 kilometres to No Name Track. The area was part of the original Gunyah Timber reserve, which had escaped clearing a century before but was never alienated from the Crown and had been classified as a 'site of botanical significance' in 1984. A strip within the Gunyah Gunyah Rainforest Reserve was also cleared.

I visited Gunyah after the logging. The experience was harrowing. A series of coupes exposed huge, old, hollowed mountain ash. Rainforest species, mountain ash regrowth following a 1914 fire, and younger regrowth following the controversial logging of 1979 had been cleared. The muddy, steep terrain was covered in stumps and bulldozed debris. I contacted the Department of Natural Resources and Environment and was informed that the forest had been vested in the Victorian Plantations Corporation (VPC) and that they were responsible for the logging. On ringing the VPC (the first I had heard of them), I was told that they were only logging plantations.[29] This didn't ring true with the stumps we had measured and photographed. The VPC said that the Department still held all the records. The Department

---

[27] 'I stress from the outset that it's the Mt. Fatigue Reforestation Plan… that's not a sleight of hand or a card trick, it definitely means that.' Ian Hemphill, 1990. Forest Production Planner, 'Tapes of a public meeting in the Mount Best Hall', 26 January. In the Mount Best Concerned Residents archive.

[28] 'Reforestation at Mt. Fatigue: Is it necessary?', *Foster Mirror*, 31 January 1990.

[29] Mr Guillfoyle, VPC accountant at Melbourne office, Personal telephone conversation with author, 7 July 1997.

insisted maps and records had been given to the corporation. A representative from the VPC said I could ask for maps and plans but that the VPC wasn't bound to release them.[30] It emerged in this conversation that this state-owned entity, created in 1993 to manage state softwood plantations, was likely to be fully privatised. In the Strzeleckis, 40,000 hectares had been vested, although only 13,000 hectares of it was pine plantation.

Our suspicions aroused, fellow campaigner Kim Devenish and I borrowed local histories, the LCC reports, botanical reports and other material. We produced relatively quickly, with an impending privatisation breathing down our necks, *Strzelecki Blues: Mucking around with a State Forest*.[31] This booklet outlined government policies and the Reforestation Scheme and questioned the selling off of a public forest. From our reading of the annual logging regimes and the aims of the reforestation scheme, we voiced our early suspicions that regenerated forest after logging was being incorrectly claimed as plantation. Witnessing what had been called 'plantation' at Gunyah, coupled with the secrecy surrounding the maps, fuelled our concerns, and later inspired various attempts to reveal the truth through freedom of information requests.

The lack of public consultation in the appropriation of the Strzelecki State Forest meant that '[l]ocal people felt left out of the picture and deeply uneasy'.[32] A loose network of groups and individuals evolved and campaigned over the following years to fight the privatisation and push for more reserves. Members of the Mount Best and South Gippsland groups, many of whom had campaigned in the 1980s, emerged, older but keen to take a part. Friends of Gippsland Bush, active on the northern fall, made contact with the southern groups and monitored logging coupes and mapped rainforest in the College Creek, Agnes and Franklin catchments. Public meetings were held across the region and local media covered the story regularly and prominently.[33] Local conservation groups and individuals wrote to the

---

[30] Mr Guillfoyle, VPC accountant at Melbourne office, Personal telephone conversation with author, 14 July 1997.

[31] Kim Devenish and Julie Constable, 1997. *Strzelecki Blues: Mucking Around with a State Forest*, Foster: self-published.

[32] Collett, *Wednesdays Closest to the Full Moon*.

[33] In 1998 meetings were held at Mirboo North, Gormandale and Monash University, Churchill. The Latrobe Valley Field Naturalists held a meeting about the Strzelecki forest, as did the South Gippsland Conservation Society at the Bunurong Environment Centre – 'Crisis in Strzeleckis'. In 1998 newspapers covering the campaign included; the *Foster Mirror, Latrobe Valley Express, Sentinel Times, Southern Star,* and *Yarram Standard*.

government asking for further public consultation and assessment of the area.

The state government continued to pave the way towards privatisation and the conservation minister refused to meet with a local delegation. In response to an amendment bill to the Victorian Plantations Corporation Act, which would allow for a sell-off to private interests, a demonstration was quickly organised outside parliament, opposing the privatisation and calling for a national park. The *Foster Mirror* reported, '[w]ith a few days notice people from Gippsland considered the issue of such importance that 70 locals attended the protest scene'.[34] The amendment was passed, but only marginally.

The Gippsland RFA process was due to start and preliminary documentation was available. Drawing on the RFA criteria for bioregional conservation reserves, a group of us wrote *A Proposal for a 30,000 hectare National Park in the Strzelecki State Forest*.[35] This was endorsed by the South Gippsland Conservation Society and gained widespread approval from conservation, community and tourism groups throughout Gippsland and beyond.[36] Susan Davies, an Independent member for Gippsland West, tabled a petition of 7,000 signatures in favour of the park in the state parliament. Greens Senator Bob Brown and international conservation expert Dr David Bellamy lent their support. The park proposal in some ways cut through the complexity and gave voice to decades of frustration with the treatment of the forest.

The Strzelecki State Forest fell under the governance of three shires, and councils were taking notice of the public reaction to the vesting of the forest. The South Gippsland Shire sent a delegation to treasurer Alan Stockdale to press for a national or state park to preserve the native forest in the Strzeleckis. The Latrobe Shire implemented a process to have high

---

[34] 'Strzelecki Sell Off To Go Ahead', *Foster Mirror*, 6 May 1998.

[35] Julie Constable, Kim Devenish, Kim and Allan Standering, 1998. 'A Proposal for a 30,000 hectare National Park in the Strzelecki State Forest'. Foster: self-published.

[36] Groups supporting the national park include Society for Growing Australian Plants, the South Gippsland Conservation Society, the Mount Best Concerned Residents Association, the Latrobe Valley Field Naturalists, the Strzelecki Hills Branch ALP, Wonthaggi/Bass Branch ALP, Friends of the Tarra-Bulga National Park, Friends of Morwell National Park, Friends of the Gippsland Bush, Prom Coast Tourism, Foster Community Association, Mount Eliza Association for Environmental Care, and on a state level Environment Victoria, Greening Australia, Greens Party and the Victorian National Parks Association. The proposal may be viewed at http://members.dcsi.net.au/kimjulie/aaa.html.

conservation areas in the Strzeleckis removed from the Victorian Plantations Corporation tenure.

The injustice felt about the vesting of the Strzelecki State Forest spilled over into the Gippsland Regional Forest proceedings that got under way in May 1998. Events and actions that occurred during the LCC investigation have some noticeable parallels in the Gippsland RFA process.

It became apparent that the Strzelecki forest had been marginalised and isolated prior to the public process. In 1995 the Strzelecki State Forest was declared a Deferred Forest Area (DFA), excluding it from logging pending the outcomes of the agreement. However, in 1997 the VPC logged the Gunyah forest prior to the RFA process beginning, echoing the pre-emptive logging at Gunyah and Macks Creek in the late 1970s. A further distressing sign was a map at the first public meeting showing the Strzelecki State Forest (including the Gunyah Rainforest Reserve) coloured purple, indicating that it was available for plantation development.

Despite the logging and the pending privatisation, groups and individuals prepared their submissions and attended workshops and meetings. The LCC investigation had initially held out some promise of conservation outcomes and local groups responded with submissions. Likewise there was some hope that the RFAs would protect the forest, which by the RFA criteria was woefully under reserved. Reams of paper complained about the lease arrangements, the lack of reserves, forest management, and the whereabouts of the 7000 hectares of plantation. Many supported the national park proposal.[37] State environment groups had become aware of the Strzelecki issues, some supporting the park and some writing submissions.

However, in November 1998 the Strzelecki forest was sidelined further when the state government leased the 7000 hectares of eucalypt regeneration and reforestation as plantation to HVP.[38] The deal allowed HVP to

---

[37] Groups writing submissions in support of forest protection in the Strzeleckis included Friends of the Gippsland Bush, Concerned Residents of 'Sustain the Strzeleckis', Friends of the Earth Forest Network and Recreational Fisherman, Mt. Best Concerned Residents, South Gippsland Conservation Society, South Gippsland Conservation Society (Foster Branch), Friends of Tarra Bulga National Park, Victorian National Parks Association, and Field Naturalists Club of Victoria.

[38] It wasn't until February 1999 that the first maps of the hardwood plantation were made available to the public, confirming fears that reforestation and native forest regeneration were being claimed by HVP as plantations. The Gunyah coupes of 1997, situated within a site of botanical significance, had been replanted with the non-indigenous shining gum and were marked as plantations. The matter of a government selling a resource without making mapping available prior to the sale for public scrutiny was raised in Parliament by Susan Davies and Sherryl Garbutt.

manage a further 20,000 hectares of public native forest (including sites of botanical and zoological significance), rainforest, and even the Big Tree at Gunyah. Industrial security beyond that of the forest agreements was taking precedence over conservation in the Strzeleckis. The revised DFA, the vesting and the privatisation seemed to have circumvented the debate as to how much of the Strzeleckis should be available to the timber/pulp industry and what should be available for conservation, tourism and water needs. Privatisation notwithstanding, the Gippsland Local Government Network signed a memorandum of understanding supporting the creation of a 'major tourism asset in the form of an enlarged park or reserve system in the Strzelecki Ranges' and to 'seek funds for buy back of Hancock timber rights on public land'.[39]

The LCC's ineffectiveness in achieving conservation outcomes in the Strzeleckis was echoed by the RFA process. The *Social Assessment Report* acknowledged that the Yarram community's vision for a major national park in the Strzeleckis, yet apart from small additions to the Morwell and Tarra-Bulga national parks, the Strzelecki forests had been excluded.[40] The RFA Independent Panel's summation demonstrated a limited poignancy (unusual in the scientific and statistical style of government reportage) outlining the initial inclusion of the Strzelecki State Forest, subsequent alienation from the process after the lease to HVP, and acknowledging that despite this they had received:

> … numerous submissions, from community groups and individuals, many in great detail, expressing their concern for the flora and fauna of the native forests of the Strzeleckis. A common theme of their submissions is the need for Government intervention to protect the biodiversity values of this unique area.[41]

That was the closest the public came to a bureaucratic apology for their efforts participating in a process that couldn't even delay a privatisation during the proceedings.

---

[39] Gippsland Local Government Network, 1999. 'Memorandum of Understanding, Action Plan 1999–2000'. 9 April 1999. The seven member councils signed the document.

[40] Joint Commonwealth and Victorian Regional Forest Agreement Steering Committee, 2000. *Gippsland Comprehensive Regional Assessment: Social Assessment Report*, East Melbourne: Department of Natural Resources and Environment, 110.

[41] Gippsland Independent Panel, 2000. *Public consultation issues report: a report to the Joint Commonwealth/Victorian RFA Steering Committee on issues raised in public submissions and hearings during the consultation process on the Gippsland RFA consultation paper*. The Panel, 15–16.

The Labor Party replaced the coalition Victorian Government towards the end of the RFA process. Bolstered by Labor condemnation of the privatisation and a promise to refer the Strzelecki forest to the 'Victorian Environmental Assessment Council to examine future opportunities for protection of native forests in the region', the campaigns continued.[42] The Strzelecki forest continued to feature in the local media and in conservation group newsletters. The grassroots 30,000-hectare park proposal gained scientific credibility with the release of authoritative studies including the *Nature Conservation Review 2001* and the *Draft Native Vegetation Plan (2000)*, both of which confirmed the need for further reservation in the Strzeleckis. The Greens backed the 30,000-hectare national park proposal. At federal and state elections the Strzelecki forest featured in local media as a topical issue.

The momentum of the campaign for a national park was to some extent sidetracked by the machinations of the Strzelecki Working Group. Initiated by the South Gippsland Shire in 1999, it had offered a forum for local councils, conservationists and citizens to discuss issues with HVP.[43] Given the make-up of the group, with its varied interests, negotiations weren't always smooth. The group commissioned *The Strzelecki Ranges Biodiversity Study*, which identified 8400 hectares of high-conservation-value forest and corridors linking Gunyah and Tarra-Bulga national parks. College Creek was identified as a site of national significance for rainforest and included as a core area, resulting in Australian Paper joining the group as leaseholders of that part of the state forest.[44] Surprisingly, the group agreed to ask the government to reserve the 'Cores and Links' as an 'urgent priority', and HVP agreed to a moratorium on logging.[45]

---

[42] Australian Labor Party, Victorian Branch 1999. *Our Natural Assets: Valuing Victoria's Natural Environment*: Melbourne: The Party, 10–11.

[43] The group included a member from Friends of the Gippsland Bush, South Gippsland Conservation Society, the La Trobe, Wellington and South Gippsland Shires. The West Gippsland Catchment Management Authority and the Department of Sustainability and Environment often sent representatives and representatives from other groups were sometimes invited. Kim Devenish and Julie Constable were citizen members. The citizens and conservationists were the unpaid members of the group.

[44] See Andrew Hill, Chris Timewell, Sally McCormick, and Stephen Mueck, 2001. *The Strzelecki Ranges Biodiversity Study*. Port Melbourne: Biosis Research. The purchase of Australian Paper Plantations by HVP in 2001 blurred the distinction between the plantations of a private company and the state forest leases, adding to the complexity of the story.

[45] Letter to Minister Garbutt, 31 January 2002. Signed David Lewis, Chairperson, Strzelecki Working Group.

After the state election the state government procrastinated on the Strzelecki forest's referral to the Victorian Environment Assessment Council (VEAC) and exploited the presence of the working group and its negotiations to deflect and delay decisions about the national park, the misclassification of native forest, lease arrangements and logging practices. When the Strzelecki Working Group agreed on the reservation of the Cores and Links, the buy-back process between HVP and the government went behind closed doors and stalled.

Disturbed by the lack of progress, conservation groups banded together to hold the SOS Save our Strzeleckis rally at Parliament House in August 2003.[46] More than 300 people gathered to show their support and listen to various speakers, including then senator Bob Brown, and Robbie Thorpe of the Gunai-Kurnai nation. By 2006, with no apparent breakthrough, HVP announced it was breaking the moratorium on logging in the proposed reserve. This prompted a public meeting in Churchill attended by 150 people. Brendan Jenkins, Labor member for Morwell, said that HVP was 'inflating the value of the logging rights', thereby hindering the buy-back.[47] In early 2008, HVP announced that a shortfall of wood in their private plantations would be sourced from within the Cores and Links to service their contract to supply the Maryvale mill.[48] Six years after the Strzelecki Working Group asked the government to reserve the Cores and Links as an urgent priority, the environment minister, Gavin Jennings, signed a deal with Hancock Victorian Plantations. The deal allowed for logging in the proposed reserve and postponed protection for another 20 years. At

---

[46] Groups who assisted with the rally included The Greens, Environment Victoria, Wilderness Society, Victorian National Parks Association, Friends of the Earth, Friends of Gippsland Bush and South Gippsland Conservation Society.

[47] 'Fate Hangs in the Balance: Tug O'war Over Key Conservation Areas' *Latrobe Valley Express*, 19 January 2006.

[48] South Gippsland Conservation Society's Newsletter (see South Gippsland Conservation Society, 2008. *Newsletter*, May) said that Hancock's decision to log in the Cores and Links was the 'consequence of massive company incompetence – the result of seriously over-estimating their timber supply and entering into risky binding contracts'. South Gippsland Shire Council Minutes, 19 March 2008, report that the shortfall increased from 600,000 cubic metres of pulpwood to 835,000 cubic metres. It is ironic and tragic that a shortfall in APM's private plantations, which should have decreased the burden on state forests, are now owned by HVP, who can top up the shortfall by further fragmentation and logging from the leases they hold on public forest in the Strzeleckis.

present, the company has been paid $5.5 million, but the public do not have a park.[49]

The *Review of Victorian Plantations Corporation (VPC) Vested Lands (1993)* appeared in the press in 2008. Successive governments have ignored this report, which confirmed that native forest and reforestation had been sold as plantation. Calls for an inquiry have been denied. In November 2011, the VEAC completed its *Investigation on Remnant Native Vegetation* recommending that six priority regions (including the Strzelecki Ranges) undergo VEAC investigations into 'protected area establishment'. Consistent with the reluctance of Victorian governments to permit a full investigation of the Strzeleckis, the Baillieu state government rejected this recommendation.

The history of forest campaigns in the Strzelecki Ranges has been characterised by a strong community presence often including support and advocacy from local government. While public meetings, rallies and occasional demonstrations have highlighted concerns, Gippslanders have inclined towards writing submissions, reports, proposals, and participating in public processes. Achieving conservation outcomes has been undermined by the forest's proximity to Australia's largest paper mill. Although the industry utilises large areas of private plantation in the Strzeleckis, the pressure on the public forest has not abated. Governments have been reluctant to create a viable park system in the remnant public forest of the 'Great Forest of South Gippsland', despite scientific backing and grassroots support. Efforts to remove the Strzelecki forest from due process are evident. In addition to contending with a lack of public consultation and the withholding of information, campaigners struggled with the complexity of the story itself: changes in land tenure, lease arrangements and plans and schemes that have been misdirected on the ground. The use of freedom of information processes and the gradual release of material to the Public Records Office may provide further resources for examination in the future.

---

[49] The *Strzelecki Agreement*, signed August 11 2008, bypassed the Strzelecki Working Group and is not available to the public. Press releases indicate the deal allows a one-off logging of 1500 hectares of the Cores and Links and stalls reservation by another 20 years. The type of reserve is unknown. Protection of 15,000 hectares of other native vegetation was announced, but without details of how it would be protected. Much of it is between areas that would still be subject to highly intensive forestry, or on steep slopes and close to waterways and out of bounds in any case. It wasn't the cohesive national park which the public supported.

References

The Age, 19 November 1979.
Australian Labor Party, Victorian Branch, 1999. *Our Natural Assets: Valuing Victoria's Natural Environment*: Melbourne: The Party.
Anderson, Esther, 2000. *Victoria's National Parks, A Centenary History*. Melbourne: State Library of Victoria.
The Bird Observers Club, 1982. 'Submission to Land Conservation Council regarding Proposed Recommendations for South Gippsland District 2.' Nunawading: The Bird Observers Club.
Collett, Barry, 2009. *Wednesdays Closest to the Full Moon: A History of South Gippsland*, 2nd edition. Melbourne: Melbourne University Press.
Constable, Julie, Devenish, Kim and Standering, Allan, 1998. *A Proposal for a 30,000 hectare National Park in the Strzelecki State Forest*. Foster: self-published.
Devenish, Kim and Constable, Julie, 1997. *Strzelecki Blues: Mucking Around with a State Forest*. Foster: self-published.
Devenish, Kim and Constable, Julie, 1998. The Strzelecki Website. URL <http://members.dcsi.net.au/kimjulie/aaa.html>, accessed 7 August 2014.
Dorrington, Susan and Percival, Paddy, 1986. 'Submission on the Future of the Regrowth Forest on the Southern Face of the Mount Fatigue-Woomerra Creek Ridge'.
Foster Mirror, 11 April 1979.
Foster Mirror, 9 May 1979.
Foster Mirror, 15 October 1986.
Foster Mirror, 1998.
Foster Mirror, 6 May 1998.
Foster Mirror, 31 January 1990, 'Reforestation at Mt. Fatigue —Is it necessary?'.
Gippsland Independent Panel, 2000. *Public consultation issues report: a report to the Joint Commonwealth/ Victorian RFA Steering Committee on issues raised in public submissions and hearings during the consultation process on the Gippsland RFA consultation paper*. The Panel.
Gippsland Local Government Network, 1999. 'Memorandum of Understanding, Action Plan 1999–2000'.
Guillfoyle, Mr. VPC accountant at Melbourne office, Personal telephone conversation.
Gullan, P.K., Earl G.E., Forbes S.J., Barley R.H. and Walsh N.G., 1984. *Sites of Botanical Significance in Central Victoria*. Melbourne: Department of Conservation, Forests and Lands.
Joint Commonwealth and Victorian Regional Forest Agreement Steering Committee, 2000. *Gippsland Comprehensive Regional Assessment: Social Assessment Report*. East Melbourne: Department of Natural Resources and Environment.
Hancock Victorian Plantations, 1999. 'Woorarra Plantation Block 10'.
Hemphill, Ian. 1990. Forest Production Planner, 'Tapes of a public meeting in the Mount Best Hall', 26 January. In the Mount Best Concerned Residents archive.
Hill, Andrew, Timewell, Chris, McCormick, Sally and Mueck, Stephen, 2001. *The Strzelecki Ranges Biodiversity Study*. Port Melbourne: Biosis Research.
Latrobe Valley Express, 1979–1980.
Latrobe Valley Express, 19 January 2006.
Land Conservation Council, Victoria, 1980. *Report on the South Gippsland Study Area, District 2*. Melbourne: Land Conservation Council.
Land Conservation Council, Victoria, 1982. *Proposed recommendations, South Gippsland Area, District 2*. Melbourne: Land Conservation Council.

Land Conservation Council, 1982. *South Gippsland Area, District 2: Final recommendations*. Melbourne: Land Conservation Council.
Land Conservation Council, 1993. Review of Victorian Plantations Corporation (VPC) Vested Lands. Requested by the Minister for Planning, August 1993.
Letter from Balook & District Residents Association to Mr Hamer, 2 October 1978, signed Ms Margaret Long, President.
Letter to Minister Garbutt, 31 January 2002. Signed David Lewis, Chairperson, Strzelecki Working Group.
Mansergh Ian M. and Norris K.C., 1982. *Sites of Zoological Significance in Central Gippsland Volume 1 and 2*. Melbourne: Ministry for Conservation.
Morgan, Patrick, 1997. *The Settling of Gippsland: A Regional History*. Leongatha: Gippsland Municipalities Association.
Pridgin, Ross, 1989. 'Mt. Fatigue Reforestation Plan'. Yarram: Department of Conservation, Forest and Lands.
Sentinel Times, 1998.
South Gippsland Conservation Society, 1980. 'Submission to the LCC South Gippsland Area District 2'.
South Gippsland Conservation Society, 2008. *Newsletter*, May.
South Gippsland Conservation Society Sub-committee, 1982. 'Recommendations for consideration by the Land Conservation Council before preparation of Final Recommendations'. South Gippsland Conservation Society.
South Gippsland Shire Council Minutes, 19 March 2008.
Southern Star, 1998.
Traill, Barry and Porter, Christine, 2001. *Nature Conservation Review Victoria 2001*. Melbourne: Victorian National Parks Association.
Victorian Government, 1986. *Timber Industry Strategy Victoria*. Melbourne: Government Printer.
Weekly Times, 27 July 1983.
Yarram & District Conservation Group, n.d. 'Submission to LCC on Proposed Recommendations for South Gippsland 2 Study Area'.
Yarram Standard, 1998.

Chapter 13

'Nothing we liked better'
An original homestead in a changing landscape

Jillian Durance

Moyarra is 120 kilometres south-east of Melbourne. It lies in the heart of rural south Gippsland where today, from each side of the winding roads, stretch green rolling hills dotted with dairy and beef cows and the occasional house fringed with trees.

Framed by a row of gaunt, twisted, almost spectral pine trees and at the end of a long driveway sits a grand asymmetrical Victorian timber farmhouse, still wearing the charm of something original. Out in the paddock stand a few slightly dilapidated corrugated iron outbuildings painted in that distinctive red oxide colour of the nineteenth century. The garden is still as appealing and beckoning as it was when we (the current owners) first stepped into its welcome shade in February 1997, 100 years after William Rainbow first built his Moyarra homestead at The Pines.

In its beautiful sweep around a rare, perfectly level hilltop the garden holds the remnants of an old orchard, an ancient fruiting feijoa, a giant pear tree, a spreading walnut tree and a huge apple tree, all easily 100 years old and still bearing abundant, delicious fruit. There is a mature grove of leafy elms that frame the view of farmland to the north. To the south you can just see Andersons Inlet and Bass Strait.

Until recently this small property was surrounded on three sides by a well-established dairy farm that occupied the former Rainbow property. Our small remnant parcel of land of 7 acres was, we were to discover, all that was left of the 320-acre selection of Crown Allotment No. 50 in the Parish of Jumbunna East, County of Mornington. It was first selected by Mr John Gannon, an accountant from Drouin and then, in 1895, bought by Mr William Rainbow, also an early Moyarra pioneer.

Figure 13.1: "The Pines", as it was in 1997.
Copy in possession of author.

William Rainbow was among those early settlers who, in 1889, decided to change the name of their district from Jumbunna East to Moyarra.[1] He was also the founder and first trustee of the Moyarra Leased State School, a shareholder in the short-lived Moyarra Co-operative Butter Factory Limited, an elder of the Korumburra Presbyterian Church, and contributor to a well-known local history, *The Land of the Lyrebird*, published in 1920. Rainbow was father to Miss Mabel Rainbow who, until only nine years before our arrival, had owned and lived in the house. The house is still known locally as Rainbows', and has changed little over the decades, beyond slipping into a genteel deterioration. The landscape, on the other hand, has changed almost beyond recognition from what it had once been: a part of the Great Forest of South Gippsland.

The first selectors, including William Rainbow and John Gannon before him, cleared the slopes and steep valleys. The bushfires of 1898 ravaged

---

[1] 'Local News', *South Gippsland Express*, 27 August 1889; Miss C. Elms cited in *The Land of the Lyrebird: A Story of the Great Forest of South Gippsland*. 1998. Korumburra: Korumburra & District Historical Society, 369.

the remaining forest, while the demands of the nearby coal industry, the development of a road infrastructure and the progressive fragmentation of landholdings before and after the Great War quickly changed forever the face of those rolling hills. Driving those changes to the landscape were the choices of people as they responded to environmental forces and the economic pressures, political events and personal fortunes playing out in their own individual lives.

First landscape: the Great Forest

William Rainbow came to South Gippsland in the early 1880s, some 15 years before he settled at The Pines. He came from Buninyong in central Victoria to the forest in the foothills of the Strzelecki Ranges with his three brothers, Joseph, Henry and James, and sister Matilda. Buninyong friends William John Williams and George Matheson came at the same time. William Rainbow's father Henry had first selected land at the foot of Mount Buninyong and, after his death in 1864, his wife Matilda continued to farm with assistance of her family.[2] When she sold the land and retired to Dandenong, near Melbourne, the Rainbow brothers were drawn to the new land being thrown open for selection in South Gippsland; it promised a more reliable rainfall, new fertile soil beneath the abundant trees, and a landscape upon which they could make their own mark, economically and socially.

William Rainbow first selected Crown Allotment No. 57 in the Parish of Jumbunna East. The Buninyong settlers all chose adjoining selections and helped each other to clear their land, helping to create the story that all the pioneers helped each other in their quest to forge a home in the inhospitable forest environment. Several of them, those who survived the longest, recorded their pioneering reminiscences in *The Land of the Lyrebird*, giving detailed, though sometimes triumphal and sentimental accounts of their hardships. The pioneer writers also offered wide-ranging descriptions of the landscape, climate, dairying practices, road building and revealingly, as in James Rainbow's memoir, a description of the original forest, written in 1913.

James Rainbow described the bush on those yet to be developed selections, including that of The Pines, with a sense of entrancement and a naturalist's eye for detail:

---

[2] Rainbow family records, unpublished, private collection, courtesy Mrs Jenifer Newman.

It was very interesting to walk through this great bush and observe the habits of the various animals and birds and note the various kinds of trees, shrubs, ferns, and mosses, etc. which grew in such profusive luxuriance everywhere; from small ferns and moss at your feet to creepers reaching to the tops of the trees, the tallest of which were the bluegums and the blackbutts, one hundred and fifty feet high, and as close together as they could grow. Their trunks were very straight, and would do well for ships' masts or piles. The next in height were the blackwoods or wattles, about 50 or 60 feet high. These woods are useful for cabinet making, being very pretty in the grain and taking a beautiful polish. Their blossoms in springtime were very attractive. Then came the hazel, its leaves a dull green with veins deeply marked and the musk with leaves bright green on the upper surface and silvery underneath, with a rather pleasant musk scent.[3]

Very little of this vegetation remains today along the Foster Creek at Moyarra. The majestic tree ferns ('very pretty indeed when a number of them were seen together'), are now single specimens, like the occasional bluegum, while tussocks of sword grass will still scratch and cut the passer-by. Here and there, hazel and blackwood now merely fringe the roadsides. Similarly, the animals and birds James Rainbow described ('a menagerie and aviary open for inspection and entertainment free of charge on any day or night of the week') had nearly become a thing of the past only 30 years after the white settlers had come to what they later referred to as the Great Forest.

Selection and its effect on the landscape

The seven settlers from Buninyong and their neighbour, John Gannon at The Pines, all worked hard to comply with the regulations necessary to obtain a Crown lease to their properties. They had all selected their properties under the 1869 and 1984 Land Acts, which required a commitment to reside on the property for the six-year term of the original licence. During that time they had to carry out improvements to the value of £1 per acre, to pay rent of one shilling per acre, and to secure the boundaries of the selection with fencing, which was very difficult to do in the rugged terrain.[4] After

---

[3] Mr James Rainbow cited in *The Land of the Lyrebird*, 323.
[4] See especially Marilyn Lake, 1987. *The Limits of Hope: Soldier Settlement in Victoria, 1915*–38. Melbourne: Oxford University Press, 16; Joseph White, Ken Perrett and

a number of years of compliance they could apply to pay the balance and purchase their selection outright. John Gannon's experience at The Pines illustrates the process of selection and the ensuing changes to the forest landscape.

In 1889 John Gannon applied for his Crown Lease on Allotment No. 50 Jumbunna East. In the following five years he made improvements to a total cost of £563/11s. While Gannon had not managed to fence all his 320 acres, he had cleared 90 acres of dense forest through various methods of felling timber, burning, and picking up and grubbing the remaining stumps. He had then sown down his new paddocks, now rich in potash, with 'English' grasses, ready for his dairying enterprise.[5]

John Gannon's application also noted one and a half acres sown down to potatoes, a 25 by 12 foot cottage built of timber palings with iron roof, and verandah and a separate kitchen of slightly smaller dimensions. He had sunk a well and planted a hedge of hawthorn, both of which still feature at The Pines. He had built a yard for stock. His fences consisted of 10 chains of split post and rail, 16 chains of chock and log fence, and 6 chains of picket fence around his buildings. Gannon had carved out a small farm in that unique natural landscape which James Rainbow described in his *Lyrebird* memoir.

As each selector worked from their first clearing outwards toward his neighbours' boundaries the landscape resembled a patchwork of homestead and orchard enclosures, grassed and cultivated paddocks, and fringes of forest, connected by a rough network of muddy pack-tracks with log bridges crossing the streams. Fern gullies were often preserved as favourite places for summer picnics, but much natural habitat was destroyed. But not quite all: even in the 1930s in the deep fern gullies of the Foster Creek, lyrebirds could still be heard imitating the sounds of a nearby sawmill, and platypus still swam in the water.[6]

The difficulties posed by the pioneering life took their toll. Settlers like George Matheson often underestimated the amount of capital required to develop such difficult land and had to take on other work such as labouring for others or taking on a mail run.[7] Freight costs to the coast ate into profits

---

Jillian Durance 2009. *Valley of Peace: History of Kongwak*, Kongwak: Kongwak Public Hall Committee: 9.

[5] John Gannon's Crown Lease application reproduced in Joseph White, 1983. *Valley of Peace*, Kongwak: Kongwak Centenary Committee, 11.

[6] Ray Irving, 2005. Interview.

[7] Mr George Matheson, cited in *The Land of the Lyrebird*, 283.

for butter production, until the coal mining townships, which developed from the 1890s, provided local markets. Hard physical labour also affected the men and their families and sometimes other occupations held more appeal. Henry Rainbow set up a butcher's shop in the nascent township of Jumbunna in the mid 1890s while Joseph Rainbow moved his large family to Mildura to grow fruit. The Rainbow brothers' sister Matilda Gillespie died giving birth to her daughter Tilly in 1897. Tilly's father Robert Gillespie then administered her selection until 1917.[8]

When John Gannon found himself in financial difficulties he had to mortgage his property; the mortgagees foreclosed and by the mid 1890s The Pines was up for sale. While four of the Buninyong selectors managed to hold onto their other selections, or parts of them, other blocks were snapped up and fragmented over time by land sales, annexations by neighbours, demands for road building and new townships, and the infrastructure requirements of the burgeoning coal industry.

The effects of coal mining

Coal mining continued to change the face of the landscape. Any timber still standing on the steep slopes was taken for tramways, tunnel infrastructure (pit props and shoring), mine buildings and fuel. Railway sidings, tracks, mine shafts, excavations, quarries, drilling rigs, tramways and the huge Jumbunna gorge embankment all changed the natural configurations and drainage features of the land, rapidly transforming it from a farming to an industrial landscape.[9]

The townships of Outtrim and Jumbunna sprang up around the new mines, and a railway line feeding out from the Great Southern Railway at Korumburra, and terminating at Jumbunna, was opened in 1891. Road traffic increased with bullock and horse teams transporting supplies, building materials and even coal before the Outtrim line was completed. In 1894 the Jumbunna correspondent of the *Korumburra Independent* newspaper reported on the proliferation of houses, wine shops, hairdressers and coffee palaces: 'The township of Jumbunna is pushing out in leaps and bounds,' he noted, 'and by the time mining operations are in full swing every conceivable

---

[8] Rainbow family records.
[9] This section is based on photographs in the Korumburra and District Society's photographic archive.

article of commerce may be purchased from vendors in what a few years ago was nothing but a vast wilderness'.[10]

By 1894, William Rainbow, with his young wife Mabel Earl, had survived the earliest tests of clearing and fulfilled his selection obligations, but now found he was surrounded by coal exploration and the beginnings of new mining enterprises. Black coal was first discovered in 1890 on Mr Thomas Horsley's property in Jumbunna and then, in 1892, good steaming coal was found at Outtrim. A number of mining companies were formed to develop the seam.[11]

The newly discovered black coal deposit forming the Outtrim and Jumbunna coalfield lay under Mr McLeod's selection adjoining William Rainbow's own land, which was needed for access. He was not permitted to reap the benefits of royalties on the coal and he was not able by law to farm the land while exploration took place and coal mining was conducted nearby. He managed to negotiate for some time with Mr R.B. Stamp of the Outtrim Howitt British Consolidated Coal Company. As he recalled in his memoir, 'one of the directors bought me out and I may say, that transaction put me on my feet financially'.[12]

William Rainbow's search for new land had coincided with John Gannon's troubles at Moyarra. Crown Allotment No. 50, The Pines, still adjoined his Buninyong friends and remaining family; it was ideal. By his own account, William Rainbow was delighted by his new prospect, 'there being plenty of places for sale at the time I travelled about a bit,' he wrote, 'but I saw nothing I liked better than around Moyarra, and as Mr John Gannon's selection was for sale I decided to purchase it'.[13] He could profit from the good work that Gannon had accomplished, but also take advantage of the close and growing market for his butter, cream and bacon, without the immediate, physical interference of the coal mining being carried out just two miles away.

William Rainbow took on his new selection and, by late 1897, had built a fine new homestead on one of the few level building sites in the district. He moved in with Mabel and their two young sons, William Ernest and Henry Phillip. Their first daughter, Lily Beatrice, was born soon afterwards. Within a year William had reason to further appreciate his fortunate choice.

---

[10] *The Korumburra Independent*, 29 March 1894.
[11] Mr M. Halford, cited in *The Land of the Lyrebird*, 233–234.
[12] Mr W. Rainbow cited in *The Land of the Lyrebird*, 301.
[13] Ibid.

During the first few weeks of 1898, after record heat and a long period without rain, severe bushfires ravaged much of the Strzelecki Ranges. Apart from an unknown number of fatalities, there was a great loss of vegetation, housing, livestock and much human suffering. This natural disaster became known as The Great Fire.[14]

The efects of the great fire

What later became known as Red Tuesday, 1 February 1898, lies at the midpoint of nearly four weeks of bushfire. The Rainbows' neighbour, Arthur Elms, who would later become the driving force and editor of *The Land of the Lyrebird*, describes in harrowing detail the effects of the fires that raged through much of west and south Gippsland. The Elms' property and homestead, Athelney, could be clearly seen from the eastern facing verandah at The Pines less than two miles away. Arthur Elms recounts how the fires raged around the Moyarra district and beyond in the fire-filled weeks of January and February 1898. The fires destroyed thousands of acres of grass and miles of fencing. Many houses were burnt, stock roasted alive, the landscape desolated.[15]

Settlers fought the fires by beating down the flames with wetted hessian bags; they sheltered under wetted blankets, or under the protective bodies of horses or in holes excavated by uprooted trees, or in wells. A baby was born as cinders and sparks hailed down on the roof of a cottage. The Moyarra Presbyterian Church, built only a few short years before, was burned to the ground. The Parsons family at Cloverdale carried their valuables onto a bare ploughed paddock and fled the flames. They returned to find their house spared, but their belongings destroyed.[16]

The Parsons, neighbour Mrs Wilson, George Matheson at Bonnie Vale, William John Williams at Ferndale, and Arthur Elms at Athelney all lost fences, grass, outhouses, haystacks, cows, pigs, poultry and horses. George Matheson's entire dairy herd of 35 cows was lost to the flames. But by what the settlers saw as a miracle, along with their own desperate fighting of the fires, their homesteads had been spared. Others elsewhere were not so

---

[14] 'The Great Forest' is a commonly used term in *The Land of the Lyrebird*. 'The Great Fire' distinguishes the 1898 fires from the many others that have wreaked havoc in the South Gippsland area.

[15] Mr Arthur Elms cited in *The Land of the Lyrebird*, 302–310.

[16] This information was obtained from Ray Irving, whose family later lived at *Cloverdale* 1911–1960s.

fortunate, as Arthur Elms noted: 'After years of pioneering, many were left with nothing but the clothes on their backs.'[17]

William Rainbow, newly established at The Pines, was one of the lucky few in that perilous time: 'We were more fortunate than some of our neighbours, as we did not get any of our buildings burned, but only lost some fencing and a few head of cattle. Our old homestead, from which we had only moved a few months, was burned to the ground; only one little log hut, a few fruit trees and a brick chimney were left to mark the spot.'[18]

William Rainbow's good fortune and optimistic approach is mirrored in other survivors' accounts of the time: good rains fell, the grass grew and houses lost were rebuilt. Nevertheless, the years following the fires saw an exodus of settlers, already pressured by financial constraints and the hardships of selecting on such difficult terrain. The only government relief at the time was for hay and chaff for stock in only the most pressing of cases. Those who remained on the land would benefit from the fires in that they had 'thinned out much of the thick scrub, making less of it to remove and the ashes helped increase the fertility'.[19]

The effects of the Great War

Despite the loss of his 40-year-old wife in 1913, at about the same time as he would have begun his memoir for *The Land of the Lyrebird*, William Rainbow describes his early pioneering days in a rosy light: 'We, like all the others, had our trials and sorrows, but rough and all as it was at first, we have had a very happy time.'[20] The land he chose was nothing he liked better and it continued to prosper. He had raised four children, escaped the worst effects of The Great Fire, and lived in a level of comfort not experienced by many of his neighbours. His two sons, William Ernest and Henry Phillip, had been spared the horrors of the Great War and began to purchase and establish their own farms nearby. Although they had attempted to join up in July 1915, they were both rejected as medically unfit.[21]

---

[17] Mr W. Rainbow, cited in *The Land of the Lyrebird*, 301.
[18] White. *Valley of Peace*, 53.
[19] Mr W. Rainbow, cited in *The Land of the Lyrebird*, 301.
[20] Mr Arthur Elms, cited in *The Land of the Lyrebird*, 307.
[21] Rainbow family records. Jennifer Newman is a cousin to Mabel Rainbow (William Rainbow's daughter), and has compiled much of the family history.

William Rainbow's relations and friends were not to escape so easily. William Joseph Rainbow, his nephew, was killed in 1917 at Bullecourt and Alfred George Wesley Williams, second son of his friends William and Euphemia Williams, described in *The Land of the Lyrebird* as 'the first white child born at Kongwak' was lost at Passchendaele.[22] The sons of the pioneers had enlisted along with many of their farm workers, including Richard Ives. Ives, who died in Flanders, who had only recently migrated from England and had been employed at The Pines. One quarter of those local men who enlisted did not return; they were killed in action, or died of wounds and illness and buried in the cemeteries of France and Belgium. Farming families not only lost their workforce, but also the descendants who were once destined to take over their selections.

Those soldiers who returned had the opportunity to take up farming even if they had not done so before. The Discharged Soldier Settlement Act of 1917 provided employment and land to returned servicemen.[23] The Soldier Settlement Board, responsible for the administration of the scheme, acquired and made available a number of blocks in the Moyarra district. At nearby Outtrim, part of William Rainbow's former selection Crown Allotment No. 57 was offered up for soldier settlement along with Henry Rainbow's selection. Their niece, Tilly Rainbow, daughter of their deceased sister Matilda, also sold her mother's selection, Crown Allotment No. 51, right next door to The Pines.

While most of the 320-acre selections were divided into three farms, No. 51 was divided into two, which allowed a more viable 120 acres for Bill Hair to farm. This, in part, contributed to his relative success, as well as his proximity to the rest of his extended family, the reliable rainfall, and the excellent carrying capacity of the former bluegum country. Bill Hair remained farming on his block until he was 90.[24]

To the north – and within view of The Pines, beyond Foster Creek – lay Selection No. 38, acquired during the war years by Mrs Isabella Glasgow of nearby Kilcunda Road, with the view to assisting her son Will Glasgow to set up after his return from the front. He was killed six weeks before the Armistice, and the property was offered for sale to the Soldier Settlement

---

[22] Mrs Euphemia Williams, cited in *The Land of the Lyrebird*, 351.
[23] Lake, *The Limits of Hope*, 49.
[24] Bill Hair's Soldier Settlement file in Public Record Office of Victoria, Inquest database, Soldier Settlement Land Files, PROV VPRS 5714/797; John Gow, 2009. Unpublished notes.

Board, divided into three blocks, all taken up, dairy farmed and further cleared intensively.[25]

Soldier settler George Dunstan's 100-acre block held many precarious slopes and his tiny former homestead still sits perched on one of them above a creek bed. For years, like Bill Hair to the south, he battled through poor returns and the persistent scourge of bracken fern that would colonise the bare soil after the bush had been cleared. Both settlers were inspected regularly to make sure they were keeping up the necessary improvements and performing efficiently.[26]

Nearby, Dunstan's neighbour Les Dowel, plagued with war injuries, found the steep slopes difficult. Les's small farm, annexed in 1919 from his father's selection, continued the Dowel tradition of running Jersey milkers until 1937, when the 85 head pure bred cows were sold at the clearing sale. His original homestead, Callemondah, has been clearly visible from The Pines since 1920 when it was transported there by bullock wagon and re-erected as two former miners cottages, no longer needed in nearby Jumbunna, where the coal industry declined after the Great War.[27]

The soldier settlement movement, and the ensuing closer settlement movement, created pressure on the twentieth century settlers, like the nineteenth century selectors before them, to make their farms viable and make a living while repaying their loans and fulfilling the conditions of their leases. Their families often fulfilled the role of unpaid labourers, especially on the dairy farms, while the smaller farms of the twentieth century saw more houses on the rural landscape.[28]

The pressure to produce as much as possible and to have a farm to leave to the next generation led farmers to clear all the available land, including the creek banks and almost inaccessible slopes. Dairy herds needed year round pasture and hay to supplement feed in the winter. Horses needed oaten hay for chaff. Trees, as well as weeds, became 'the enemy'. A farm with a few remaining trees and a covering of bracken was called 'dirty'.[29]

Another phase of soldier settlement after the Second World War fragmented further the grand selections of the nineteenth century. Alex Thomson's

---

[25] Lois Glasgow, 2003. Unpublished Interview; Australian War Memorial, Roll of Honour Circulars, AWM 131/5021.

[26] Keith Dunstan, 2000. Interview, Moyarra.

[27] Wilfred O'Flaherty, 2004. Unpublished notes, Moyarra.

[28] Dulcie Mitchell, 2002. Unpublished memoir, Korumburra.

[29] Jessie Tyers, 2000. Interview, Moyarra; Dulcie Mitchell, 2001. Interview, Korumburra; Mitchell. Unpublished memoir.

former selection neighbouring The Pines and RN Scott's selection were combined to make Wolonga Estate, which was then divided into four 170-acre dairy farms for soldier settlers.[30] The quest for clear rolling pasture continued well into the twentieth century.

Tragedy strikes

By 1922, only 25 years after settling on his Moyarra selection, William Rainbow held the largest reserve of relatively flat to gently undulating land in the district. Beyond his comfortable eight-room homestead, lay rolling pastures of fine English grasses and clovers now stretching over 400 acres in some of the best dairying country in South Gippsland. The forest was now felled and silent to the axe-men's stroke. The entire property, except for 25 acres, was all wire-netted (to protect against the ravages of rabbits) and subdivided into numerous paddocks, all well watered and cleaned of thistles and bracken fern.[31]

A detailed real estate advertisement for an auction at The Pines, published in June 1922, highlights the vast changes that had taken place since the days of the lush forest illuminated in James Rainbow's description of the flora and fauna three decades earlier. It also describes a very different place to that which William Rainbow had acquired from John Gannon only 25 years before.

Out on his pastures he had:

> 12 three-year-old heifers, some springing and 27 heifers rising two years, some of which were in calf, having been served by his three pure bred Ayrshire bulls, all pedigrees at sale. There were 20 heifer poddies, ten working bullocks, rising four years old and fat, 27 steers rising 2 years old, 2 draft mares, 8 years old and good in all work, a light medium draught colt, rising 3 years and still unbroken, a draught filly, a pony mare aged, but still good in all harness, a Yorkshire sow with her litter, a Tamworth sow to farrow early and a Berkshire boar who had been bred by Mr. Jenkins.[32]

The property also had many established outbuildings:

---

[30] Gow, unpublished notes; White et al. 2009. *Valley of Peace*, 100–101.
[31] *Great Southern Advocate*, 6 June 1922.
[32] Ibid.

… all of galvanised iron and in first class order: a three-stall stable, hay shed, chaff room, machinery bay, housing the latest in farm machinery and implements, a cart shed and a motor garage… a cowshed with a brick floor and 32 bails, separator and engine shed.

There were '60 exceptionally choice grade Ayrshire cows from the 2nd to the 5th calf and some to calve in September'.

William Rainbow's farm in just 30 years had become a viable, successful enterprise.

The advertisement explained that William Rainbow was at the end of his farming life, at 66 years of age; he was retiring and he had 'sufficient faith in his property to leave the bulk of his money in it' as he was selling on exceptional terms. He was selling the lot at auction or, failing that, the property would be submitted 'in four blocks of approximately 100 acres each.' The Pines was described as being 'in good heart', but William Rainbow's heart was no longer in it.

The real reason behind his decision to sell lay elsewhere; it lay in the events of 22 January 1922, only five months before. What happened then reversed his good fortune and changed his life irrevocably. And it is the story of these events that were the first story we were ever told about The Pines.

William Rainbow's sons, now aged 27 and 25, and a farm worker, Charles Scarborough, aged 19, went to bathe in the dam since there was not enough water at the house. And when they did not return for their Sunday dinner William, in his concern, went down to the dam where he found the dog waiting on the bank and two bundles of clothes, but no sign of the three young men. He raised the alarm with his neighbour, John Irving; the Korumburra police were fetched and the dam dragged. All three were found, all drowned, seemingly in the act of trying to rescue each other from the mud, two naked, one fully clothed.[33]

In his unimaginable despair William Rainbow decided to sell The Pines. Ultimately, most of the property was sold to neighbours, but the house block retained. The family, now a bereaved father and his two daughters, Lily and Mabel, maintained the remaining 100 acres in conjunction with a series of share-farmers. William Rainbow died in 1935. His share-farmer at the time, Cocky Bennett, moved out of the house to the little 'fern-cutter's cottage' which is still part of the property. Later share-farmers with families lived in the larger cottage down the hill, now a bleak wind-blown

---

[33] Public Records Office of Victoria Inquest Deposition Files, 1922, PROV VPRS 24, No.86.

pile of rubble surrounded by hawthorns and cherry plum trees. Following the death of Miss Lily Rainbow in 1960 the old farm was sold off. Mabel Rainbow retained the old house on a 7-acre parcel of land and lived there until she died in 1988.[34]

Over the past 100 years the timber homestead with its surrounding garden and orchard at The Pines has changed less than the surrounding countryside. All the neighbouring original selections have been divided up many times; a few farms have increased in size while many others have fragmented into much smaller holdings and 'hobby farms'. The loss of natural vegetation, assisted by the damage caused by successive rabbit plagues, has resulted in obvious land degradation; soil erosion and landslides in the wetter years have denuded the steeper slopes of cover. With the local decline of dairying, beef cattle and sheep are slowly replacing the dairy cows. Reforestation of the Foster Creek and the Powlett River catchment is bringing back pockets of natural vegetation and a slow increase in birdlife and the odd koala. The Pines is very much part of that larger story of the settling of South Gippsland. It continues to command a view of the rolling South Gippsland hills, and endures as a physical reminder to the people who helped change them.

References

Australian War Memorial, Roll of Honour Circulars, AWM 131/5021.
Dunstan, Keith, 2000. Interview, Moyarra.
Elms, Ron, 2007. Unpublished memoir.
Glasgow, Lois, 2003. Unpublished Interview.
Gow, John, 2009. (soldier settler) Unpublished notes.
Great Southern Advocate, 6 June 1922.
Irving, Ray, 2005. Unpublished Interview.
Korumburra and District Historical Society photographic archive, Korumburra, Victoria.
Lake, Marilyn. 1987. *The Limits of Hope: Soldier Settlement in Victoria 1915–1938*. Melbourne: Oxford University Press.
Mitchelle, Dulcie, 2001. Unpublished Interview, Korumburra.
Mitchell, Dulcie. 2002. Unpublished memoir, Korumburra.
O'Flaherty, Wilfred, 2004. Unpublished notes. Moyarra.
PROV, Inquest database, Soldier Settlement Land Files VPRS 5714/797.
PROV, Inquest Deposition Files, 1922, PROV VPRS 24, No.86.
Rainbow family records, collected by Jenifer Newman, (nee Rainbow).
South Gippsland Express, 27 August 1889.

---

[34] Mr Ron Elms, 2007. Unpublished memoir. Ron Elms, who was born in 1924, is a second cousin to Miss Mabel Rainbow and grandson of Moyarra pioneer, Arthur Elms.

The Land of the Lyrebird: A Story of the Great Forest of South Gippsland 1998. Korumburra: Korumburra & District Historical Society.
The Korumburra Independent, 29 March 1894.
Tyers, Jessie 2000. Interview, Moyarra.
White, Joseph, 1983. *Kongwak: The Valley of Peace 1883–1983*. Kongwak, Victoria: Kongwak Centenary Committee.
White, Joseph, Perrett, Ken and Durance, Jillian, 2009. *Valley of Peace: A History of Kongwak*. Kongwak, Victoria: Kongwak Hall Committee.

Chapter 14

Rural land use change and the Ridley paddock, 1897–2012

Deirdre Slattery

The Ridley paddock, in the Stockdale district of East Gippsland, is just an ordinary paddock, a rectangle of 320 acres (half a square mile), or 130 hectares. Like thousands of such paddocks in Victoria, it is a product of the nineteenth-century Land Selection Acts. It was a half-cleared grazing paddock for over 100 years. That changed when my siblings and I sold it in 2004, along with the rest of the farm. Before putting it on the market, we put a Trust for Nature Covenant on one-third of it, the Back Ridley, which was still mainly native vegetation. We pictured this grassy woodland remnant as a haven for a nature-loving idealist who could live in the bush section and run a low-key farming enterprise on the more developed Front Ridley. Instead, the Macquarie Bank bought the paddock: the Ridley is now run by Midway Plantations Pty Ltd as the land manager for the Bank's Stockdale holdings. Midway manages the Front Ridley as a blue gum plantation and the Back Ridley as a conservation site.[1]

My research was facilitated by a fairly intact collection of family records and diaries, as well as the remarkable records of the European history of individual paddocks held in the Public Records Office of Victoria. My cousins' and my own memories and contextual documentation on rural land management history enabled me to put together a continuous story spanning 115 years.

This chapter focuses on the lived experience of each of the seven owners or managers of the Ridley as they worked to achieve land use goals in the context of changing values. Each section describes the aims and purposes of

---

[1] Trust for Nature, January 2005. 'Our Latest Covenants'. *Conservation Bulletin*: 16. East Melbourne: TFN, 16.

the landholder, exploring how these affected the approaches and practices employed in the Ridley paddock in the context of wider changes in land use and management.

Settlement and selection in Gippsland, 1886–97

The land that became known as the Ridley is situated on forested sandy soils in the upper reaches of the Providence Ponds catchment, so before 1897 it was probably not used for squatting and was slow to be taken up by the selection process. Although my family, the Boyds, were not the first to select the Ridley for farming, they were already settling in Stockdale, and knew its first selectors.

Previously the area belonged to the Brayabaulung clan of the Gunai-Kurnai Aborigines, but there was no transition of ownership from Aboriginal to settler; the Boyd family knew nothing about the people whose land they occupied. Today the only evidence of this culture in the Ridley paddock is a possible scar tree and the parish name, Bow Wurrung.

This was not the first experience of land settlement in Victoria for John Boyd, my great-grandfather. Boyd and his wife, fellow Irish migrant Margaret Cusack, had made the decision in 1886 to move from densely settled land near Beechworth, where John Boyd mined from the 1850s. The couple married (1871) and had six children in Beechworth. They farmed a small block of 153 acres (62 hectares). Boyd's obituary, written by an old family friend, says that Beechworth was too settled to meet his restless energy, but his growing family also offered a compelling motivation to secure more land.[2] At Stockdale, still a pioneering frontier for European settlement, he initially purchased two 40-acre (16-hectare) blocks. Here the family lived in a bark hut, then selected 320 acres on which the family weatherboard home was built in 1888.[3] Their youngest child, my grandfather William, was born there in 1888.

Defining the Ridley Paddock, 1897–1920

Ten years later, a neighbouring block also came into the settler world of Land Acts, leases, and government files. It was previously unfarmed land

---

[2] *Gippsland Times*, 14 December 1908.
[3] Victorian Public Record Office Series (VPRS) – Land Selection and Correspondence Files, VPRS625/P0/539, VPRS 625/P0/8 & VPRS/5357/P0/751.

Rural land use change and the Ridley paddock, 1897–2012 | 259

Figure 14.1: The Ridley paddock, 1920.
Personal collection of the author.

that became available for selection in the latter years of the Grant Land Act (1869).[4] Of the four Land Acts that underpinned land use philosophy and policy from 1860 to the 1890s, the Grant Land Act was the most successful in achieving government intentions to establish thriving communities of yeoman farmers.[5] The Acts were a response to the demand for land post 1850s by gold-rush migrants, who formed a powerful political lobby arguing for access to land.

An impressive 10 million hectares were alienated for agricultural production under Land Acts in Victoria, of which the Ridley paddock was one small part.[6] The general conditions of the Act were that a selector held the land for an initial three years under licence. He paid an annual rent of two shillings an acre and had to make 'improvements' such as clearing, fencing, sowing a crop, building a house and residing there. He continued paying

---

[4] VPRS – Land Selection and Correspondence Files, 5357/P0/787.
[5] Tony Dingle 1984. *The Victorians: Settling*. McMahons Point: Fairfax, Syme and Weldon Associates, 63; J.M. Powell, 1970. *The Public Lands of Australia Felix: Settlement and Land Appraisal in Victoria 1834–91 With Special Reference to the Western Plains*. Melbourne & New York: Oxford University Press, 149.
[6] J.M. Powell, 1973. *Yeomen and Bureaucrats: the Victorian Crown Lands Commission, 1878–79*. Melbourne: Oxford University Press, 149, xx.

rent for a further seven years until the full purchase price of £1 an acre was paid. He then gained a Crown Grant of the land that gave him freehold title.[7]

Beginning in 1897, three selectors tried to settle the Ridley. The first was Chas E Lott, a hairdresser from Cassilis on the Omeo goldfield, married with four children. Correspondence in the file offers an inexorable sense that he wasn't able to maintain the 'improvements' required of him.[8] Despite his reports of building a wool shed 20 feet by 20 feet with slab sides and a bark roof, an orchard, fences and lots of 'grubbing', cutting scrub, 'ringing', 'picking up' and 'burning off', he could not make a living, and only paid off one shilling an acre towards the purchase price.

For several years Lott twisted and turned to escape the bailiff's recognition that he was an absentee farmer at best. By 1906 he admitted that he was living as a hairdresser in Stratford and that his neighbour, Charles Ruecastle Ridley, was unofficially managing the land for him. The file contains applications for transfer of the lease to Ridley in 1907, and one from James Rooke in 1913.

None of these settlers lived in the paddock or made a lasting impression on its vigorous scrub and sandy soils. They suffered from the typical problems of selectors. Many failed in such paddocks through poverty, poor land quality, lack of capital, unsuitability to the life, and lack of personal support or labour.[9] As land like this was remote, the products that could be successful were limited by lack of transport to markets, and disaster in the form of pleuro pneumonia, scab, weed invasion, fire, drought and flood was always possible.[10] Depression and drought for 10 years from the 1890s was a further problem.[11] Rabbits, another enduring threat, were in the Stockdale area by 1910.[12]

My grandfather, William Boyd, took over the lease for the Ridley, as it continued to be known, in June 1918. At that time of the transfer, he paid Ridley £900. The reason for the large sum is unspecified in his records.

---

[7] Dingle, *The Victorians*, 63–76.
[8] VPRS 5357/P0/787.
[9] Dingle, *The Victorians*; Powell, *Yeomen and Bureaucrats*.
[10] Don Garden, 1984. *Victoria: A History*. Melbourne: Nelson.
[11] Dingle, *The Victorians*.
[12] L. Hamlyn, 1985. 'The Legacies of Agricultural Development in East Gippsland.' Paper presented at the *Farming in East Gippsland: Past, Present and Future Seminar*, Bairnsdale, Vic. Melbourne: Department of Agriculture for East Gippsland Primary Industry Committee.

It may have been payment for improvements. Boyd paid the Crown half the purchase price, £160, and the remainder in 1920. Boyd's access to this capital showed that he was well off compared to the previous selectors, and indeed he employed Rooke as a casual labourer in the years to come.[13]

His financial security undoubtedly came from his membership of a cohesive family group which methodically selected blocks together, each of his three older brothers having already selected nearby land by the early 1900s.[14] As the youngest son he also benefited from their hard work in establishing Glenariff, a home and farm nearby, where he lived with his widowed mother, inheriting this land on her death.[15]

To this point, the Ridley story illustrates the common difficulties and failures of selection. However, it also shows that the experience of selection lasted well into the twentieth century rather than ending in the nineteenth. Further, the factors leading to success need to be noted as well as those involved in failure. But, as Boyd's experience will show, success did not necessarily conform to the rules. The goal of selection was to establish a family both living on, and living off, the land. Success, however, involved complex relationships and cooperative family efforts across a set of selections. Leased land was rented to others, labour, tools and products were shared, and selectors worked on land other than their own.

Freehold land, 1920–1950

Boyd kept daily diaries from 1917 until 1954 and intermittent ones after that, and his tax records from 1918 to the 1950s are virtually complete. The diaries record his land management. Their daily entries describe small farming tasks and domestic events, virtually without change for 40 years. They show that the practices of the selection years continued long after freehold tenure was achieved.[16] In 1920, Boyd was a young married man with two small children (Sheila, born in 1917, and Ewen, in 1919). The Ridley was his personal investment in his family's future. He took his mother and wife to see the

---

[13] It appears that Rooke was unsuccessful in gaining the transfer and that he worked for Ridley until Boyd took over the lease. See William Newstead Boyd, 1917–1954. 'Farm diary', vols. 1–16. Author's possession (hereafter Boyd Farm Diary).

[14] Powell, *Yeomen and Bureaucrats*, xv.

[15] D. Slattery, 2004. 'Sixty Years Hard Labour: The Continuing Legacy of Selection in Gippsland.' *Gippsland Heritage Journal*, 28, 15–26.

[16] Records consulted include the Boyd diaries, 1917–54 and William Newstead Boyd, 1918–1951. 'Taxation returns.' Author's possession.

new paddock the Sunday after he took possession: the photo (Figure 14.1) was probably taken then. It suggests the daunting task ahead of him.

Freehold tenure offered the opportunity to take a long-term approach to land management, but Boyd's hopes for the paddock may have been short lived as new factors intervened in the early 1920s. Others were leaving the land: Boyd had been managing 640 acres for his older brother, but Dave was reluctant to settle to clearing it on his return from France. By 1919, another brother, Henry, was renting his 640-acre selection to Boyd: he sold the lease to him in 1923. After his mother's death, in 1925, he was managing more than 1600 acres (648 hectares) and struggled to maintain this land, let alone improve it. Although the paddock was soon divided into the Front Ridley and the Back Ridley, change was small and piecemeal and stocking rates reflected the paddock's unimproved state, remaining constant at around 2–300 sheep (less than one to the acre) and a few cattle.[17]

Both his brothers often worked for Boyd in the 1920s, mainly 'scrubbing' and bark collecting, clearing of which seems to have been a priority, but in the Depression years Boyd's expenditure on employment for all tasks but the essential shearing dropped radically from an average of £150 in the 1920s to £20–£50 in the 1930s. Wool prices were fixed by agreement with Great Britain during World War One but were very low once that arrangement ended.[18] Boyd's taxable income varied between £767 in 1921–22 and £35 in 1929–30. He had employed Bob Miller virtually full time from 1925, but that ceased in 1930 for lack of income. For the years 1920–45 the Ridley paddock seems to have suffered from the general problems of decline in the wool price, and lack of either manpower or mechanical capacity to increase production.[19] In addition, perhaps Boyd had inadvertently become responsible for too much unproductive land through the various family processes described.

Farm work was characterised by monotonous physical labour. For years the Ridley was managed by frequent 'riding around' or walking around the stock (occasionally carrying home a sick sheep), bouts of rabbit poisoning

---

[17] Boyd Farm Diary, 1917–54.

[18] The price was fixed by imperial agreement until 1920 and the excess wool stockpiled. The British Australian Wool Realisation Association was then formed to avoid a drop in price and to arrange disposal of the stockpile. This arrangement was wound up in 1930. See Alan Barnard (ed.), 1962. *The Simple Fleece: Studies in the Australian Wool Industry*, Melbourne: Melbourne University Press, 480; S. Wadham, R. Wilson, R. Kent and J. Wood, 1964. *Land Utilization in Australia*. Melbourne: Melbourne University Press.

[19] Dingle, *The Victorians*, 179–8; Chris McRae, 1976. *Land to Pasture: Environment, Land Use and Primary Production in East Gippsland*. Bairnsdale: Department of Agriculture, 68.

Figure 14.2: Wattle bark was a big source of income from the Ridley in the 1920s.
Personal collection of the author.

using traps and baited carrots, stripping and carting wattle bark, controlling weeds, hoeing scrub, cutting firewood, fencing, and checking fences for damage. There were periodic wild dog problems necessitating use of a private 'dogger'. Many tons of household firewood and fencing posts as well as wattle bark were taken from the paddock. The latter was obviously an attempt to benefit from the necessity to clear it, but also a source of casual employment for other locals who seem to have been able to sell the bark they collected from Boyd's land. Sale of tan bark mainly from the Ridley earned more than 10 per cent of the farm's cash income in most of the 1920s as well as income for half a dozen others who were stripping the wattle bark. Cutting down, stacking up and burning timber was repeated endlessly, followed by periodic sowing of manure and clover seed. Ironically, after so much removal of shelter and shade, by 1938 Boyd was planting cypress pines around the sheep yard, and in the dry years that followed he carted water to them. One is still alive, a monument to 'improvement'.

In describing his work, Boyd uses selectors' language: words and phrases specific to the task of clearing and settling the land.[20] 'Vermin' describes

---

[20] Jay Arthur, 2003. *The Default Country: A Lexical Cartography of Twentieth-century Australia*. Sydney: University of New South Wales Press; William Ransom, 1991.

all non-domestic animals, native and otherwise. Poisoning and trapping invoke complex rituals of warfare, the native shrubs 'manuka', 'black ti-tree', 'mangrove' and 'burgan' are 'scrub' and 'weeds', as are the introduced thistles; and 'scrubbing', 'ringing', 'grubbing', 'bark cutting' 'burning off' are the means of control.

Some aspects of nature were uncontrollable, no matter how much work was done. With State Forest to the west of it, the Ridley paddock was occasionally the point where fire entered the property, a dreaded event, particularly in 1926 and 1932.

> The fires being very bad. Mustered the sheep out of the Ridley paddock, School paddock, House paddock, Old paddock and David Boyd's. The fire burning out the Ridley paddock, School paddock and part of the House paddock as well as J Boyd's paddocks.[21]

Limitations to the yeoman dream

By the 1940s some of the negative impacts of closer settlement on local ecology were beginning to show.[22] The most demanding task was clearing the persistent understorey of 'burgan' (*Kunzea ericoides*) and 'manuka' as they were colloquially called (in this case *Melaleuca parvistaminea*, but also *Leptospermum continentale*). These native species responded vigorously to disturbance, repeatedly 'taking over' the clearing efforts, even though Boyd, Miller and Rooke spent several weeks on them every year.

The land's stubborn resistance brought realisation across the district that clearing had become a matter for broader policy and action. The farmers sought a political solution, holding well-attended public meetings in Lindenow and Bairnsdale in 1948 to lobby for government assistance.[23] They were frustrated that although they had done everything reasonable to 'improve' their land, they had had inadequate support. They angrily accused *laissez faire* government policy of setting them up to fail for 40 years. Hardy Rash, a neighbour of the Ridley paddock, proposed government support for

---

'Wastelands to Wilderness: Changing Perceptions of the Environment.' In Mulvaney, John (ed.). *The Humanities and the Australian Environment*, Australian Academy of the Humanities, Occasional Paper No. 11. Canberra: Highland Press: 5–20.

[21] Boyd diary, 19 January 1932.
[22] Dingle, *The Victorians*, 179.
[23] Boyd Farm Diary, 1917–54.

increased use of bulldozers. As he was a bulldozing services contractor at the time he was understandably enthusiastic about lobbying for this solution. Others wanted these 'weeds' declared 'noxious'.

The farmers' case for government assistance involved interest-free loans, assistance from the Commonwealth Scientific Industrial Research (renamed the Commonwealth Scientific Industrial Research Organisation, CSIRO, in 1949), and access to heavy clearing machinery. However, representatives of the Country Party government (deputy premier John McDonald, and Albert Lind, member for East Gippsland), fobbed the farmers off with vague promises of a bright future.[24] McDonald's evasive response echoed the original logic of selection policy: it was a question of manpower. In predicting a future population of half a million for East Gippsland, he fudged the issue of the farmers' present problem by reiterating static government policy, offering the vision of prosperity for the district through further settlement of new areas of public land.

These days the clearing problem could be seen as a long-term 'ecological disordering', caused by the settlement process.[25] In the *Advertiser* the Supervisor of Land Settlement, ER Pemberton, offered this explanation in a polite and technical article, addressing angry farmer concerns by explaining the plant's ecology.[26] Manuka's spread was connected with removal of the tree cover that had previously inhibited growth, and poisoning it or clearing it piecemeal without addressing the question of cover would not help. Pemberton also explained that it was a native plant, whose spread was enabled by a change in 'the balance of nature', but the contrary opinion (that it is a New Zealand weed that came in with settlement) is still active in the district today.[27]

Hence, the activities in the paddock for its first 60 years, under both selection and freehold tenure, were overwhelmingly directed at controlling the very problems that had arisen from the settlement process itself: regrowth of native 'scrub', 'vermin', 'weeds', damage from frequent uncontrolled fire, loss of shade. These activities may also have been instrumental in developing long-lasting attitudes: my cousin suggests that maybe they shaped a generation that thought in 40-acre increments of 'improvements', a requirement under

---

[24] *Bairnsdale Advertiser* October 1948; Slattery 'Sixty Years Hard Labour'.
[25] G. Main, 2005. *Heartland. The Regeneration of Rural Place*. Sydney: UNSW Press, 51–2.
[26] E.J. Pemberton, 'More Effective Utilisation of Land: Suggested as Solution to Combating Burgans Scrub', *Bairnsdale Advertiser*, 15 October, 1948, 3.
[27] J. Boyd, personal communication, January 2006.

the Land Acts.²⁸ Perhaps the low technology, self-sufficient experience of farming had shaped the farmers' practices. As my cousin points out, they used hand tools rather than machines, they walked everywhere, even when they owned utilities, and so took a lot of time to go around their stock.²⁹

Until the 1940s the settlers' achievements were measured in clean land with no shelter for 'vermin', cultivated land where a new balance was achieved between farmer and ecology, and ordered pastures where the bush no longer threatened to 'come back'. The security of a family, home and community enabled the land to be worked. Family farm enterprises both supported and relied on communities that had been through similar experiences of settlement.

Families like the Boyds and others who survived this process remain the backbone of farming in 'deep countryside' districts.³⁰ Their success is hard earned, achieved through intensive labour and values such as self-sufficiency and cooperation, austerity and practical innovation.³¹ Pride is part of this culture: in place, in family, in community, in farming skill and achievement, in the aesthetics of landscape which have developed from their successful, if precarious, control of land through farming. The culture that was built around these personal and social relationships has been long enduring, but the dependency of such farms on declining prices and demand for their products makes them vulnerable to decline or change.³²

Ridley paddock in the era of scientific agriculture, 1950s to the 1980s

Until about 1946, the world defined by Boyd's diaries was small. He recorded most daily visits to and from the farm: nearly all of these are local

---

[28] J Boyd, personal communication, January 2006.

[29] Ibid.

[30] Stephen Dovers, 1992. 'The History of Natural Resource Use in Rural Australia: Practicalities and Ideologies.' In *Agriculture, Environment and Society. Contemporary Issues for Australia*. Geoff Lawrence, Frank Vanclay, and Brian Furze (eds). South Melbourne: Macmillan; Don Aitkin, 2005. 'Return to "Countrymindedness".' In *Struggle Country. The Rural Ideal in Twentieth Century Australia*. Graeme Davison and Marc Brodie (eds). Clayton: Monash University ePress: 11.01–11.06.

[31] J.M. Powell, 1988. 'Patrimony of the People: The Role of Government in Land Settlement.' In R.L. Heathcote (ed). *The Australian Experience. Land Settlement and Resource Management*. Melbourne: Longman Cheshire, 15–25.

[32] Neil Barr, 2009. *The House on the Hill. The Transformation of Australia's Farming Communities*, Canberra: Land and Water Australia, 132–6.

or personal, whether they involved family, friends, employees, local suppliers and businesses or community activity. But after 1946 change gathered pace. The diaries record the arrival of the telephone and electricity. These changes are confirmed by my own memories. What had been a seemingly unchanging culture at Stockdale was revolutionised: domestically, the farm became less self-sufficient in dairy foods, meat, fruit and vegetables. Washing and cooking were transformed, and outside the house tractors and utilities replaced horses, mobility increased, activities were more specialised in wool production. Thanks to the Korean War, the wool price was at an all time high, Boyd's income more than quadrupling from 1945–46 to 1949–50, and my grandparents bought a new car and went to the district centre of Sale to shop weekly rather than monthly.[33] I remember school holidays in the 1950s where, instead of the slow routine of walking around sheep, we spent the time on the back of Uncle Ewen's ute, driving the eight miles to and from the railway siding in Munro carting superphosphate imported from a mysterious place called Nauru, where, in passing, we were told Dave Boyd's son Ivan worked for a body called the South Pacific Commission.

As the increased use of superphosphate showed, agricultural science began to dominate perceptions of land with new values about farming life. 'Land settlement' was in the past and land productivity was the prospect. 'Mastery' and conquest 'of the earth for man's use' were the aims of the CSIRO and agricultural scientists.[34] A 'biography' of the agricultural district, *Land to pasture. Environment, land use and primary production in East Gippsland*, or the Department of Agriculture *Journal*, describe the context in which the Ridley and other recalcitrant paddocks could now be seen.[35] These publications were full of belief in technical possibilities, relatively problem free, and confident.

New visitors mentioned in the diaries included representatives of the Australian Paper Mill and of Pivot and Cresco Fertiliser Companies, the latter a South Australian-based superphosphate company recently established at Geelong. New outings also appeared. My grandfather notes that he and his now adult son Ewen participated in field days, through a very active Department of Agriculture extension program. Trial plots and experiments on model farms encouraged farmers to update their methods. Once the land was cleared, the emphasis was on pasture improvement

---

[33] Boyd, Taxation Returns. 1918–50.
[34] Main, *Heartland*, 48–9.
[35] McRae, *Land to Pasture*.

through use of 'super', and introduced pasture plants. In East Gippsland the Bengworden Stocking Rate Trial (1957) was a great success, showing 'that stocking rates on improved pastures could be doubled compared to ruling district practice'.[36]

The lack of permanent water and regrowth in the Ridley could now be dealt with through the use of heavy machinery. Some new dams were installed, one in the 1970s, probably replacing earlier handmade ones. Boyd records the application of greatly increased amounts of superphosphate to his paddocks from 1947–51, spreading 61 bags (of 1 cwt each, or 51 kilos) in 1947, 132 bags in 1948, 144 bags in 1949 and 316 bags in 1951. Although superphosphate had been used as fertiliser since the 1920s, and there had been superphosphate bounties previously, its use had been constrained by the wartime occupation of Nauru by Japan, and by cost and lack of machinery to cart and spread it. Now it was spread in the Front Ridley, and subterranean clover planted. In 1949 the Boyds bought Cobran rams to improve their specialisation in fine-wooled Merinos.

The Australian Paper Mill began to buy land, as my grandfather recorded: 'A representative of the Paper Pulp people called'.[37] Beginning with one of the original Boyd paddocks opposite the Ridley, in the 1950s and 1960s Australian Paper Mill bought hundreds of hectares of formerly selected land. Soon, the Ridley was surrounded on three sides by young pine trees whose dense rows paid no heed to watercourses, contours or previous occupancy. Their needles formed a thick silent blanket, through which an occasional wattle or manuka daringly sprouted, as well as blackberries. This introduction of agribusiness replaced land owned by Boyd, Rash and Jensen, families working to create and maintain a way of life, with a company operating for profit. This depopulation was sufficient to result in the closure of the school in the 1960s, leaving a major hole in the landscape and in community life. New houses sprang up close to the nearby settlement of Munro for Australian Paper Mill employees who did not live in Stockdale.

But the Back Ridley escaped most of these changes. I went fox shooting there with my father, a ritual that involved an eternity of crouching behind him in the dense manuka, being vigorously 'shushed' for swatting mosquitoes as he imitated the shrill scream of an injured rabbit on his fox whistle. But this time I recall looking at the myxomatosis-glazed eyes of a

---

[36] M. J. Lee, 1965. 'Land Use in East Gippsland.' *The Journal of Agriculture*, 63(6): 259–262; McRae *Land to Pasture:*, 13.

[37] Boyd diary, 5 April 1951.

rabbit he shot; a new wonder scourge of vermin, I was told. The subsequent decrease in rabbits could have contributed to the increase in wool yield that followed.

Diverging values, 1960s to the 1980s

By the 1960s, new personal factors also affected the nature and extent of land use. My grandfather's approach to subduing the recalcitrant scrub through personal toil supplemented by the hoe and the horse was being replaced by huge technological and scientific changes, but he was conservative, and growing old, and could no longer sustain his efforts. His diaries were patchy after 1954, although he lived until 1965.

The time for generational change had arrived, but that was a slow and indecisive process. A partnership, 'Boyd and Son', was not established until 1947, when my uncle was 28. Until then he had apparently worked for an allowance, and although he increasingly performed the harder physical work, there is little evidence that he had much managerial autonomy even after the partnership was set up. When he inherited the property in 1965, he was unmarried and there was little incentive and too much land for him to continue to improve productivity other than marginally, although he maintained the commitment to good animal husbandry with which he had been brought up. He sold a third of the farmland to a cousin and continued with much the same methods as his father on the rest. My grandfather's will left some money but no land or managerial role for the other main interested party, his daughter.

Unlike my uncle, who had begun working on the farm at 14, my mother had been a primary teacher and had married a teacher. Her life and relationship to the farm, where she had not lived permanently since going to boarding school in the 1930s, had developed in very different ways from her father's and brother's. By the 1960s, busy with her own growing family in distant parts of Victoria, she visited her childhood home a few times a year, but maintained a deep emotional commitment and loyalty to it. As the absent and now disempowered daughter, she was frustrated by the lack of 'progress' at Glenariff, particularly exemplified by the Ridley paddock. Succession planning was unheard of then, but there was no attempt to encourage the next generation, my siblings and myself, to think of ourselves as farmers or potential owners.

Despite this we remained very attached to the place, partly out of affection for family and places, partly as a romantic part of our heritage. By the 1970s, we were irregular visitors from the city, now with our own small children with whom we were keen to share the essence of the place. This was best experienced in the long fly-swatting walk over to the Ridley with Uncle Ewen and the dogs to 'go around' the sheep. That didn't change. The dusty lane retained its yellow box and wattles, and the Back Ridley was still the best place to spot some wildlife or wildflowers, the kind of experience I normally enjoyed through visiting and camping in national parks in other parts of the state.

Nevertheless, in the 1970s – as a result of cooperative work with my now-adult second cousin – part of the Back Ridley was bulldozed and sown with oats. My uncle's erratic notes indicate that stocking rates increased markedly in both the Front and Back Ridley paddocks through the 1970s and 1980s to a regular 400–500 sheep. His production of wool increased from 23–25 bales in the 1940s to 35 bales in the 1970s and 1980s, although he farmed less land. As well, he stocked more cattle there, presumably because they needed less detailed care than sheep. Occasional jottings of his wool sales results and his tax returns show that his income from wool alone was around $15,000 to $20,000 per annum, more than enough for a single person.[38] This was clearly assisted by the wool Reserve Price Scheme, which ensured a fixed price for wool from 1974 until 1991, after his death.

New owners and new relationships to land: 1986–2005

My Uncle died in 1985, leaving the farm as a life income to my mother as a trustee for her children to inherit on her death. Too old and lacking practical experience, she leased it to a series of tenants but took little responsibility for it, treating is as a place to go for a walk on her increasingly infrequent visits. Her fiercely defensive and nostalgically romantic attitude towards her childhood home seemed to paralyse her ability to move forward in any particular direction. Her sense of place for the Ridley and Glenariff in general was fixed in a stable secure past, not a decaying present.

The problems of agricultural land with which we are all too familiar today became my responsibility in the early 1990s. A series of short-term tenants argued for reductions in the rent each time the lease was renewed on the grounds that 'times were bad'. The Ridley was the least profitable of all the paddocks, with its decaying fences, multiplying kangaroos, foxes, rabbits and wombats, burgeoning blackberry patches, overgrown dams and rapidly regenerating 'scrub'. It made me recognise that as absentee

---

[38] John Ewen Boyd, 1948–1951; 1970–83. 'Taxation returns.' Author's possession.

Figure 14.3: Melaleuca and Leptospermum species colonising previously grazed woodland, Ridley paddock 2013.
Personal collection of the author.

landholders we were no better than farmers I sometimes criticised for poor land management. A party of kangaroo shooters drove through the padlocked gate one night, leaving a trail of dying and butchered corpses and beer cans littering the paddock. The family 'ancestral home' sat empty and neglected in a nearby paddock, like many other sad relics around the landscape of Victoria.

From rental property to plantation and nature reserve: 2000–2006

To resolve these problems of land management and personal responsibility, my siblings and I put the Ridley on the market along with the other land titles. One cold windy day in 2004, a group of us visited it to appraise its worth. The stock and station agent looked over the dense bush and shook his head: 'It's gone back terribly since your uncle died. It would cost about $100 an acre to bring it back'. A neighbour wanted to buy it and convert it to pasture, and nodded agreement. Further down the hill, the Trust for

Nature assessor smiled happily: 'Look, here's a patch of greenhoods! This leptospermum and melaleuca combination is really unusual around here. There's virtually none left. This paddock is coming on really well.' My siblings hovered between these two attitudes to land value: will we get more for it if it's covenanted, or less?

Coming on, going back! The language of our appraisals was at odds, showing our conflicting perceptions and purposes. We were not farmers. As urban people who love the bush, we favoured the 'coming on' view, and regarded the native vegetation and wildlife in the paddock with pride and interest. But our cousin, who worked here to make it more productive and was then clearing remnant bush off his own place, probably shared the latter view more than the former: he wisely didn't say. We went ahead with the covenant, but the anticipated 'green' buyer didn't turn up, and a year later we accepted an offer of $600 per acre from Macquarie Bank. The purchase price was more than local farmers would pay for 'unimproved' land.

The new land manager, a trained forester, lives in Stratford. He is proud of his professionalism, believing that managing trees is no more damaging than some past agricultural practices. He still calls the paddock 'the Ridley', as it's important to those locals with whom he wants a good relationship.[39] Since then, Macquarie Bank has bought at least half a dozen whole farms in the district, including a large property owned by my grandfather's older brother and his descendants since the 1890s. A financially secure retirement became possible for the owner, another cousin. The district has changed markedly. Long-standing views of farmhouses, hills, patches of bush, patterns of gullies and creeks and roads are hidden by the uniformly growing trees. Such radical changes in farmers' physical and social lives go against their deep-seated belief in the value of the land and its product, and of the lifestyle choice that they have often made, which is more important to them than mere financial success.[40] John Boyd would 'like the price of wool to go up, then no one would sell'.[41]

The Front Ridley is just a small example of a huge change in land use in the Stockdale district, driven through owner investors who will never see the land they have so altered, or take any interest in the product from which they hope to profit. Although Macquarie Bank's investments have not collapsed

---

[39] M. Schofield, personal communication, January 2006.
[40] M. Tonts and A. Black, 2003. 'The Social and Economic Implications of Farm Plantation Forestry: A Review of Some Key Issues.' *Rural Society*, 13(2): 174–192.
[41] J. Boyd, personal communication, January 2006.

– as those of Great Southern Plantations and Timbercorp did in 2008–09 – yields may well be less than expected in the Ridley as was the case for some of these companies' plantations.[42] As the Australian Tax Office (ATO) puts it, companies have focused on selling tax-free incentives, not on profitable commodity production: the outcome has put the future of agribusiness as a land use in serious doubt. Some commentators are contemplating the huge cost of restoring timbered land to farmland; such land has certainly lost the biodiversity value it had retained over the previous 120 years. In a welter of parliamentary inquiries, ATO guidelines and receiverships it is clear that whatever difficulties traditional farmers have faced, agribusiness schemes cannot claim a more convincing approach to land management.

Reflecting on the Ridley story

The story of the Ridley paddock under eight different landholders offers both an historical and a sociological angle. It illustrates the interactions between broad general land use policy and detailed personal response. The direct, determined aims of the Land Acts of 100 years ago – with their values of colonisation, productivity and community stability – have long disappeared from government policy, although they shaped a landscape that looked stable from the 1920s to the 1980s and these values remain alive in some visions of rural life.

But this stability was short term when the relationship between productive capacity, community security and ecological health of land is considered seriously.[43] Taken one by one, the choices of the landholders seem accidental and random, but the story of the Ridley actually reflects well-known social factors in land use. The size and gender distribution of a family, education (or lack of it), access to a life partner (or lack of it), the influence of history and tradition, and of personal style all act as an invisible backdrop against which land use goes on, or changes. The influence of the wider culture cannot be sifted out: the role of second-, third- and fourth-generation families does matter, as does the influence of rural or urban backgrounds and experiences of landowners.

---

[42] Wikipedia entry for the 'Great Southern Group 2012'. URL <http://en.wikipedia.org/wiki/Great_Southern_Group> accessed 1 June 2013.

[43] Dingle, *The Victorians*; Garden, *Victoria*; Dovers, 'The History of Natural Resource Use'; Powell, *Yeomen and Bureaucrats*; Powell, 'Patrimony of the People'; McCrae, *Land to Pasture*; Wadham, Wilson and Wood, *Land Utlization in Australia*.

Ecological health is a relatively new value in the equation, and is currently driving major federal government funding for rural land. The emergence of Landcare, with its emphasis on community capacity and personal stewardship, underpins government aspirations which place individuals at the forefront of policy implementation.[44] But policies for a landscape that integrates biodiversity and productivity look very tentative and prone to uncertain outcomes when compared with the Land Acts.

Shaped by both historical and current trends, the Ridley has experienced three extreme styles of use in the last 10 years: conventional farming, nature conservation, and plantation forestry. Of these, family farming proved not to be sustainable; in part because this light grazing land is not adaptable to diversification when commodity prices fall, but also because of personal factors. Plantation forestry looks likely to be the most short lived of the various scenarios for the long-term use of the Ridley, and is certainly the least 'balanced' from the point of view of land health. The covenant has the advantage of being permanent, but greening the landscape through significant protection, as in the case of the Back Ridley, offers only a partial solution to the challenge to live better with the land. As a model of land management it relies on income from outside the land holding itself and so its occupiers, if there are any, actually depend on products from land elsewhere.

The Ridley paddock is an ordinary place in the Victorian landscape, but its small story throws some light on Australians' continuing struggles to achieve a stable and mature relationship with land. The changes this paddock has undergone over the last 120 years do not add up to a coherent culture of land use. The shifting interactions of policy demands and inducements and changing personal and family experience have led to a clumsy outcome. Land selection, family farming, agribusiness schemes and biodiversity conservation initiatives are all government driven, but no one set of practices and values comes to grips with both the individual values and needs of farmers and rural communities and the broader society's need for an integrated healthy landscape.

---

[44] Deirdre Slattery, 2010. 'Community Participation in Land Restoration.' *Proceedings of the Royal Society of Victoria*, 122 (2): 123–129.

References

Aitkin, Don, 2005. 'Return to "Countrymindedness"'. In Graeme Davison and Marc Brodie (eds), *Struggle Country. The Rural Ideal in Twentieth Century Australia*. Clayton: Monash University ePress: 11.01–11.06.
Arthur, Jay, 2003. *The Default Country: A Lexical Cartography of Twentieth-century Australia*. Sydney: University of New South Wales Press.
Bairnsdale Advertiser, October 1948.
Barnard, Alan (ed.), 1962. *The Simple Fleece: Studies in the Australian Wool Industry*. Melbourne: Melbourne University Press.
Barr, Neil, 2009. *The House on the Hill. The Transformation of Australia's Farming Communities*. Canberra: Land and Water Australia.
Boyd, J. 2006. personal communication, January.
Boyd, John Ewen, 1948–1951; 1970–83. 'Taxation returns.' Author's possession.
Boyd, William Newstead, 1917–1954. 'Farm diary', vols. 1–16. Author's possession.
Boyd, William Newstead, 1918–1951. 'Taxation returns.' Author's possession.
Department of Primary Industry, 1999. *Plantations 2020 Vision Strategy*. Accessed 25 January, 2006. Available from: http://www.plantations2020.com.au
Dingle, Tony, 1984. *The Victorians: Settling*. McMahons Point: Fairfax, Syme and Weldon Associates.
Dovers, Stephen, 1992. 'The History of Natural Resource Use in Rural Australia: Practicalities and Ideologies'. In *Agriculture, Environment and Society. Contemporary Issues for Australia*. Geoff Lawrence, Frank Vanclay, and Brian Furze (eds). South Melbourne: Macmillan.
Garden, Don, 1984. *Victoria: A History*. Melbourne: Nelson.
Gippsland Times, 14 December 1908.
Hamlyn, Laurie, 1985. 'The Legacies of Agricultural Development in East Gippsland'. Paper presented at the *Farming in East Gippsland: Past, present and future Seminar*, Bairnsdale, Victoria, 1 May. Melbourne: Department of Agriculture for East Gippsland Primary Industry Committee.
Lee, M. J., 1965. 'Land Use in East Gippsland'. *The Journal of Agriculture*, 63(6): 259–262.
Main, George, 2005. *Heartland. The Regeneration of Rural Place*. Sydney: UNSW Press.
McRae, Chris, 1976. *Land to Pasture: Environment, Land Use and Primary Production in East Gippsland*. Bairnsdale: Department of Agriculture.
Powell, J.M. 1970. *The Public Lands of Australia Felix*. Melbourne: Oxford University Press.
Powell, J.M., 1988. 'Patrimony of the People: The Role of Government in Land Settlement'. In *The Australian Experience. Land Settlement and Resource Management*. Edited by Heathcote, Ronald Leslie. Melbourne: Longman Cheshire, 15–25.
Powell, J.M. (ed.), 1973. *Yeomen and Bureaucrats. The Victorian Crown Lands Commission 1878–79*. Melbourne: Oxford University Press.
PROV. Land Selection Files. VPRS 5357/P0/787; VPRS 625/P0/539; VPRS 625/P0/8; VPRS/5357/P0/751.
Ransom, William, 1991. 'Wastelands to Wilderness: Changing Perceptions of the Environment'. In John Mulvaney (ed.), *The Humanities and the Australian Environment*. Australian Academy of the Humanities, Occasional Paper No. 11. Canberra: Highland Press: 5–20.
Schofield, M. 2006. personal communication, January.
Slattery, Deirdre, 2004. 'Sixty Years Hard Labour: The Continuing Legacy of Selection in Gippsland'. *Gippsland Heritage Journal*, 28: 15–26.

Slattery, Deirdre, 2010. 'Community participation in land restoration.' *Proceedings of the Royal Society of Victoria*, 122 (2): 123–129.
Tonts, Matthew; Black, Alan, 2003. 'The Social and Economic Implications of Farm Plantation Forestry: A Review of Some Key Issues'. *Rural Society*, 13(2): 174–192.
Trust for Nature, January 2005. 'Our Latest Covenants'. *Conservation Bulletin*: 16. East Melbourne: TFN.
Victorian Public Record Office Series (VPRS) – Land Selection and Correspondence Files, VPRS625/P0/539, VPRS 625/P0/8 & VPRS/5357/P0/751.
Wadham, Samuel, Wilson, R. Kent and Wood, Joyce, 1964. *Land Utilization in Australia*. Melbourne: Melbourne University Press.
Wikipedia, 'Great Southern Group.' Accessed 30 January 2012. Available from: http://en.wikipedia.org/wiki/Great_Southern_Group.

Chapter 15

People and forests in East Gippsland
Change and continuity in forest industries

Helen Martin

This chapter draws extensively on oral history interviews conducted in 2008 with people who had worked in East Gippsland's timber industry or in Victorian Government agencies responsible for forest management, or who had knowledge of the forests and forest industries through the experiences of family members.[1] The interviews were commissioned by Gippsland Timber Development Inc. (GTD), which was formed in 1996 to encourage value adding in the Gippsland timber industry by maximising the use of timber and minimising waste.[2] The interviews contributed to an exhibition at the East Gippsland Institute of TAFE (now Federation Training) Forestech campus near Lakes Entrance which opened in 1997, but are also available for use by researchers or local historians.

These interviews provide insights into the people who spent their lives working in the forests of East Gippsland and in the industries that depend on them. They reveal the way in which individuals responded to the challenges presented by the forests and the economic opportunities they offered. East Gippsland's forests are very diverse, requiring adjustments in techniques and approaches, but they also provide a range of 'niches'

---

[1] The original recordings – listed in Appendix 1 – are held by the East Gippsland Shire Library and are available for research purposes.

[2] Gippsland Timber Development Inc. (GTD) was formed in 1996 under the auspices of Gippsland Development Limited (GDL), the regional development organisation. It attracted funding to establish the East Gippsland Institute of TAFE (now Federation Training) Forestech campus near Lakes Entrance and developed an interpretative display and a self-guided forest drive. The board was made up of representatives of GDL, East Gippsland TAFE, Monash University, state agencies, local government, and the relevant industry body and union. GTD's projects were mainly funded by Commonwealth grants. The author acted as executive officer from 1995–2001 (through East Gippsland Shire) and was a board member thereafter. GTD was disbanded in 2011.

to produce different outputs, often with a lot of hard work but relatively simple technologies. The forests and the economy also delivered a series of blows to participants in the industry, through the dangers of working in the bush or the mills, bushfires, diseases affecting the timber resource, market fluctuations and changes of government policy. As a result, the story of forest industries in East Gippsland demonstrates a high level of adaptability as well as a degree of continuity.

Utilisation of forests has been one of the backbones of East Gippsland's economy.[3] In 1974 the Land Conservation Council (LCC) reported that the 25 mills in the East Gippsland Forest Management Area (FMA) – equivalent to the former Orbost Shire, plus the eastern part of the former Tambo Shire – produced 27 per cent of Victoria's total hardwood output and employed 700 people. Including the families of timber industry workers, 44 per cent of Orbost Shire's population depended on the timber industry.[4] By 1985, East Gippsland FMA produced 25 per cent of Victoria's hardwood output. An estimated 25 per cent of Orbost Shire's workforce was directly employed in the timber industry and another 10–15 per cent in industries wholly or partially dependent on it. The 23 mills supplied from public land employed nearly 500 people.[5] The amount of mills and number of people employed in the timber industry in East Gippsland has since contracted significantly, as a result of reductions in timber supply, market forces and industry rationalisation. In 2014 the industry seems to be at a critical juncture which may see it become largely irrelevant or may lead to the development of a native forest timber industry that is sustainable for the long term, albeit on a smaller scale than in the twentieth century.

Diversity

The forests of East Gippsland reflect the wide climatic, geomorphic and topographic variations within the region. They range from pure stands of

---

[3] The forests of East Gippsland are located in East Gippsland Shire and encompass the Tambo and East Gippsland Forest Management Areas (FMAs). The Tambo FMA covers the western part of the Shire and includes the forests between Bairnsdale and Omeo, while the East Gippsland FMA is the area generally east of a line running north from Lakes Entrance to the western end of the straight line border between Victoria and New South Wales.

[4] Land Conservation Council (LCC), Victoria. 1974. 'Report on the East Gippsland Study Area'. Melbourne: Land Conservation Council.

[5] LCC, Victoria. 1985. 'Report on the East Gippsland Study Area Review'. Melbourne: Land Conservation Council.

alpine ash and shining gum in the cool, moist environments of the mountains to mixed species forests in the foothills and coastal plains. Pockets of remnant cool temperate rainforest along rivers and streams contrast with dry callitris forests in rainshadow areas in the middle reaches of the Snowy River. East Gippsland accounts for only 10 per cent of Victoria in area, but over one-third of the state's total plant species occur there.[6]

Forests are the dominant land cover in East Gippsland. Seddon quotes Alfred Howitt's 1890 paper 'The Eucalypts of Gippsland' on the effects of early pastoral settlement on the nature and distribution of forests. Howitt noted that some areas that were forested by the late nineteenth century had been grassy flats when the first settlers arrived and the forest itself had been much more open. The Aboriginal people maintained a grassy forest understorey by regular burning; grazing by native animals did not make much impact on it. Sheep and cattle, on the other hand, put much more pressure on the grass and settlers excluded fire to protect their properties from damage. This resulted in a greater regeneration of trees and shrubs, leading to more widespread and denser forests.[7] Keith Gidley, a retired forester, told how his father started in the forests in 1927, when a horse and gig could be driven through most areas, unimpeded by scrub.[8] He contrasted this with scrub in some areas that is now so thick that 'a dog couldn't open his mouth to bark in it'. However, Seddon points out that Howitt's picture of open forests was not universally true: explorers found some areas of East Gippsland — particularly the more humid parts — very difficult to penetrate because of thick forests.[9]

In East Gippsland Shire as a whole, less than 25 per cent of the area is freehold land and in East Gippsland FMA only 12 per cent.[10] Most public land is forested and there are substantial areas of natural bush on adjoining private land.[11] The usual concept in Victoria of remnant native vegetation

---

[6] LCC. 1974. 'Report on the East Gippsland Study Area'.

[7] G. Seddon, 1994. *Searching for the Snowy: An Environmental History*. St Leonards: Allen & Unwin, 296.

[8] Where full names are used in this paper, without a footnote, the reference is to the GTD interviews.

[9] Seddon, *Searching for the Snowy*, 298–299.

[10] East Gippsland Shire. Website. Accessed 24 August 2014 at: http://www.eastgippsland.vic.gov.au/About_Us/Our_Shire; LCC. 1985. 'Report on the East Gippsland Study Area'.

[11] East Gippsland Catchment Management Authority. 2005. 'East Gippsland Regional Catchment Strategy'. Bairnsdale: East Gippsland Catchment Management Authority.

being endangered 'islands' in a sea of cleared land is reversed in most of East Gippsland. Instead, it is the cleared land that exists as discontinuous pockets, separated by wide swathes of forest. This has a profound effect on local attitudes to forests, forestry and land clearing.

The range of products won from the forests mirrors the diversity of the forest environments. The early graziers felled timber for huts, cattle yards and firewood; the mining industry also made demands on the forests and so did agricultural settlers at a later date.[12] Many of the areas cleared for pastoral activities, mining and dairying in the nineteenth and early twentieth centuries have since reverted to bush — what might be called 'the triumph of the forests'.[13] Sawmills were established on the Tambo River in the 1860s and other mills opened soon after at Sarsfield and Lindenow.[14] A mill set up by the Richardson family on the Brodribb River near Orbost in 1881 played a significant role in the development of Orbost, Marlo and surrounding farming districts.[15] Other early sawmill locations in East Gippsland included Splitters Range and the upper Livingstone Creek near Omeo, Orbost, Lake Tyers and the Bark Sheds area on the Omeo Highway north of Bruthen.[16] Most of these operations were relatively short-lived.

In 1914, the railway was extended from Bairnsdale to Nowa Nowa and in 1916 to Orbost, opening up forests suitable for production of railway sleepers and hewn beams for heavy wooden construction for wharves and bridges.[17] By 1920, the area was producing around 60 per cent of the state's requirements for hewn beams and 25 per cent of sleepers.[18] The hewing industry became even more important after the Depression. Between 1950 and 1960, the East Gippsland FMA supplied up to half of Victoria's

---

[12] M. Fletcher, 2005. 'East Gippsland: Environmental history'. Vol. 3 of Context Pty Ltd *East Gippsland Shire: Heritage Gaps Study*. Brunswick: Context Pty Ltd for East Gippsland Shire, 21.

[13] The study of old growth forest in East Gippsland contains a series of maps (Maps 6–8) that show the locations over time of grazing leases, agricultural selections and historic mining sites and tracks. Many of these areas are now covered by forest and some freehold land has reverted to the Crown. P.W. Woodgate, W.D. Peel, K.T. Ritman, J.E. Coram, A. Brady, A.J. Rule, and JCG. Banks, 1994. *A Study of the Old Growth Forests of East Gippsland*. Melbourne: Department of Conservation and Natural Resources.

[14] Fletcher, 'East Gippsland: Environmental History', 22.

[15] M. Leatch, 2008. *Curlip: The Life and Times of a Snowy River Paddle Steamer and the Family Who Built Her*. Orbost: Thandwalla, 13–28.

[16] See the Victorian Heritage Database at http://vhd.heritage.vic.gov.au/vhd/heritagevic.

[17] LCC. 1974. 'Report on the East Gippsland Study Area'.

[18] Ibid.

railway sleepers, as well as heavy construction timbers and a substantial numbers of poles to support telephone cables and electric wires.[19]

Brian Donchi is the fourth generation of his family to work in East Gippsland's forests. His great-grandfather, grandfather and father all cut sleepers in the forests around Orbost, along with many other Italian workers. His grandfather started with broad axes, crosscut saws and a bag of steel wedges as his tools of trade — but with no safety equipment — and was still working into the era of machine-driven swing saws. They were piece-workers, paid only for what they produced, so if the weather was bad or the cutters were ill, they and their families had no income. Skilled workers could get 20 to 30 sleepers out of a single tree — and up to 120 – and the sleeper cutters learned quickly to recognise the right types of tree for their purpose:

> There were some species of trees in East Gippsland which would not be suitable for cutting sleepers. You might look at apple box, you might look at silvertop, yertchuk and some of those, which would be inferior and would rot out quickly. A lot of the sleeper cutters sort of knew which species they had to cut, which might have been white stringy[bark], yellow stringy[bark], blue gum, box trees, ironbark or red box.
>
> The areas that they chose were across a vast area in the flatter hills. They didn't move right up into the mountains where it was sort of wetter, redder soil and bigger timber, but they stayed in the coastal areas where the trees were slower growing and of the right species.

The diverse use of the East Gippsland forests is also evident in the wattle bark industry, which began as early as 1877 and continued until the 1960s.[20] Bark was stripped from black wattle trees and sold to tanneries in Bairnsdale and Melbourne – and even Britain.[21] Around Lakes Entrance, bark was also used to tan sails and season fishing nets to increase their resistance to rotting in salt water and to make them less visible to fishes.[22] Bakeries valued the wattle timber, a by-product of bark stripping, because of the hot fire it

---

[19] Ibid.; D.G. Buntine, P.C. Fagg, and E.K.Gidley, 1976. 'Forestry'. In *'Land to Pasture: Environment, Land Use and Primary Production in East Gippsland'*, edited by C. McRae, Bairnsdale: Department of Agriculture, 292–297.

[20] Fletcher, 'East Gippsland: Environmental History', 21.

[21] C. Dow, 2004. 'Tatungalung Country: An Environmental History of the Gippsland Lakes'. Ph.D. thesis, Churchill: Monash University, 105.

[22] Ibid, 107; Fletcher, 'East Gippsland: Environmental History', 21.

produced.[23] According to Jack Whadcoat, when the fire was lit in the bakery oven it could get up to temperature very quickly. The wattle timber produced a very fine ash, which was blown out and the bread cooked in the residual heat.

Jack grew up at Lake Tyers and was involved in the wattle bark industry from an early age. He recounted how he and his father stripped bark to supplement their income from a very small dairy farm.

> Wattle bark was a saviour for those people in a lot of places, including my Dad. The price that they paid for wattle bark was quite reasonable for a day's living, but they were long hours. You'd be up at four o'clock in the morning milking cows for a couple of hours in the dairy, then you'd go out during the day stripping the bark and be back about four o'clock and milk your cows in the evening, so they were very long days, very hard work…
>
> In order to get the bark off the tree itself, and some of them were 30' high, when it was stripping well in the springtime you could quite often run the bark off right to the limbs but in most cases you had to cut the tree down.

During the Second World War there was a significant industry producing charcoal as an alternative fuel for road transport vehicles and farm machinery. The government promoted the use of the so-called 'producer gas', based on burning charcoal in an enclosed environment to produce carbon monoxide. The gas producer unit was fitted to the vehicle and the gas fed to the engine. It did not pay to be in a hurry, as the apparatus had to be 'lit up' 15 minutes before the engine could be started.[24] Up to 13 charcoal kilns operated in the Nowa Nowa area during the war.[25] Josephine (Josie) Jacobi is the daughter of Alma May (Maisie) and Joe Byrne, who lived in a tent in the Colquhoun Forest in 1942–43 while Joe worked as a kiln operator. In her interview for GTD, Josie talked of her parents' recollections of the time and read from surviving written accounts (poems and other fragments) that Maisie had penned over the years. Josie explained:

---

[23] Ibid.

[24] The description of the apparatus and the time taken to activate it is derived from notes of a 2008 talk by Don Bartlett, Chairman, Engineering Heritage Australia: 'Producer Gas and the Australian Motorist – The "Alternative Fuel" of the 1939–45 Fuel Crisis'. Address delivered to Engineering Heritage Victoria, 21 February 2008. Accessed 21 September 2011 at: http://home.vicnet.net.au/~engherit/08_producer_gas.htm

[25] Fletcher, 'East Gippsland: Environmental History', 22.

The kilns were like big ovens that they could put even sized pieces of timber into and start the fires going. Once the fires were going well they would shut down the oxygen supply. It's like a reduction firing that you do with pottery, reduce the oxygen and the wood doesn't burn away, it just turns to charcoal, so the carbon is left... There were two shifts, my Mum & Dad and another, I think just a single man or two men who camped there together. The kilns were kept going 24 hours a day.

As well as stoking the kilns, the operators had to cut wood to feed them. Following the 1939 fires that destroyed much of the mountain ash resource east of Melbourne, sawmillers such as Jack Ezard from Erica sought to move into the East Gippsland forests.[26] Ezards, as the business was known, shifted to Swifts Creek progressively from 1943 to 1946 and set up an extensive operation that included drying and seasoning kilns and 17 houses for employees.[27] Ezards also developed a high lead/skyline logging system, a technique that used a winch and aerial cables supported on tall 'spar' trees to transport logs from where they were felled to the loading areas.[28] The system was common in the central highlands but unusual in East Gippsland.

Until relatively recently the main output of the East Gippsland sawmills was 'green scantling', unseasoned timber cut into specified sizes and lengths and packaged in 'house lots' for construction purposes in Melbourne. As Brian Donchi explained:

I can remember even as a kid when we used to... come in from the bush with Dad to go down to unload the sleepers [from the logging trucks]. You would see this train sitting in this rail station at Orbost and have 50 to 60 carriages of green scantling that would come in from all those mills across East Gippsland to be pulled out every night to Melbourne bound.

Fence palings were an even lower-value product, which some mills turned to when supplies of better quality logs were tight. Other commercial uses of the East Gippsland forests during the last 100 years have included firewood, apiculture, horse safaris and, more recently, other ecotourism ventures.

The forests also provide 'ecosystem services' such as clean water and highly valued opportunities for recreation and inspiration. The first national parks

---

[26] Woodgate et. al. *A Study of the Old Growth Forests of East Gippsland*, 49.

[27] Details of the Ezards mill at Swifts Creek are derived from the recollections of Fred Lawrence, recorded by his son Terry Lawrence, in a 1999 letter to Graeme Dunstall of Bairnsdale (transcribed copy in author's possession).

[28] See the Victorian Heritage Database at http://vhd.heritage.vic.gov.au/vhd/heritagevic.

in East Gippsland were reserved in 1909 for their natural beauty, flora and fauna.[29] Other small parks were declared in the 1920s and 1960s. Following the LCC investigations in the 1970s and 1980s, major new national parks were created in the region.[30] Many stories can be told about the campaigns to have these parks declared and to exclude exploitative uses from the forests. Some GTD interviews touched on these issues, but space does not allow them to be explored here.

Before leaving this theme, I might mention that there is a downside to diversity, at least for sawmillers trying to value-add to East Gippsland mixed species timber. There is enormous variation between species and even within the same species from different areas or different sites in terms of density and moisture content, which makes kiln drying a much more delicate art than is the case for very consistent species such as alpine ash.[31]

Adversity

Adversity, my next theme, has been experienced in East Gippsland through the inherent dangers of working in the forests or forest industries, as well as through fires and floods, pests and diseases, market conditions and the fluctuations of government policy concerning access to timber resources. Many of the GTD interviewees talked about the dangers involved in working in the bush. Brian Donchi described the risks of cutting down trees in the days of selective logging:

> The trees had to be placed in a perfect position into the clearest spot, which might have meant that the canopy or crown of the tree would brush past another tree. And when you get that effect in trees going past each other you'll always get what they call these hang-ups or 'widow makers'… as canopies go past each other, limbs will break and then they will either come back towards the faller or be left broken up there swinging around, that they have to work under when they're cutting sleepers.

---

[29] Fletcher, 'East Gippsland: Environmental History', 32.

[30] D. Clode, 2006. *As If for a Thousand Years: A History of Victoria's Land Conservation and Environment Conservation Councils*. Melbourne: Victorian Environmental Assessment Council, 99.

[31] A PhD student from Monash University's Civil Engineering Department was based at Forestech in the late 1990s, researching species identification and customised drying techniques for mixed species timber.

Other risks in the bush included branches that shed spontaneously during still weather, old stag trees falling when their roots failed, and logs rolling off stockpiles. The interviewees also spoke of the perils of operating machinery, such as tractor rollovers or bulldozers pinning men against trees. Those who worked in the mills had a similar catalogue of accidents with saws, chains and chippers. Leona Lavell, the first woman to work in a sawmill in Buchan, recounted a close shave:

> I nearly had a bad accident because... the docker timber jammed... Another boy and I were sent up there to release this block... We'd just got down to a few of these small pieces that were blocking [it] and someone turned the chain on. I was on a slope and my feet gave way and I went down in this channel. As I fell I threw my hands up and he grabbed my hands but I had the biggest rubber burn on my bottom because it took the seat out of my shorts.

Another major risk for these forests and for the workers who gained their livelihoods from them was bushfire. East Gippsland was less affected than other parts of Victoria by the 1939 fires, but there was significant damage caused to forests (and buildings in and near them) in the Omeo district and other parts of the Tambo FMA. There were major fires in the western part of East Gippsland in 1965 and in the far east in 1980 and 1983.[32] The huge fires of 2002–03 and 2006–07 affected nearly all the forests along the Great Divide, as well as much of the bush west of the Great Alpine Road above Bruthen.

Several GTD interviewees talked about past and present practices concerning fire management in East Gippsland forests. They also pointed out the challenges of matching the frequency of fires to the needs of particular ecosystems. Keith Gidley advocated learning from the way the Aboriginal people managed the forests, with frequent relatively cool burns, and increasing the amount of fuel reduction that is undertaken in strategic areas. Dennis Matthews spoke of the tensions between fire management for forest health and the need to protect settlements and other assets in and adjacent to forests. He pointed out that the object of current fire management is towards prevention and away from suppression, in order to lessen the chances of 'megafires' or 'campaign fires' like those of the last decade. Since weather conditions in East Gippsland mean that most of the fuel reduction activity has to be carried out in autumn, Matthews predicted

---

[32] Woodgate et. al. *A Study of the Old Growth Forests of East Gippsland*, 50.

that community tolerance might be stretched in future by the amount of smoke in the atmosphere at this time.

Adversity has also come in the shape of pests and diseases, such as *Phytophthora cinnamomi*, a water mould (previously described as a root rot fungus) prevalent in the 1970s. Measures were introduced to stop its spread on plant and machinery and some affected sites were rehabilitated with disease-resistant stock. The disease does not seem to have expanded significantly since the 1970s in East Gippsland, although a return to wetter seasons may see it become a problem again.[33] Dennis Matthews, in his GTD interview, also mentioned that there was a period when there were serious attacks on shining gum stands by the ambrosia beetle.

To participants in the timber industry, government policy has also been seen as a source of adversity as it has led to progressive reductions over the past 45 years in the areas of land available for timber harvesting and frequent changes in commercial arrangements for access to timber supplies. Sawmill owner Bob Humphreys summed it up:

> In the early 1960s the industry had access to nearly all the State Forest in East Gippsland. And it wasn't until about the mid-70s the Land Conservation Council did a study in East Gippsland and made recommendations that certain areas of it be alienated from production and from commercial use and be set aside into National Parks and other reserves which precluded industry's participation or operation in them. And that was really the thin edge of the wedge…
>
> [Forest protests on the Errinundra Plateau in the early 1980s] kicked off another lot of studies… Every time we had another study we had another loss of resource, to a stage now where we are probably operating in about 15 per cent of the forest, maybe even a little less than that when you look at some of the regulatory and prescription exclusions…
>
> We are getting beaten further and further down. One of the last nails in our coffin I think has been the change where Vic Forests took over the commercial arm of the State Government's interests in forests and then chose to do their own harvesting and haulage… For the first time in the 45 years I have been in the game we've run out of logs a couple of times in the last six months.

---

[33] Victorian Government, Department of Sustainability and Environment (VDSE). 2008. 'Victoria's Public Land *Phytophthora cinnamomi* Management Strategy'. Melbourne: Department of Sustainability and Environment.

The number of sawmills in East Gippsland had expanded significantly after the Second World War and growth continued until the early 1970s (Table 15.1). In 1977, the Victorian Government accepted the LCC's recommendations for extensive new national parks in East Gippsland.[34] Further reductions in available resources followed the 1982 LCC Gippsland Lakes Hinterland study (which covered Tambo FMA) and the 1985 LCC East Gippsland review.[35]

Table 15.1: Sawmills in East Gippsland

Date	Area	No of mills	Locations	Source
1942	East Gippsland (EG & Tambo FMAs)	9	Bendoc (3), Bonang, Bruthen, Cabbage Tree, Omeo, Orbost & Nowa Nowa	Buntine et al. 1976
1951	East Gippsland	54	[Locations not recorded]	Buntine et al. 1976
1970	East Gippsland	63	[Locations not recorded]	Buntine et al. 1976
1974	EG FMA	24 (+18 outside area, supplied from FMA)	Bemm River (2), Bendoc (2), Bonang, Cabbage Tree (2), Cann River (5), Club Terrace (4), Combienbar, Maramingo, Newmerella, Noorinbee North, Orbost (3) Waygara	LCC 1974
1979–80	Tambo FMA	11 (+ 3 supplied from private land)	Bruthen (2), Buchan South, Ensay North, Nowa Nowa (6), Swifts Creek (2)	LCC 1982
1984	EG FMA	13 (+ 6 supplied from private land; 3 outside area were also supplied)	Bemm River (2), Bendoc (3), Brodribb River (2), Cabbage Tree (1), Cann River (5), Club Terrace (3), Combienbar, Maramingo (2), Newmerella (4), Noorinbee North, Waygara	LCC 1985
2001	East Gippsland (EG & Tambo FMAs)	19	Bairnsdale, Bendoc (2), Brodribb (2), Buchan, Cann River (3), Club Terrace, Genoa, Marlo Plains, Mt Taylor, Newmerella, Noorinbee Nth, Nowa Nowa (3), Waygara	Dept of Natural Resources & Environment Victoria – information to East Gippsland Shire

---

[34] Clode, *As If for a Thousand Years*, 99.
[35] LCC, Victoria. 1982. 'Report on the Gippsland Lakes Hinterland Study Area'. Melbourne: Land Conservation Council; LCC, Victoria. 1985. 'Report on the East Gippsland Study Area Review'.

The Regional Forest Agreements (RFAs) between the Commonwealth and Victorian governments were concluded in 1996 for East Gippsland FMA and in 2000 for the remainder of the Gippsland region (including Tambo FMA). The agreements were intended to provide certainty to the industry and to conservation interests. They specified which areas were to be included in the reserve system and which were available for timber harvesting 'for the purpose of providing long-term stability of forests and forest industries'.[36] Through government structural adjustment packages, some logging contractors left the industry at this time and some sawmillers accepted compensation for reduced allocations. A number of smaller mills closed or were taken over by larger ones.[37] Some timber from East Gippsland was transported out of the region for milling at Heyfield. After political pressure, the Swifts Creek mill site (formerly Ezards, then Neville Smith Timber Industries) was taken over by Dormit, a manufacturer from Dandenong. It is now used to process low-grade sawlogs for pallets — not necessarily a desirable outcome in terms of 'highest and best use', but a strong benefit for local economic development.[38]

For the industry, any illusion of stability as an outcome of the RFAs was soon shattered when a new forest resource inventory showed that the previous sustainable yield calculations (the basis for assessing allowable harvesting volumes) were severely flawed. Processors faced further substantial decreases in supply. A socio-economic study of the timber industry in Gippsland noted that considerable rationalisation occurred in the industry in East Gippsland after 2002.[39] The remaining processors had to grow in order to survive, demonstrated by the fact that the seven largest sawmills in the total Gippsland region in 2005 accounted for over 81 per cent of the input of hardwood logs.[40] Of the four East Gippsland mills amongst these seven,

---

[36] Commonwealth of Australia and State of Victoria. 2000. 'Gippsland Regional Forest Agreement'. In the agreement, the Commonwealth and Victorian Governments confirmed their commitment to developing and implementing ecologically sustainable forest management; establishing a comprehensive, adequate and representative reserve system; and facilitating the development of an internationally competitive wood production and wood products industry. The agreement listed the volume of sawlogs expected to be available from the Tambo FMA, but also noted timber supply levels in Victoria were subject to change, based on periodic review of sustainable yield.

[37] Gippsland Private Forestry Inc. 2005. 'The Timber Industry in Gippsland: A Socio-economic Assessment'. (Original report: MBAC Consulting Pty Ltd.) Bairnsdale: Gippsland Private Forestry Inc.

[38] Ibid.

[39] Ibid.

[40] Ibid.

only three were still in operation by 2014: Dormit at Swifts Creek, Auswest at Brodribb (plus their value-adding site outside Bairnsdale) and Hallmark Oaks at Cann River, operating on a reduced scale from the former Sumberg mill. By 2005, the area available for timber production in the two FMAs was reduced to 4410 square kilometres or about 20 per cent of the total area of public land in East Gippsland.[41] Of this, approximately 0.7 per cent or 3150 hectares was cut each year.

The closure of mills had significant impacts on some of the smaller townships. Bob Humphreys, owner of Hallmark Oaks and a senior office bearer in the Victorian Association of Forest Industries (VAFI), described his satisfaction at the contributions his business – by far the largest private sector employer in Cann River – had made to the community. He spoke quite movingly about the sense of responsibility he felt to his employees and the township. However, he foresaw poor outcomes for the school, the volunteer emergency services organisations and the economic base of the town if he was obliged to close up his operations or was unable to find a buyer when he wanted to retire.[42] Leona Lavell talked of the effect on Buchan when the largest mill closed in 2003 with the loss of 32 jobs. Most of the men and their families eventually left and Leona said it 'took the whole inside out of Buchan' and the town had not recovered.

Other changes in Government policy from the 1960s caused the timber industry in East Gippsland to transition from selective logging – where only the better trees were removed from each area – to clear falling in relatively small coupes (typically about 40 hectares).[43] Clear falling involves clearing most of the trees within a coupe, other than those retained as seed trees or for habitat. Logs are transported out of the forest, with those that are not suitable for sawmilling being converted to woodchips and ultimately pulp. The upper parts of the trees and the stumps are burnt on-site to provide a good seed bed to aid regeneration.

The GTD interviewees were divided in their views on the merits of clear falling versus selective logging. Several saw environmental as well as safety

---

[41] Information supplied by DSE for the Forest Discovery Exhibition at Forestech.

[42] In an April 2011 interview with ABC Rural Radio, Bob Humphreys stated that his workforce had declined from over 70 at its peak to only 18. Bob Humphreys, 2011. 'Timber Industry Faces Further Squeeze'. Interview with Cath McAloon on ABC Rural Radio, 29 April 2011. Accessed 24 August 2014 at: http://www.abc.net.au/site-archive/rural/vic/content/2011/04/s3203695.htm.

[43] This logging method is called 'clear felling' in most public documents and the scientific literature, but as foresters and logging contractors – including those interviewed for GTD – all call it 'clear falling', this term has been used here to avoid confusion.

benefits in clear falling, provided proper measures were taken to ensure good regeneration. Bob Humphreys maintained:

> Clear falling is the only way to operate. We go in and knock out nearly everything now, we leave some fauna habitat and we leave some seed trees… we get rid of nearly all the canopy, the soil gets disturbed from the logging operation, from the machinery, they have a slash burn afterwards that opens all the seed pods that drop into the potash and stirred up soil… and they'll grow up your trouser legs.

On the other hand, Brian Donchi opposed clear falling:

> I think the management of the forests, whether it be Vic Forests or DSE, has become better over the years, but there's other areas where I believe that they've sort of fell down too. And I believe that some of the things they do, like completely clear-falling an area, is wrong and that the technology and the practice will change again in time. I believe in the future, and more so as a result of these fires that went through the Great Divide, that we'll produce a lot more trees. There will be a lot more trees for sawmills and for use, they'll be smaller and there will be a lot more thinnings of the forest and you'll see a lot of this clear falling phased out, which would be a good thing I believe.

Many of the older respondents, including Stan Collins, Jack Whadcoat and Maurice 'Maurie' Killeen felt that the industry was no longer driven by the demand for sawlogs but had become a captive of large operators with short-term commercial interests. For example, Jack argued that 'management today seems to be more towards developing [the forests] for commercial purposes rather than keeping it for posterity. And that's a pity, because if you lose that forest you've lost life'.

Stan Collins was also concerned:

> I feel that the forest resource still has not been taken to its highest and best end use. I don't think it is being run profitably, for a number of reasons. The allocation of wood should be towards sawlogs. The industry, back in the early days when the [Timber Industry] Strategy was implemented, our native forest had to be sawlog driven… Now it's being pulpwood driven for two reasons: one I say because of the unskilled management; and two, because of the volume of wood that has become available because of the bad fires.

Bob Humphreys, however, believed that sawlog demand was still driving the industry:

> Victoria still has a sawlog-driven industry and if we, the sawlog operators, stopped processing I think… apart from thinnings, it would probably be the end of the pulpwood industry as well. Because the pulpwood at the moment is a by-product of, is the utilisation of, the residues from sawlog harvest.

Keith Gidley, while supporting clear falling, was concerned about the increasing size of coupes. Dennis Matthews acknowledged that the community wanted to move away from clear falling and discussed potential ways in which this might be accomplished, including the use of machinery similar to that employed for thinning operations in the coastal forests. The ongoing debates about the most environmentally sustainable ways of producing timber from native forests mean that the logging industry will potentially be faced with the need to invest in new types of machinery and to develop new skills in removal of logs in order to minimise environmental impacts.

The discussion above illustrates some of the difficulties encountered by the timber industries in East Gippsland, as seen by the GTD interviewees. These included the inherent risks faced by workers and the need to respond to changes brought about by government policies that sought to conserve important areas and improve the sustainability of forest management.

Opportunity and adaptability

Opportunity and Adaptability, my next two themes, are strongly linked with each other and also with creative responses to adversity. Opportunity came in many forms: improved technology in the bush (through the use of chainsaws, crawler tractors, bulldozers and excavators) and in the mill, as well as improved transport infrastructure and the ability to access new markets and export opportunities. Forest industry operators, in the bush and in processing industries, have always been very adaptable and far more skilled than has been recognised generally. In earlier times, much of the machinery was homemade, including swing saws and portable sawmills, or adapted, like the use of farm tractors by the sleeper cutters. Some of the infrastructure erected for fire surveillance in the forests is a marvel of ingenuity in its own right, such as the observation tower built in 1941 on Splitters Knob that consists of two 60 foot stringybark logs scarfed together, with a small cabin on top.[44] Max Reynolds recalled one of the tower's builders saying that the hardest part was 'how to get off the roof when the last sheet of iron was nailed on'.

Partly as a response to declining timber availability, there has been a move towards more value adding to the available wood. Some would say that the industry in East Gippsland has been slow to embrace the need to ensure that all timber harvested is utilised to the maximum extent possible. This was due partly to the problems of kiln drying mixed species and partly to the small scale of many operations, which could not easily access the capital required. However, some forward-thinking interviewees tried to give a lead to their colleagues. One of these was Stan Collins (a former president of VAFI), who undertook an amazing array of value-adding activities in association with his Bairnsdale mill. These included a wooden toy factory employing 27 people (some with disabilities) and producing 167 types of toys. He also installed a research kiln and established a subsidiary business that made furniture components for Moran Furniture in Melbourne. Stan recalled:

> I just delighted in exploring the opportunities to value add, not just to improve the product but in the value… We did all sorts of things just to keep my total staff usefully employed… [The toy factory] came about by trying to get the maximum return from the resource that I had and employing some physically and mentally handicapped people and that went in tandem, and that was a great venture. There wasn't much profit in it, or very little, but there was a lot of satisfaction in doing that.

Stan aimed to go beyond the first stage of value adding, which was supplying kiln-dried flooring and structural beams. He explained that 'we wanted to gather up other markets and displace some of the imports… because a lot of people didn't realise the quality of our hardwoods'. Max Reynolds, who holds a licence for speciality timbers such as blackwood, sassafras, and red ironbark, described the more demanding process of sawing these timbers and producing a range of value-added goods from them. Along similar lines, Bob Humphreys re-equipped his mill to process smaller logs, which would not previously have been used for sawn timber. He put his investment by 2008 at $2.3 million in plant and equipment and

---

[44] See the Victorian Heritage Database at http://vhd.heritage.vic.gov.au/vhd/heritagevic.

$6 million for the value of timber being air-dried before further processing. Brian Donchi reflected on the multi-step and highly mechanised operation at Fenning's mill in Bairnsdale, where he was working in 2008, which processed alpine ash for furniture grade uses and structural beams.

Relationships with the forests and communities

One of the consistent themes that came through the GTD interviews was the strong relationship the participants had with the forests and with timber. While differing to some extent amongst themselves about current management practices, there appeared to be a general acceptance that a sustainably managed industry represents an appropriate use of available East Gippsland forests in the future.

Despite the physical hardships of their working lives, most speakers expressed a great love for and understanding of the bush. For example, Stan Collins stated 'I love it [the bush], you can say I'm a really strong conservationist because I've grown up in it since I was a kid, I understand it'. Ron Schrader had a slightly different take on his attachment to the forests, reflecting that he couldn't travel anywhere 'unless I'm looking out the window of a car looking at the trees and working out what's in each tree... It doesn't matter where you go, you're looking at the bush'. Max Reynolds also said that he loved the bush, but like Ron he admitted to always sizing up the logs as he drove by. Brian Donchi talked of some of the practices that were common in his early career as a timber worker, such as allowing trees to fall into rainforest gullies, and took pride in the fact that current management conserves buffer strips along streams and other areas of high conservation value.

Leona Lavell spoke nostalgically about the smell of the timber and the other smells of the mill (even the diesel). Max Reynolds extolled the beauty and unique characteristics of the different speciality timbers, which come in 'all the colours of the rainbow' and described the various patterns of grain and features in the timber as 'nature's oil painting'. Jack Whadcoat talked of the freedom, independence and self-reliance developed by the bush children in an environment that was friendly, but sometimes savage (often through their own fault) and contrasted it with the sheltered, urban lives of modern children disconnected from nature. He also described the ways in which bush people learned to read the environment and recognise signs of disturbance or weather patterns that might affect the safety of their activities.

Although Maisie Byrne's wartime life in the Colquhoun Forest was relatively isolated and lacking in most amenities, her accounts of the time she spent in the forest –recounted by her daughter Josie Jacobi – described it as the happiest period of her life. Her recollections included stories about the wildlife of the forest, how the kangaroos would huddle round the warm kilns on frosty nights until the dingo packs started to take advantage of their torpor. She also told how the old horse feared the dingoes and sought refuge in the tent, while the dog tried to protect all its owners' possessions, including the horse.

The interviews reflect a day-to-day familiarity with the bush and its seasons that appears to be disappearing as a result of mechanisation in the timber industry and, to some degree, professionalisation of forest management.

Several interviewees lived in very spartan circumstances in their childhood and young adulthood. But looking back – often 50 years after this period – nearly everyone talked of the comradeship of the bush and the mills and the strength and supportiveness of communities in the small country towns. Leona Lavell recounted how the Buchan sawmill company donated timber and workers' time to build a house for a young mother widowed in an accident in the bush.

Change and continuity

The title of this paper refers to change and continuity in forest industries in East Gippsland. Change there has certainly been, and this is unlikely to alter. Climate change will be a major challenge for the forests and the industries that rely on them, including ecotourism. However, we might question whether there will be continuity. As Bob Humphreys stated in 2008, 'unfortunately I don't [see a future for the East Gippsland timber industry], I don't think the Government wants us here. That's fairly obvious in their actions in recent times'.

The then current Coalition Government released a Timber Industry Action Plan in December 2011 which sought to provide the conditions for a productive, competitive and sustainable Victorian timber industry, involving both native timber and plantations. The plan was welcomed by forest industry representatives but opposed by conservation groups.

I have been a committed conservationist for over 40 years, but nonetheless I believe that there is a place for an environmentally sustainable native forest timber industry in East Gippsland. The alternative, for sawlogs at least, is continued reliance on imported timbers from forests that are almost certainly not managed with the same attention to environmental quality and biodiversity that Victoria could achieve. However, despite recent audits showing a very high level of compliance with the regulatory framework governing the planning of timber harvesting, harvesting operations and forest regeneration, I do not think we are there yet.[45]

Achieving this aim, in my view, involves further improvements in the environmental sustainability of forest operations, which means moving out of old growth forests and (potentially) restricting clear falling to forest types that require fire for regeneration and developing new harvesting methods for other ecosystems. In the mills, there is considerable scope for increasing value adding to forest products, providing timber supplies can be guaranteed and competitiveness in the marketplace maintained.

The challenge for the immediate future will be for forest managers and timber industry operators to develop the leadership skills, wisdom and resilience to adapt to new circumstances and seize opportunities. I believe there is hope that this can be achieved.

Appendix 1

Gippsland Timber Development Inc.: Oral History Project, 2008

Name	Topic
Josephine Jacobi	Charcoal production WWII at Colquhoun Forest – recollections of her parents, Alma May (Maisie) & Joe Byrne
Stan Collins	Early value adding in the East Gippsland industry
Ron Schrader (died March 2011)	Life of the bush workers and their families/changes in the timber industry/timber haulage
Keith Gidley	History of forest management in East Gippsland
Brian Donchi	Sleeper cutting – experiences of his father, Alan – and his own life in the timber industry
Jack Whadcoat	Wattle bark industry/sourcing the product
Leona Lavell	Women's involvement in the timber industry (sawmills)
Dennis Matthews	Fire management history and forest management
Max Reynolds	Specialty timbers of East Gippsland – their use and history
Bob Humphreys	Value-adding in the East Gippsland timber industry
Maurice 'Maurie' Killeen	Sawmilling
Andrew Rule	Re his father, Keith Rule, a noted axeman and timber-getter in the Maffra district

---

[45] Press release from Ryan Smith, MP, Minister for Environment and Climate Change, 3 December 2012.

The interviews, recorded by David Huxtable and Bryce Grunden (LookEar Pty Ltd) for Gippsland Timber Development Inc. (GTD), are available to researchers through the Local Collection of the East Gippsland Shire Library, Service Street, Bairnsdale, Victoria.

All interviewees signed release forms granting permission for GTD to use the material from the interviews and for them to be lodged in East Gippsland Shire Library's local collection and made available to the public for use for research, broadcast or publication (with acknowledgement to the interviewee). Note that the interview by Andrew Rule has not been used in this chapter.

References

Bartlett, Don. 2008. 'Producer Gas and the Australian Motorist – The "Alternative Fuel" of the 1939–45 Fuel Crisis'. Paper delivered to Engineering Heritage Victoria, 21 February 2008. Accessed 21 September 2011 at: http://home.vicnet.net.au/~engherit/08_producer_gas.htm.

Buntine, D.G., Fagg, P.C., & Gidley, E.K. 1976. 'Forestry'. In *Land to Pasture: Environment, Land Use and Primary Production in East Gippsland*, edited by C. McRae, Bairnsdale: Department of Agriculture: 292–297.

Clode, D. 2006. *As If for a Thousand Years: A History of Victoria's Land Conservation and Environment Conservation Councils*. Melbourne: Victorian Environmental Assessment Council.

Commonwealth of Australia and State of Victoria. 2000. 'Gippsland Regional Forest Agreement'. Signed 31 March 2000.

Department of Sustainability and Environment. 2012. Forest Explorer. Accessed 6 August 2012 at: http://nremap-sc.nre.vic.gov.au/MapShare.v2/imf.jsp?site=forestexplorer.

Dow, C. 2004. 'Tatungalung Country: An Environmental History of the Gippsland Lakes'. Ph.D. thesis, Churchill: Monash University.

East Gippsland Shire. Website. Accessed 24 August 2014 at: 'http://www.eastgippsland.vic.gov.au/About_Us/Our_Shire'.

East Gippsland Catchment Management Authority. 2005. 'East Gippsland Regional Catchment Strategy'. Bairnsdale: East Gippsland Catchment Management Authority.

Fletcher, M. 2005. 'East Gippsland: Environmental history'. Vol. 3 of Context Pty Ltd *East Gippsland Shire: Heritage Gaps Study*. Brunswick: Context Pty Ltd for East Gippsland Shire.

Humphreys, Bob 2011. 'Timber Industry Faces Further Squeeze.' Interview with Cath McAloon on ABC Rural Radio, 29 April 2011. Accessed 24 August 2014 at: http://www.abc.net.au/site-archive/rural/vic/content/2011/04/s3203695.htm

Gippsland Private Forestry Inc. 2005. 'The Timber Industry in Gippsland: A Socio-economic Assessment'. (Original report: MBAC Consulting Pty Ltd.) Bairnsdale: Gippsland Private Forestry Inc.

Land Conservation Council, Victoria. 1974. 'Report on the East Gippsland Study Area'. Melbourne: Land Conservation Council.

Land Conservation Council, Victoria. 1982. 'Report on the Gippsland Lakes Hinterland Study Area'. Melbourne: Land Conservation Council.
Land Conservation Council, Victoria. 1985. 'Report on the East Gippsland Study Area Review'. Melbourne: Land Conservation Council.
Lawrence, Fred. Recollections cited in Terry Lawrence's letter to Graeme Dunstall, Bairnsdale. c.1999. (Transcribed copy in author's possession).
Leatch, M. 2008. *Curlip: The Life and Times of a Snowy River Paddle Steamer and the Family Who Built Her*. Orbost: Thandwalla.
Seddon, G. 1994. *Searching for the Snowy: An Environmental History*. St Leonards: Allen & Unwin.
Smith, Ryan M.P. Press release, Minister for Environment and Climate Change, 3 December 2012.
Victorian Government, Department of Sustainability and Environment (VDSE). 2008. 'Victoria's Public Land Phytophthora cinnamomi Management Strategy'. Melbourne: Department of Sustainability and Environment.
Woodgate, P.W., Peel, W.D., Ritman, K.T., Coram, J.E., Brady, A., Rule, A.J., Banks, J.C.G. 1994. *A Study of the Old Growth Forests of East Gippsland*. Melbourne: Department of Conservation and Natural Resources.
Victorian Heritage Database: http://vhd.heritage.vic.gov.au/vhd/heritagevic.

Chapter 16

'The campers' powerful headlights are shining on the water'

Motor tourism and the environment in far East Gippsland

Sarah Mirams

Tom Mutch first visited Mallacoota in 1909 with the poet Henry Lawson. This was a rescue mission, an attempt to get Lawson out of Sydney, into the bush and off the booze.[1] They were to stay with their friend and fellow writer EJ Brady, who had spent the previous year living under canvas overlooking the beautiful Mallacoota Inlet. This was the most easterly farming and fishing settlement in Victoria, a favoured holiday destination for the more adventurous nature lover, angler, poet and hunter.[2]

Mutch and Lawson travelled by steamer from Sydney to the port town of Eden and then along the rutted tracks that wound through the bush to Genoa. There they met Brady, who rowed them the 27 kilometres down the Genoa River to Gipsy Point and Top Lake, through the Narrows and into the Bottom Lake. Here they entered the inlet, with its 320-kilometre coastline of sandy beaches, coves and bays, surrounded by lowland sclerophyll forest. The visitors had left Sydney on the Friday and arrived at Mallacoota late Sunday night.

During his visit Lawson captured something of the region's beauty and isolation in letters to his son. 'The scenery,' he wrote, 'is wild and grand… we are out of the world.'[3] He was not alone in his admiration. Zoologist Le Soeuf, describing his first view of the inlet in *Wild Life in Australia*, wrote,

---

[1] John Le Gay Brereton to Henry Lawson, 2 January 1910, in Colin Roderick, *Henry Lawson Letters:1890–1922*, Sydney: Angus and Robertson, 1970, 460.

[2] The borders of the sub-region of East Gippsland have changed over time. In this article East Gippsland encompasses the former Shire of Orbost. This was East Gippsland as Brady recognised it. Council amalgamations in 1996 created the new Shire of East Gippsland, extending the borders to include the Gippsland Lakes and Lakes Entrance.

[3] Tom Mutch diary cited in Roderick, *Henry Lawson Letters*, 462.

'its beauties far exceeded our expectations'.[4] He went on to describe the waters teeming with fish, the lyrebird mimicking a cross-cut saw in the fern gullies, and lace monitors sunning themselves on logs.

The influential progressive reformer and national park advocate James Barrett lobbied successfully to have a national park temporarily reserved at Mallacoota in 1909, shortly before Lawson's arrival. The 4560-hectare national park consisted of Crown Land situated within a distance of 60 chains from the high water mark of Mallacoota Inlet. The reservation enclosed the existing selections and freehold land, most of which was located in the fertile creek valleys and on land overlooking the inlet.

Twenty years later Henry Lawson was dead and Tom Mutch a seasoned politician in the NSW Parliament. Brady now owned the Mallacoota House guesthouse. His business catered for motor tourists who could reach the inlet from Melbourne or Sydney in half the time it had taken Mutch and Lawson all those years ago. Despite Mallacoota's greater accessibility Mutch doubted he would visit again. He'd heard Mallacoota had become too civilised. 'The motor-car,' he wrote, was 'ruining Australia.' He observed that once a good spot gets opened up 'it begins to look like any other place'.[5]

Historian Paul Sutter explores the impact of the automobile on nature tourism in America in *Driven Wild: How the fight against the automobile launched the modern wilderness movement*.[6] He argues that the car made it possible for more Americans to explore nature, and had a profound influence on the way nature was experienced and marketed in the interwar years. Sutter describes an America where the popularity of motor tourism saw national parks overwhelmed by cars and campers and subsumed by a tide of consumerism and commercialism. There was a growing fear that the American wilderness was being lost when increasingly uncultivated and natural public lands, such as forests, were targeted for highway development. A nature-centred, rather than utilitarian, view of wilderness emerged amongst preservationists and the Wilderness Society was formed in 1930.[7] The society's lobbying eventually resulted in the passing of the 1964 *Wilderness Act*, which banned the construction of roads in wilderness areas.

---

[4] W.H. Dudley Le Soeuf, 1907. *Wild Life in Australia*, Melbourne: Whitcomb and Tombs, 94.

[5] Tom Mutch to E.J. Brady, 21 February 1921, in Edwin James Brady collection Ms 206/1/352, National Library of Australia (hereafter NLA)

[6] Paul S. Sutter, 2005. *Driven Wild: How The Fight Against The Automobile Launched the Modern Wilderness Movement*. Seattle: University of Washington Press.

[7] Ibid., 51, 111, 9.

This article considers the impact of motor tourism in East Gippsland during the interwar years and considers if any parallels can be made with the American experience as described by Sutter. My study is far more contained and intimate than Sutter's, focusing on one bioregion, coastal East Gippsland, and one tourist community, Mallacoota. Whereas Sutter explores the careers of significant figures in the broad American environment movement, such as Aldo Leopold, I explore the lives of the one concerned Victorian citizen and the local farming families who ran tourism enterprises. While there were significant differences between the broad American and the local East Gippsland experience, there were also some similarities. In both instances the automobile was to reshape both the tourist experience and natural environment.

The charms of Mallacoota

The old Gippsland saying 'It is wise to grow nothing that cannot walk out' was particularly apt in East Gippsland.[8] East Gippsland extended over 9521 square kilometres. It had a population of 3826; of these, 2000 lived in the regional centre of Orbost.[9] When Mutch visited East Gippsland there were no all-weather roads through the forests, only remnant mining and cattle tracks and bridle paths linking the scattered sparsely populated farming settlements. The Old Gippsland Road, the major land road land route from Melbourne, was impassable in winter months beyond Orbost. The stories of horses sinking up to their necks in mud and carts disappearing into morasses gave the eastern stretches of the road an 'almost legendary notoriety'.[10]

The lack of reliable transport routes into East Gippsland was explained by its environment and its political and social history. The region was characterised by daunting environmental barriers, including forests, steep hills, poor soils, deep gullies, rivers, creeks and a shipwreck coast. To build roads and railways into East Gippsland would require political will, sophisticated engineering and money. Much of East Gippsland, with its steeply forested hills and poor soils, had a history of failed land selection. There was little economic incentive for the Victorian Government to invest in East Gippsland.

While there is no doubt that this environment was modified by farming it had not experienced the wholesale transformation experienced in more closely settled parts of Victoria. It was then, as it is now, scenic, with a mild

---

[8] 'Far Mallacoota', *Argus*, 19 November 1911.
[9] *Victorian Municipal Directory*, 1910. Melbourne: Arnell and Jackson, 550.
[10] W.K. Anderson, 1994. *Roads for the People: A History of Victoria's Roads*. Melbourne, Hyland House, 225.

climate and rich in biodiversity. This attracted city tourists, and from the 1890s local farming families offered services such as accommodation, food and boat hire to the visitors. James Barrett was one such visitor and it was his persistent lobbying which secured some temporary reservation of the national park in 1909.

'National' parks

Australian and US visions of national parks were very different during this period. The preservation of American landscapes such as Yellowstone and Yosemite were seen as a federal responsibility, as was the provision of natural space and outdoor recreation for all citizens. Successive American presidents supported the national park ideal and took a personal interest in its development. Federal money was invested in building roads into national parks and providing tourism infrastructure. A dedicated body, the National Parks Service, was created in 1916 to administer national parks.

The Australian story of national parks is far more fragmented. There was little 'national' about national parks in Australia in 1909. National parks were largely state concerns, gazetted by state governments on a case-by-case basis, usually after concerted local campaigns by concerned citizens including public intellectuals, scientists, field naturalists and urban reformers.[11] An Australian national park in the early twentieth century could be anything from a pleasure park in easy reach of the city by rail or a remote patch of forest largely inaccessible except by packhorse.

In Victoria, national parks – a term rarely defined during this period – came under the mantle of the various Land Acts, and were managed under similar conditions as other types of public land reserved for community use, such as schools, cemeteries and Mechanics' Institutes.[12] Of the five areas set aside for national parks by 1908 only two, Fern Tree Gully and Wilsons Promontory, had a committee of management, employed rangers and had a budget.

Mallacoota National Park existed, but was little more than (temporary) lines drawn on the revised parish plan in a Lands Department file. It was the most remote national park in Victoria. There was no budget for the park, and no ranger appointed to enforce regulations. There was no government

---

[11] Drew Hutton and Libby Connors, 1999. *A History of the Australian Environmental Movement*. Cambridge: Cambridge University Press, 31–40.

[12] Sandra Bardwell, 1974. 'National Parks in Victoria 1866–1956 "For All The People For All Time"', Melbourne: PhD thesis, Monash University, 344.

investment in tourism infrastructure. The existence of the national park was not used in advertising promoting the region as a tourist destination.

In fighting for the establishment of the national park at Mallacoota, Barrett argued the land being excised was useless for farming and that tourism was a more economic use of the land.[13] The locals disputed this and sent a petition opposing Barrett's proposal. They argued the ban on hunting in national parks would discourage tourists and that the land should be given over to selection to encourage more farmers into the region.[14] While tourism in 1909 provided a welcome supplementary income, they believed that the future in Mallacoota lay in farming. This reflected the belief shared in Australia across the political spectrum that the farming frontier could extend into forests and deserts, as yet untouched, and that environmental barriers to settlement could be overcome by technology and transport.[15]

E.J. Brady's most well known work, *Australia Unlimited*, written in Mallacoota, promoted this ideal.[16] The book celebrated both Australian nature while arguing that Australia's natural resources were inexhaustible. Brady made the transition from tourist to local in 1917 when he took up a selection on land overlooking the inlet. He championed East Gippsland's development in the press. Railways, ports and roads were amongst the schemes he promoted to connect East Gippsland's farmers to the outside world.[17]

'Charms of Mallacoota Inlet'

It was only with the establishment of the Victorian Country Roads Board (CRB) in 1914 that a reliable transport route linking Melbourne and Sydney via a coastal route became possible. After extensive work, the Old Gippsland Road was renamed the Princes Highway and officially opened in 1922. In Mallacoota the locals came together to build a gravel road connecting the Princes Highway to the western side of the inlet.[18] This was paid in part from funds set aside by the Victorian Government to build

---

[13] National Parks Association pamphlet, 1908. Melbourne.
[14] Mallacoota and District Progress Association, letter from Association to Honorable James Cameron MLA, 16 June 1909.
[15] David Walker, 1999. *Anxious Nation: Australia and the Rise of Asia, 1850-1939*. St Lucia: Queensland University Press, 99–112.
[16] E.J. Brady 1918. *Australia Unlimited*, Melbourne: George Robertson.
[17] 'East Coast Railway – Nowra to Orbost – Making Australia Settled and Safe', news clipping in Edwin James Brady Collection, Ms. 206/10/3B54, NLA.
[18] Tourist Committee to Brady, 15 April 1926, Tourist Committee Minute Book, VPRS 10344, PROV.

roads to 'places of interest'.[19] In Australia, as in America, tourism was a major impetus for road building.[20]

The new Mallacoota West road and the vehicles that used it were to reshape the spatial dynamics of the hamlet. Prior to the road being built the locals used horses and boats as the main form of transport for goods and people and there was no distinct township. The post office operated from Lake View Hotel on the eastern side of the inlet, which was only accessible by water, and declined in popularity with the growth of motor tourism. The road into Mallacoota led to the township reserve, located on land excised from the national park. The Lake View liquor licence was transferred to the new Mallacoota Hotel in 1922. This became the first substantial building on the township reserve.

E.J. Brady lobbied hard to have Mallacoota as a stop on the Royal Automobile Club's 1000 Mile Alpine Reliability Trial in 1922. This boosted the region's tourist profile and linked it to motoring as an adventurous, modern sport. The trial was a vast military-style undertaking with 56 cars taking part and involving over 2000 participants and organisers. The drivers competed in a number of endurance events including a hill climb, a time trail and a road race. At Brady's guesthouse petrol consumption was tested in the large garage to decide first, second and third place.[21] The motorists then spent two days by the lake fishing, swimming, walking and shooting. This event was widely reported in the press.

Most motor tourists were not seeking such a quasi-military excursion, but rather a more democratic tourist experience. Only the Americans rivalled Australians in their passion for car ownership. Falling prices and prosperity meant that the car was affordable to the more comfortable middle classes, and by 1929 one in five Australian families owned a car.[22] The car was praised as a 'liberator and leveller' in America.[23] Australians also enjoyed the freedom travel by car allowed. Tourists moved through the countryside at their own pace, decide which roads they wanted to take and camp where they wished.

---

[19] Anderson, *Roads for the People*, 60.
[20] Sutter, *Driven Wild*, 25; Richard White, 2005. *On Holidays: A History of Getting Away in Australia*. North Melbourne: Pluto Press, 98.
[21] 'Scene at Mallacoota: Photographers Busy', *Argus*, 20 November 1922; 'Motoring Through Gippsland East', *Argus*, 21 November 1922; 'Petrol Consumption Tests', *Argus*, 30 November 1922, (Photograph).
[22] Justine Greenwood, 2011. 'Driving Through History: The car, the Open Road and the Making of History Tourism in Australia 1920–1940', *Journal of Tourism History*. 3(1): 27.
[23] Sutter, *Driven Wild*, 31.

The discomfort that came with driving long distances on poor roads added to the modern experience of nature tourism. You could truly believe you were close to nature negotiating the final miles that led from Gipsy Point to Mallacoota in your sporting car, dodging bushfires. Bogged cars, swinging your axe to remove fallen trees and even taking shelter from the rain all were memorialised in photographs and became part of holiday memories in East Gippsland.[24]

This was a period in Australia in which, as in America, the motorcar and increased consumerism led to an expansion of tourism. Tourists sought out holidays in places different from their everyday life.[25] Brady had worked in advertising and his marketing emphasised Mallacoota's distance from the world of the city with its noise, work and routine. It was where visitors could retreat from 'the strident cares of modern living'.[26] Mallacoota House was advertised as a rest house where guests could 'reinvigorate their mind'.[27] Brady also listed the services available to visitors, including a dance hall, a gramophone, tennis courts, motorboat hire and, most importantly, a garage.

Sutter argues the desire to visit nature was a modern desire but one that was 'driven by a deep ambivalence about modernity'.[28] Brady's marketing of Mallacoota as a tourist destination identified, perhaps unconsciously, that paradox. In *The Overlander: Princes Highway* his travel book recounting a coastal trip from Melbourne to Sydney, he suggested that in the emancipated mood that accompanied the 1920s, the motor tour appealed to 'the younger generation anxious to escape for a season from all conventionalities'.[29] He also tapped into the growing interest in history tourism, describing how motor tourists were following the route taken by pioneers who first ventured into the forests of East Gippsland.[30] It was the car, that very symbol of modernity and consumerism which allowed those who could afford it the means to visit nature as yet unspoilt by modern life.

---

[24] See for example Frank Tate, *Record of a Camping Trip 1918*, Frank Tate Papers, ASSEC no. 90/158, Melbourne University Archives, University of Melbourne; Photographs of motor tourists held by the Mallacoota and District Historical Society.

[25] Paul Thorton, 1997. 'Coastal Tourism in Cornwall since 1900', in Stephen Fisher (ed.) *Recreation and the Sea*, Exeter: University of Exeter Press, 66.

[26] Mallacoota House Tourist Brochure, Ms 206/Box 54 Folder 12, NLA

[27] Mallacoota House Tourist Brochure c.1925, Ms 205/Box 54/Folder 12, NLA.

[28] Sutter, *Driven Wild*, 27.

[29] E.J. Brady, 1927. *The Overlander: The Princes Highway*, Melbourne: Ramsay Publishing, 147.

[30] Greenwood, 'Driving Through History', 26.

By the mid 1920s tourism had become significant part of Mallacoota's economic and social world.[31] Local farmers sold milk, bread and vegetables to campers. They acted as intermediaries between these tourists and their pursuit of sport, taking them on trips by horse or boat to various fishing spots on the lake, advising on bait and rods. The locals added another layer to the holiday experience, with visitors commenting upon their bush skills and familiarity with boats and their environment.[32] Tourism was very much a family occupation with the sons of landholders spending the summer acting as guides and daughters and wives running guesthouses, setting nets to catch fish and preparing meals. Some fathers set their sons up in business. Frank Buckland built Karbethong guesthouse for his son Henry, who had returned from World War I with an English bride. Brady's son Hugh bought a Buick and met visitors at the Orbost Railway Station, which opened in 1917. He then drove them along the 125 kilometres of gravel highway and road into Mallacoota.

An anglers' paradise

In the US, the popularity of auto-camping or motor touring boomed during the same period, with the *New York Times* reporting that five million vehicles were used for auto-camping in the mid 1920s.[33] As more roads were built, more people ventured further and the demand for services grew. Municipal auto-camps catered for these travellers initially, but by the late 1920s commercial camps, motels and inns became lucrative commercial enterprises along American highways en route to national and state parks. They changed the human and natural landscape and, in some cases, obscured the very nature the motor tourist sought.

Tourism also made an imprint on Mallacoota's landscape. The empty township allotments were hired out to visitors as campsites over the tourist season and the money raised used to improve tourist facilities.[34] A committee of management was elected in 1925 to develop the recreation reserve.[35] A clay pigeon shooting range was set up over the summer and the golf course was

---

[31] Sarah Mirams, 2010. 'Dreams and Realities: E.J. Brady and Mallacoota.' Churchill: PhD thesis, Monash University, 147,151.

[32] 'Record of a Camping Holiday 1918', Frank Tate Papers, Melbourne University Archives, Melbourne, 25.

[33] Sutter, *Driven Wild*, 35.

[34] 'Mallacoota Easter Holiday Prospects', *Argus*, 26 March 1929.

[35] 'Country Districts', *Argus*, 8 August 1925; 'Holiday Resorts', *Argus*, 12 December 1922.

almost completed by 1929. Tennis courts and a croquet lawn were attached to the Mallacoota Hotel and there were three public telephone lines for tourists to use.[36]

Mallacoota was no longer host to only the most adventurous nature lovers who could afford the time and expense to retreat to such a remote location. In 1925 the *Argus* Holiday Report described hundreds of tents along the inlet stretching for more than a quarter of a mile. At night 'the campers' powerful headlights are reflected in the water of the Inlet and the landmarks which the resort is noted appear very picturesque'.[37] As in America the car made the nature tourism experience more inclusive, with families and couples now making the long trip. Mr Godding, an auctioneer from Melbourne, set up his tent with his family by the lake in 1924. He wanted, he wrote to Brady, 'to potter about on the lake in a small rowing boat esteeming ourselves explorers and pioneers on a small scale'.[38]

Not all Australians could drive, or indeed relished the prospect of navigating the rough country roads and highways. Guided motor tours, run by the Victorian Railways or private operators, offered an alternative to independent driving. Mallacoota became a destination on the Alps, Lakes and Grampians Tour through southern New South Wales and eastern Victoria. Advertised as 'Holidays Without A Care', these scenic tours were intent on 'embracing lake, river, ocean, forest and mountain scenery, in addition to the Cave country'.[39] The visitor book for Karbethong guesthouse suggested the market for these trips was older couples and middle class women travelling in small groups.[40] This trend mirrored, in a more genteel way, the greater participation of women in bushwalking and skiing clubs during the period and spoke perhaps to the greater freedoms women enjoyed after the Great War.[41]

As more cars came down the Mallacoota West Road tourism profits increased but problems emerged. The locals discovered, as was the case in America, that motor tourism reshaped the very nature that tourist travelled

---

[36] *Argus*, 30 December 1926.
[37] 'Holiday Resorts', *Argus*, 9 January 1925.
[38] JE Godding to EJ Brady, 1 October 1924 in Edwin James Brady Collection, Ms. 206/5/1128, NLA.
[39] Tourist Brochure in Edwin James Brady Collection, Ms.206/53/13, NLA.
[40] Karbethong Guest Book 1926–32. Photocopy of original owner Ian Senior.
[41] Hutton and Connors. *A History of the Australian Environmental Movement*, 68.

so far to visit. This is what Sutter describes as the internal contradiction of modern tourism, which sees the tourist travel to find the new, the authentic, the unspoilt, and then move onto 'virgin' territories. Once a place is discovered it changes, and tourists find themselves surrounded by other tourists, or in a landscape denuded by the tourist presence.[42] Roads played an important role in hastening this process of discovery and disillusionment in the early years of the twentieth century.

Rubber wheels tore up dirt roads, causing erosion and soil damage. Campers cut down trees for fires and saplings to build the cross-poles of tents, and left smouldering camp fires and rubbish. Many tourists rejected the guesthouse model and sought out what Sutter called a 'more primitive relationship with basic resources', utilising nature to recreate domesticity, but at the same time causing environmental damage.[43] Sometimes campers set their tents on farming land without permission. In Mallacoota the locals complained in 1926 of the destruction of tea-tree, which helped support the fragile lake bank, by campers along the foreshore and demanded a police presence in Mallacoota to control tourists.[44] Several bushfires during the 1920s were blamed on campers abandoning campfires. In 1927, Brady wrote to the local paper of tourists who 'come down with a car and about a thousand cartridges and blazed at everything in sight, out of season, in season, protected, not protected.'[45] The visitors also put a strain on existing facilities. Supplies of fresh drinking water and sanitary facilities were a particular problem over summer when the visitor numbers rose and the Orbost council were forced to step in and provide additional services.[46]

There was a growing realisation that the resources of nature that attracted tourists could be under threat. The resources of the inlet were not in-exhaustible. Fishing had remained a significant drawcard with Mallacoota described in the press as the 'Anglers Paradise'.[47] Holiday photos show groups of men proudly lined up and equipped for a day's fishing, hats on, ties loosened and water up to their ankles. The vast quantities of fish caught by anglers were a feature of newspaper reports that typically read 'three benzine

---

[42] Sutter, *Driven Wild*, 29.
[43] Ibid., 35.
[44] 'Destruction of Tea Tree: Complaints at Mallacoota', *Argus*, 9 February 1928.
[45] *Snowy River Mail*, 16 August 1926.
[46] 'Country News', *Argus*, 16 June 1925.
[47] *Argus*, 2 December 1927.

cases filed with bream in two hours' and 'excellent catches of snapper… 70 lb. for 30 minutes fishing'.[48]

The roads brought in more amateur anglers, but also gave commercial fishermen the means to transport fish from the inlet to city markets quickly. The inlet's low draught and sandbar at the entrance made it difficult for fishermen to quickly transfer the commercial catch onto coastal steamers bound for the Melbourne markets. Declining fish stocks in the Gippsland Lakes saw more commercial fleets travelling to Mallacoota Inlet from 1925. They were now able to transport their catch by lorry along the Mallacoota West Road and the Princes Highway to the railhead at Orbost.

The CRB banned lorries with excessive loads of fish using those roads in 1925, arguing they were cutting up the road and impeding tourist traffic.[49] This was a particular problem during heavy winter when the Princes Highway was often transformed into a quagmire during heavy rains. The CRB argued that heavy loads such as fish and wattlebark should only be carried out by sea. While there was local concern that the transportation of wattlebark was being impeded, the banning of fishermen's lorries was supported, as commercial over-fishing was increasingly seen as a threat to Mallacoota's tourism industry.[50]

This saw the locals who had opposed the national park in 1908 enter into an alliance with Barrett. He had long argued that an expanded national park and aquatic sanctuary would make Mallacoota unique amongst tourist destinations and bring long-term economic benefits to the region. The locals were not entirely convinced by his expansive vision, which included extending the borders of the park to the NSW border, but they did support the banning of commercial net fishing as a means of securing fish stocks for visiting anglers.[51] Their claims of overfishing were vigorously disputed by commercial fisherman and the fisheries department.[52] Thirty-three residents signed a petition addressed to Tom Hogan, the Labor Premier, requesting that the Lands Department immediately close the tidal and fresh waters of Mallacoota Inlet to seine fishing and establish a nature sanctuary for the preservation of

---

[48] *Argus*, 18 September 1926.
[49] *Argus*, 2 April 1926.
[50] *Gippsland Times*, 3 June 1926.
[51] Brady, 'Wonderful Gippsland: Sidelights on the Fishing Industry', newspaper article, in Edwin James Brady Collection, Ms. 206/Box 54/Folder 9, NLA.
[52] Lewis, Inspector of Fisheries to E.J. Brady, undated [1926?] in Edwin James Brady Collection, Ms 206/Box 54, NLA.

bird and animal life. Angling should continue on the lake and thus the 'Gem of Victoria' could maintain its position as a premier tourist resort.[53]

Barrett failed in his bid to ban commercial fishing in the inlet and establish an aquatic sanctuary. He did however succeed in having both the Mallacoota Inlet and Wingan Inlet National Parks gazetted. The growing popularity of motor tourism strengthened Barrett's argument for the establishment of national parks in Victoria. He looked to the national parks in Canada, America and South Africa for inspiration. In those other New World nations motorists drove along highways into the parks where they stayed in well-appointed hotels, and there they could commune with nature.[54] He argued that such government investment in national park tourism, scientifically managed to avoid exploitation, would bring economic benefit to East Gippsland.

His expansive vision was not realised in East Gippsland. Another two national parks were created in the region in 1925 but they were modest in scale. The Alfred National Park, 1357 hectares on the slopes of Mount Drummer, featured subtropical rainforests and stands of rare ferns. Lind National Park was made up of dense rain forest; it was home to the rare Gippsland waratah. These were designed as highway parks, with the Princes Highway running through them.[55] Visitors could view the towering forest from the comfort of the automobile. No rangers were appointed to manage the national parks, no committee of management appointed and no tourist services provided.

Dreams of a prosperous tourist industry in Mallacoota had faded by 1931. Barrett's campaign to have net fishing banned played out against worsening economic conditions in Australia and the world. In an attempt to raise revenue the Victorian Government imposed heavy taxes on petrol in 1931 and the number of car registrations in Victoria dropped dramatically. As a consequence, journeys to remote regions became less popular.[56] Tourist numbers dwindled and a holiday in Mallacoota became too distant and too expensive for many. It reverted to being an isolated farming settlement. It was in tourist resorts closer to Melbourne, such as Dromana, Warburton and Healesville, that local shires used unemployed relief funding to build

---

[53] Petition to T. Hogan Premier, 30 January 1930, Mallacoota Inlet National Park file, Rs 1176, Bainsdale: Parks Victoria.

[54] 'Attracting the Tourists: Canada's National Parks', *Argus*, 8 November 1936.

[55] H.J. Newson, 1975. 'National Parks in Victoria.' *Victorian Year 1975*, Melbourne: Government Printer, 15.

[56] Russel Ward, 1977. *A Nation For A Continent: A History of Australia 1901–1975*, Richmond: Heinemann, 189.

roads and tracks to improve tourist access.[57] The unemployed in remote East Gippsland were more likely to be put to work building forestry tracks on public land, or roads and tracks giving farmers better access to markets.[58]

With the Depression easing in the late 1930s, tourists again headed east. Around this time the caravan became a feature of the tourist trail. This was technology particularly suited for long family camping vacations. The *Argus* reported in 1938 that caravans 'provided the comfort of civilisation amid the quietude of far away places'.[59] Caravan parks catered for these travellers and Mallacoota was amongst one of the earliest coastal resorts to advertise this service in 1935.[60]

Conclusion

It was easy for Mutch to deplore the downside of 'civilisation', sitting as he was in his Sydney flat with all its comforts, looking back with nostalgia to the days of his youth. For the people of Mallacoota the remoteness and isolation, and its distance from civilised comforts, was endured, not enjoyed. The road and the car conferred a freedom of movement to both tourists and farmers.

The cars that brought the tourists in also helped turn their settlement into small town and provided additional income for farmers. The road into Mallacoota saw the development of a modest tourist infrastructure, which coexisted with the business of farming (until the crash). There is no evidence to support the view that Mallacoota was 'ruined' by the car; however, the belated support for the national park by locals suggests they were aware of the long-term dangers of commercial exploitation and acted to control it. Interestingly though it was not tourists that were seen as the threat, but fellow Gippslanders making a living from the aquatic richness of the inlet.

The management of all Victorian national parks came under the control of a single specialised government body in 1956. The National Parks Service took a proactive role in both protecting and promoting national parks, and its officers visited Mallacoota in 1957 to rediscover the largely forgotten park.[61]

---

[57] 'Road work for the unemployed', *Healesville and Yarra Glen Guardian*, 9 May 1931; 'Unemployment. St Kilda Baths. 1,000 men seek work', *Argus*, 8 July 1930.

[58] 'Road construction 150 men for East Gippsland', *Gippsland Times*, 17 June, 1935.

[59] *Argus*, 18 May 1927.

[60] Jim Davidson, 2000. *Holiday Business: Tourism in Australia since 1870*. Carlton: Miengunyah Press, 67.

[61] Geoff Thomson to National Parks Authority, 7 October 1957, Department of National Resources and Environment, National Parks Service, Box 166, File 7/1/21, Melbourne.

Its borders were extended and tourist facilities built to encourage tourists. Post-war prosperity saw the car become a common household purchase, and Mallacoota began attracting large numbers of tourists seeking leisure in nature.

East Gippsland's remote location, lack of government investment in tourism and national parks development, and the smaller scale of motor tourism in Australia generally, protected East Gippsland from the long-term ravages of motor tourism in the inter-war years. While East Gippsland saw changes wrought by motor tourism, they weren't of the scale experienced in America. What the car did was lay down the foundations for Mallacoota's eventual future in the second half of the twentieth century as a town which derives its main income from nature tourism.[62]

Today Mallacoota has a permanent population of roughly 1000 people, over half of whom are retirees. Over summer the tourists come and the population swells to over 8000.[63] The holiday town is now ringed by the Croajingalong National Park, which stretches 100 kilometres along the Victorian coast. A drawcard for tourists, it is recognised as one of the great reservoirs of biodiversity in Australia.[64]

Over Christmas, hundreds of caravans and cars line the Mallacoota foreshore looking out over the inlet towards the eastern arm and the national park, a view not dissimilar to that enjoyed by Lawson and Mutch all those years ago when they visited Brady's camp. Today, camping is strictly regulated in the national park and the last commercial fishing licence was revoked in 2003. If visitors look towards the town they can see caravans, tents, car parks, supermarkets and shops. Holiday houses and bitumen roads now occupy the old farming land overlooking the inlet. Since the 1960s debates have raged in Mallacoota over the place of commercial fishing, boat ramps, casinos, hotel developments and creeping suburbia in the district.[65] A hundred years on, the spectre of 'ruination' travelling into Mallacoota, persists.

---

[62] Australian Bureau of Statistics, 2006. Mallacoota Quick Stats. URL <www.abs.gov.au>, accessed 6 August 2014; note that these statistics have since been archived and may be accessed at <http://archive.today.dUT7>.

[63] Ibid.

[64] DNR, 2009. 'Victoria's Biodiversity: East Gippsland'. URL <http://www.nre.vic.gov.au/plntanml/biodiversity/directions/eastgipp.htm> accessed 6 August 2014.

[65] See for example 'Retreat Faces Threat', *The Sun*, c.1976 and the Save Bastion Point website.

References

Anderson, W.K., 1994. *Roads for the People: A History of Victoria's Roads*. Melbourne: Hyland House.
Argus, 19 November 1911.
Argus, 20 November 1922.
Argus, 21 November 1922.
Argus, 30 November 1922 (Photograph).
Argus, 12 December 1922.
Argus, 9 January 1925.
Argus, 16 June 1925.
Argus, 8 August 1925.
Argus, 2 April 1926.
Argus, 18 September 1926.
Argus, 30 December 1926.
Argus, 18 May 1927.
Argus, 2 December 1927.
Argus, 9 February 1928.
Argus, 26 March 1929.
Argus, 8 July 1930.
Argus, 8 November 1936.
Australian Bureau of Statistics, 2006. Mallacoota Quick Stats. www.abs.gov.au, accessed 6 August 2014; note that these statistics have since been archived and may be accessed at http://archive.today.dUT7.
Bardwell, Sandra, 1974. 'National Parks in Victoria 1866–1956 "For All The People For All Time"', Melbourne: PhD thesis, Monash University.
Brady, E.J., 1918. *Australia Unlimited*. Melbourne: George Robertson.
Brady, E.J., 1927. *The Overlander: The Princes Highway*. Melbourne: Ramsay Publishing.
Davidson, Jim, 2000. *Holiday Business: Tourism in Australia since 1870*. Carlton: Miengunyah Press.
Department of Natural Resources, 2009. 'Victoria's Biodiversity: East Gippsland'. http://www.nre.vic.gov.au/plntanml/biodiversity/directions/eastgipp.htm, accessed 6 August 2014.
'East Gippsland Shire Council Bastion Point Ocean Access Boat Ramp Mallacoota, Final Submissions 2008', East Gippsland Shire. www.egipps.vic.gov.au Strategic Planning, received 20 November 2009.
Edwin James Brady Collection, Ms 206, National Library of Australia, Canberra.
Frank Tate Papers, ASSEC NO.90/158. University of Melbourne Archives, Melbourne.
Geoff Thomson to National Parks Authority, 7 October 1957, Department of National Resources and Environment, National Parks Service, Box 166, File 7/1/21, Melbourne.
Gippsland Times, 3 June 1926.
Gippsland Times, 17 June 1935.
Greenwood, Justine, 'Driving Through History: The Car, the Open Road and the Making of History Tourism in Australia 1920–1940', *Journal of Tourism History*. 3(1): 21–37.
Healesville and Yarra Glen Guardian, 9 May 1931.
Hutton, Drew and Conners, Libby, 1999. *A History of the Australian Environmental Movement*. Cambridge: Cambridge University Press.
Karbethong Guest Book 1926–1932. Photocopy of original owner Ian Senior.

Le Soeuf, W.H. Dudley, 1907. *Wild Life in Australia*. Melbourne: Whitcomb and Tombs.
Mallacoota and District Progress Association, letter from Association to Honorable James Cameron MLA, 16 June 1909.
Mallacoota Inlet National Park file, Rs 1176, Bairnsdale: Parks Victoria.
Mallacoota House Tourist Brochure, Ms 206/Box 54 Folder 12, NLA.
Mirams, Sarah, 2010. 'Dreams and Realities: E.J. Brady and Mallacoota.' Churchill: PhD thesis. Monash University.
National Parks Association pamphlet, 1908. Melbourne.
Newson, H.J., 1975. 'National Parks in Victoria'. *Victorian Year 1975*. Melbourne: Government Printer, pp. 1–35.
Roderick, Colin, 1970. *Henry Lawson: Letters 1890–1920*. Sydney: Angus and Robertson.
Snowy River Mail, 16 August 1926.
Sutter, Paul S, 2005. *Driven Wild: How The Fight Against The Automobile Launched the Modern Wilderness Movement*. Seattle: University of Washington Press.
Thorton, Paul, 1997. 'Coastal Tourism in Cornwall since 1900', in Stephen Fisher (ed.) *Recreation and the Sea*, Exeter: University of Exeter Press, pp.57–83.
Tourist Committee Minute Book, VPRS 10344, PROV.
Victoria's Biodiversity; East Gippsland', Department of Natural Resources, www.nre.vic.gov.au/plntanml/biodiversity/…/eastgipp.htm received November 2009.
Victorian Municipal Directory, 1910. Melbourne: Arnell and Jackson.
Walker, David, 1999. *Anxious Nation: Australia and the Rise of Asia, 1850–1939*. St Lucia: Queensland University Press.
Ward, Russel, 1977, *A Nation For A Continent: A History of Australia 1901–1975*. Richmond: Heinemann.
White, Richard, 2005. *On Holidays: A History of Getting Away in Australia*. North Melbourne: Pluto Press.

Afterword

Erik Eklund and Julie Fenley

The challenge, when writing environmental history, is to determine how the connection between humans and nature should be depicted. Often, humankind is portrayed as a destructive force. Emphasis is placed on the intensive land-use practices of non-Indigenous people in particular, and their ability to dramatically transform the landscape – sometimes in the space of one generation. Instead of attempting to understand and live in harmony with the environment, natural ecosystems are viewed as a resource to be harnessed for profit.

Looked at this way, the rivers of Gippsland can be seen as a barometer of the power of industry. A number of rivers flow north and east from the Strzelecki Ranges and also south and east from the Baw Baw and Snowy Mountain ranges. These rivers have shaped the landscape of the Latrobe Valley, but the Latrobe and Morwell rivers have been diverted multiple times from the 1970s onwards to make way for the open-cut mining of the valley's brown coal reserves. Even here, however, the dramatic power to remake the landscape and change the course of rivers is not unassailable. In 2012, floodwater rushing down from the ranges and scouring the man-made barriers resulted in major failures of containment. The water flowed from heavy rain, the earth weakened and broke, and then the industry stopped – at least for a time. This was not the first time that these man-made structures have failed with spectacular results, and current politicians have carefully refused to rule out the possibility of future failures.[1]

What could be termed the 'revenge' of the river showed an ongoing relationship between environment and society, not a tamed subservience. Should it be any surprise then that a recent bushfire in the Latrobe Valley could also be seen to make light of plans for dominion over the land? The fire which spread into the Hazelwood open-cut mine in early 2014 cast a pall of thick, acrid smoke over Morwell and the surrounding areas for a number of weeks. The fire began on 9 February and burned for 45 days. The fire was

---

[1] Keith Seddon, 'Review of the Failure of Morwell River Diversion', Department of Primary Industries, Victoria, 13 May 2013, <http://www.energyandresources.vic.gov.au/earth-resources/exploration-and-mining/issues/yallourn-coal/review-of-failure-of-morwell-river-diversion> [URL accessed 1 October, 2013].

caused by embers from a nearby bush fire, which ignited exposed coal seams in disused parts of the mine. The 2014 fire was part of a long line of disasters caused by wild fires or faulty mine equipment that has threatened mining activities as well as the townships surrounding them.

Such stories tap into the myth about the wilful human destruction of the earth. The earth is characterised in largely positive terms, although there is recognition that it can be a powerful and hostile force. Industry is regarded in negative terms, as exploitative and unconnected from nature. The implication here is that 'earth' and 'industry' are binary opposites. Humankind is aligned with industry; nature is an opposing force. But the stories told in this book demonstrate that humans have appreciated nature, attempting to work with and preserve it.

This is not to insist that we should not discuss human modification of the landscape or the consequences of our actions but instead that we should recognise that there has been a range of ways of engaging with the environment. The chapters show the region as a mosaic of smaller sub-regions with distinctive patterns of land use which changed over time. Authors reveal changing patterns of land utilisation but also new ways of seeing the land and viewing its characteristic features; views which dissented from the dominant approach, or found ways where settler dominion over the land was held back, modified, or even contested.

As Ruth Ford found when considering the letters written by farming women in Gippsland, many celebrated the clearing of 'scrub' and evinced pride in the progress of their farms, but they also expressed a sadness for the loss of bird life and favourite habitats. Similarly, Jillian Durance described how selectors in South Gippsland were forced to clear extensive portions of their land grants to make a marginal business viable but still developed deep connections to place and lamented the loss of quiet gullies where ferns and vegetation survived and lyrebirds might still be heard. And even in this aggressively individualist policy context, selectors were still heavily reliant on neighbours and kin to assist in managing the farm.

In the varied contributions to this collection we see a very healthy snapshot of diverse approaches and different emphases. In the theme on depictions of the environment, Lennon, Legg, Ford, and Fenley and Lothian demonstrated that there can be differing understandings of nature and how humans might interact with it. From Legg's portrayal of the prominent newspaper editors who stood against prevailing opinion about unremitting development and an endless supply of forest resources, to Lennon's analysis of the conflict that can occur over the creation of pristine wilderness zones, we build a picture of competing accounts about the environment and the human impact on it. Similarly, Fenley and Lothian revealed differences across time and between individuals and groups in views on the Australian bush and how walkers might engage with it.

The theme on politics and the environment included chapters by Glowrey, Harris, Nixon and Fahey on government policy and the impact this had on settlement patterns, the income and occupation of settlers, and their approach to the land. Through Nixon we saw the ways in which selectors adopted legislators' views about independence and enterprise, while Fahey revealed the dire long-term effect of land-settlement policies on the profitability of Gippsland farms. Harris questioned the motivation of those calling for the regulation or otherwise of commercial fishing on the Gippsland Lakes, while Glowrey reminded us that the machinations of pastoralists and merchants, as well as the administrators in Melbourne and Sydney, helped shape the region we now call Gippsland.

This theme also included chapters on political activism and diplomacy. Lawrence found that the Yaitmathang people sought to benefit from the European presence in northern Gippsland, while also attempting to assist the newcomers in understanding their social and political structures and connection to the land. A similarly inclusive approach was adopted by Jean Galbraith, with Fletcher arguing that she sought the creation of reserves alongside the brown-coal mines and power-generating facilities, pine plantations, and newly developed farms of the Latrobe Valley. In contrast, Foskey and Constable highlighted the difficulties in negotiating conservation outcomes in the East Gippsland and Strzelecki forests because of the different approaches of activist groups and the intransigence of government respectively.

The themes on material environmental history yielded chapters by Durance, Slattery, Martin and Mirams that tracked changes in industries such as farming and logging, and new technologies, such as the automobile and the opening up of car touring. For Mirams, her motor tourists, and the road to Melbourne, brought Mallacoota into an important and ongoing relationship with the tourist trade and its attendant interests. For Martin and Slattery the introduction of more 'efficient' machinery and methods had not necessarily signaled an improvement in land or resource use or management. But these themes were not considered in isolation and a number of chapters in the collection shifted between these sub-genres, with most authors taking a cautiously optimistic approach to the history of the Gippsland environment.

Some chapters, such as those by Durance, Slattery, and Nixon focused on a specific place and followed its history through successive waves of new policies and land use practices. Such work - while also pointing to broader issues at a regional or state level - allows us to drill deep into the specifics of the houses, farms, tracks, hills and creeks in the one locale. Others – such as Ford, Harris, Martin and Lawrence – considered whole classes and groups of people or industries and tracked change over time in the region or parts of it. Glowrey in particular moved beyond place-based biography to consider connections across the landscape. She explored the history of Corner Inlet and saw it in a wider regional context as signalling bigger shifts in the orientation of the area towards Sydney and Melbourne. Therefore it was not just the scale of empirical focus that varied across contributions but the scale of analysis that either looked deep into a specific biographical history of place or cast a wider net following connections and relationships with the surrounding regions.

Gippsland's environment has been a vital resource for Aboriginal and non-Aboriginal people. The landscape nourished the Gunai-Kurnai, Yaitmathang and Bunurong, providing enough sustenance to allow time each day for leisure and cultural activities. The region also provided for European and other migrants: it was a source of pastoral land and dairying country; forest and timber products; fish and maritime resources; and gas, oil and of course brown coal. In the future the region's remnant biodiversity could become an important resource for mitigating climate change as methods are devised for fixing carbon in old growth and regrowth forests. Various projects to capture and store carbon as well as minimise carbon emissions have been put forward.[2] Some of these proposals involve the identification of natural, cultural and industrial resources in the region to mitigate climate change. With the region so often identified as part of the carbon problem through the relatively inefficient production of energy through burning brown coal, we can speculate that these approaches have the potential to turn this perception on its head. The case for forest conservation in Victoria now includes carbon capture and reduced carbon emissions along with more familiar arguments about protecting biodiversity and preserving natural and cultural values.[3]

---

[2] These include projects from ClimateWorks Australia, a partnership between Monash University and the Myer Foundation, and CarbonNet, which is backed by the Victorian Government.

[3] See, for example, the work of Dr Heather Keith from the ANU: 'Protecting Natives Forests More Valuable Than Logging', 11 November, 2014, ANU Research News, http://phys.org/news/2014-11-native-forests-valuable.html#jCp, accessed 20 November 2014.

These new uses and arguments, with a focus on climate adaptation and sustainability, may well mark us as different to those past generations discussed by Legg who identified the region as a vast economic storehouse of apparently endless resources.

But, as we have seen, this is only one theme in Gippsland's environmental history. There are competing narratives and alternate visions. In the myriad of stories about Gippsland's earth and industry we can also see adaptation and flexibility, productivity and profitably, as well as concern for conservation and protection. It may well be in these types of stories we might find inspiration for tackling the region's biggest challenge yet: the transition to a low-carbon future. Undoubtedly this challenge, if we embrace it, will usher in new ways of thinking about the environment, but there may also be pre-existing traditions lying dormant which will see us through too.

Index

A

Aboriginal people xv, xix, xx, xxiii, xxv, 3, 4, 8, 17, 44, 82, 83, 101, 102, 103, 104, 105, 107, 112, 115, 116, 119, 120, 126, 138, 158, 258, 317
    Bunurong (Kulin) xv, xvi, xxiv, 158, 317
    Gunai-Kurnai xv, xvi, xxiv-xxv, 82, 101, 102, 104, 115, 158, 238, 258, 317
    Land use/modification xv, xx, xxi, 3, 114, 117, 119–121, 279, 285, 317
    Yaitmathang xv, xvi, xxiv, xxv, 100–121, 316, 317

Andrew, James Moore,
    Approach to walking in the bush 56, 57–58, 63–69, 72–73, 74, 75–79
    Engagement with bush xxix, 56, 57, 69–70, 75–76, 78, 79
    Skiing trips and involvement in Rover Scouts 57, 75, 77, 78
    Walking trips/routes 55, 63, 64, 65, 68, 70, 71, 72–73, 76, 77

Australian Conservation Foundation (ACF), 195, 200, 203, 204, 209, 210, 211, 212

Australian Paper Manufacturers (APM) *see* timber and paper/pulp mills

B

Bairnsdale 125, 131, 133, 135, 264, 280, 281, 287, 289, 292, 293, 296

Balook and District Residents' Association (BADRA) 227, 229

Bass Strait xxiii, xxx, 4, 82, 83, 94, 95, 242

Bogong High Plains *see* mountains/ranges

Bonang 200, 203, 207, 214, 287

Bruthen 82, 101, 280, 285, 287

Buchan 44, 285, 287, 289, 294

Bushfire and fire management xxxi, 6, 7, 8, 9, 11, 12, 13, 15, 22, 48, 49, 52, 59, 81, 105, 119–121, 153–155, 168, 187, 196, 198, 201, 232, 243, 249–250, 260, 264, 265, 278, 279, 283, 284, 285–286, 290, 291, 295, 304, 307, 314–315

Bushwalking
    Approaches to walking in the bush xxix, 56–57, 58–63, 65–66, 68–69, 70, 71–72, 73–75, 79
    Walking clubs 56–57, 58, 60–62, 63, 65–66, 71–72, 74, 75, 78, 79, 306

C

Car touring 62, 76, 115, 207, 213, 267, 293, 299–300, 302–304, 306, 307, 309, 310, 311, 316
    And campaigns for national parks 299, 301, 302, 308–309
    Impact of motor tourism xxix, 299–300, 302–303, 305, 306–308, 310, 311
    Tourist consumption xxix, 299–300, 302, 303–304, 305–307, 309, 310, 311

Central Gippsland 22, 44, 77, 88, 96, 97

Closer settlement xxvi, xxvii, 2, 21, 23, 26, 30, 37, 39, 40, 52, 164, 165, 173–176, 177, 252, 264

Coal mining xv, xvi, xxiii, xxix, xxx, xxxi, 57, 187, 247, 248, 314, 315
    see also energy production

Coastal areas, xvi, xxii, 2, 5, 6, 16, 81, 96, 127
    Coastal change 1–3, 7, 13–16
    Coastal ports, xvi, xxiv, xxvi, xxix, 1, 13, 14, 16, 84, 85, 86, 92
    Maritime activity 4, 5, 13–14, 16, 82–83, 84, 88, 94–95
    see also Corner Inlet, Gippsland Lakes, Port Albert

Cobungra 105, 111, 114

Colonisation
    Biological xix-xx
    European xv, xxiv, xxv, xxvii, 51, 52, 81, 124, 125, 126, 127, 129, 130, 133, 134, 135, 137, 139, 141, 164, 166, 173, 273

Concerned Residents of East Gippsland (CROEG) 199, 200, 201, 202, 203, 204, 207, 209, 211, 212, 213, 214, 215, 216

Conservation Council of Victoria (CCV) 195, 200, 201, 203, 204, 209, 211, 212, 228, 229

Corner Inlet xxiv, 4, 13–15, 82, 83, 84, 86, 91, 94, 95, 96, 317

Currie family xxvii, xxviii, 146–147, 152–158, 159, 160–161, 164–165, 166–167
    And modification of property xxvii, 147, 152–156, 157-8, 159, 160, 168–169, 171, 172, 174
    And dairying income 169–171, 174
    And yeoman ideal xxvii, 148–151, 156, 159–160, 164
    Requirements under Victorian Land Acts xxvii, 146, 147–148, 150, 151, 156, 158–159, 164, 165–166, 171–172

D

Desalination plant xxx
Drouin xvi, xxvii, 146, 151, 164

E

East Gippsland xxiv, xxv, xxvi, xxviii, xxix, xxx, xxxi, 81, 195, 196, 198, 199, 200, 201, 202, 203, 204, 206, 209, 211, 212, 213, 215, 257, 265, 268, 277, 278, 279, 280, 281, 283, 284, 285, 286, 287, 288, 289, 291, 292, 293, 294, 295, 298, 300, 302, 304, 309, 310, 311, 316
Environment East Gippsland (EEG) 201, 215
East Gippsland Forest Management Area 278, 279, 280, 288
Energy production/resources xvi, xxii, xxiii, xxviii, xxix
    Electricity generation xv, xvi, xviii, xxiii, xxix, xxx, xxxi, 57, 187,
    State Electricity Commission of Victoria (SECV) xxx, 57, 187, 188
Ensay 105, 106, 107, 287
Environment
    And industry, definition of xv
    Definition of xv, xviii, xxii, xxiii,
    Historical scholarship on xviii-xxi
    Modification of xv, xvi, xviii, xix, xx, xxii, xxv, xxvi, xxvii, xxviii, xxix, xxx, 8, 21–22, 26, 37–38, 40–41, 44, 45, 46, 48–49, 51–53, 147, 152–156, 157–8, 159, 160, 243–244, 246, 252, 253, 260, 261, 262–266, 267–268, 269, 270–272, 273–274, 294, 306–309, 310–311, 317, 318
    Perceptions of xv, xvi, xvii, xviii, xxi, xxiv, xxv, xxvi, xxvii, xxii, xxix, xxx, 20, 21, 22–23, 25–27, 29–31 35–36, 37–39, 43–53, 56, 57, 69–70, 75–76, 78, 79, 187–194, 316
Environment East Gippsland (EEG), 201, 215, 216
Errinundra xxix, 196, 197, 198, 203, 204, 205, 206, 209, 210, 212, 213, 214, 215, 286
Exploration xxiv, 2, 8, 17, 58, 82–84, 102–105, 115, 120, 121, 279, 306
    see also McMillan, Angus and Strzelecki, Paul Edmund de

F

Farming and pastoralism xv, xxiv, xxvi, xxvii-xxviii, 21, 39, 50, 146, 147, 148, 149, 151, 153, 155, 156, 157, 159, 160, 161, 163–177
Field Naturalists' Clubs 7, 183, 190, 193, 225
Fish/fishing EE&JF, xv, xxvi, 5, 6, 31, 76, 106, 115, 124–142

    Angling and fish acclimatisation clubs and societies xxvi, 126, 127, 133, 135–139, 142
    Fishing inquiries 125, 127–132, 133, 135, 139, 141
    Victorian Fisheries Department and Inspector of Fisheries 127, 134, 137, 140, 142, 308
Forests Commission Victoria (FCV) 198, 199, 205
Forests/forestry xix, xxi, xxiii, xxv, xxvi, xxviii, xxix, 6, 8, 9, 15, 20–36, 38, 43, 44, 45, 48, 49, 52, 53, 82, 91, 112, 129, 151, 160, 164, 165, 166, 187, 188, 177, 190, 195–207, 208, 209–16, 219–228, 230–239, 243, 244, 245, 246, 253, 274, 277–281, 283–287, 288–291, 293–295, 310
    Campaigns xxix, 20, 26, 29, 31, 32, 195–196, 199–204, 205–216, 219, 222, 225–235, 236, 237–239, 284
    Clearing and timber/pulp harvesting xv, xxiii, xxv, xxvi, xxix, xxix, 3, 9, 15, 16, 21, 22, 37, 38, 41, 44, 48, 49, 51, 52, 95, 147, 152, 153, 155, 166, 168, 169, 171, 177, 187, 198, 199, 200, 202, 204, 220, 221, 223, 224, 230, 231, 232, 236, 246, 253, 289, 290, 291
    see also environment – perceptions of

G

Galbraith, Jean
    Campaigns for conservation and reserves xxx, 187–194, 316
    Nature writing 179, 180, 183–187, 189–193
    Response to industrialisation xxiv, 187, 188–190, 192, 316
Gippsland
    Boundaries of xvi, xxi, 81, 87, 88, 89–91, 92, 97
Gippsland Lakes, xv, xvi, xxvi, xxviii, 15, 96, 124, 125, 126, 127, 132, 133, 139, 140, 141, 142, 180, 187, 308, 316
Gippsland Mercury 23, 26, 28, 29, 33, 36
    see also environment – perceptions of
Gippsland Times 23, 26, 28, 29, 30, 32, 33, 35, 36
    see also environment – perceptions of
gold fields/mining xv, xxv, 1, 2, 5, 8, 9, 10, 11, 21, 29, 31, 32, 33, 34, 64, 95, 109, 110, 111, 112, 113, 115, 120, 121, 136, 137, 150, 158, 166, 167, 259, 260
    see also Omeo, Walhalla
Goongerah Environment Centre (GEC) 215
Government
    Federal government xxix, 30, 195, 232, 237, 274

Index | 321

Gippsland parishes/counties 147, 164, 165, 167, 171, 172, 174, 175, 177, 242, 244, 258
Gippsland local parties/government xxi, xvi, xxvi, xxx, 204
  Alberton, Shire of 225, 227
  Bass Coast, Shire xvi
  Baw Baw, shire xvi, 9, 12
  Buln Buln, shire of 164
  Cardinia, Shire xvi
  East Gippsland, shire xvi, 279, 287
  Latrobe xvi, 234
  Morwell, Shire of 225
  Narracan, Shire of 9, 10
  Orbost Shire 35, 199, 200, 202, 213, 278, 307
  South Gippsland, Shire of 30, 164, 234, 237
  Tambo Shire 278
  Wellington, Shire xvi
  State government xxii, xxvii, xxix, xxx, xxxi, 22, 95, 96, 127, 137, 139, 141, 147, 148, 151, 156, 173, 199, 200, 203, 204, 210, 211, 212, 222, 234, 237, 238, 239, 265, 277, 287, 288, 294, 300, 302, 308, 309
Great Dividing Range *see* mountains/ranges
Guest, John 30–31, 32, 34, 35
Gunai-Kurnai, *see* Aboriginal people

H
Hancock Victorian Plantations (HVP), 222, 223, 224, 232, 235, 236, 237, 238

I
Indigenous *see* Aboriginal people

K
Koo Wee Rup xvi, xxiv, 151
Korumburra 20, 243, 247, 254

L
Lake Tyers
Land Conservation Council (LCC) 10, 13, 215, 219, 221, 222, 225, 227, 228, 229, 230, 231, 233, 235, 236, 278, 284, 286, 287
Lardner 147, 151, 153
Latrobe Valley xv, xvi, xviii, xxii, xxiv, xxx, xxxi, 1, 8, 9, 22, 56, 90, 91, 96, 179, 180, 181, 187, 193, 314, 316
Leongatha 6, 226, 230
Luke, Henry 26, 29, 31–32, 33, 34
Luke, Henry Alfred 26, 29–30, 31–32, 33, 34–35

M
Maffra 20, 28, 30, 34, 36, 295
Maffra Spectator, The 26, 28, 30, 33, 34
  see also environment – perceptions of
Mallacoota xvi, 298, 299, 300, 302, 303, 304, 305, 306, 307, 308, 309, 310, 311
McMillan, Angus xxiv, 81, 82, 83, 84, 94, 103–105, 116
Melbourne xvi, xxii, xxiv, xxvi, 2, 30, 32, 42, 43, 50, 58, 64, 70, 77, 81, 83, 84, 85, 87, 88, 89, 90, 91, 92, 95, 96, 97, 112, 125, 128, 133, 134, 135, 136, 137, 138, 139, 169, 171, 173, 179, 189, 196, 204, 212, 228, 242, 244, 282, 283, 292, 299, 300, 303, 305, 306, 308, 309, 316, 317
Mining xv, xxv, xxx, 5, 280, 300
  Brown/black coal, *see* coal mining
  Gold *see* gold fields/mining
Moe xxiv, xxx, 56, 173, 187
Monaro xxiv, 3, 20, 82, 84, 85, 88, 90, 95, 101, 102, 103, 104, 105, 107, 108, 116
Morwell xxx, xxxi, 20, 187, 314
Mount Best Concerned Residents Association 230, 231, 232, 233
Mountains/ranges
  Bogong High Plains 78, 105, 115
  Great Dividing Range xvi, 78, 180, 187, 285, 290, 314
  Strzelecki Ranges/South Gippsland hills xvi, xxix, xxxi, 40, 180, 181, 182, 187, 190, 219, 220, 221, 222, 223, 224, 225, 226–27, 228, 230, 231, 232, 233, 234, 235, 236, 237, 239, 244, 249, 255, 314
Moyarra, xxviii, 242, 243, 245, 248, 249, 251, 253

N
National Parks, *see* parks and reserves
Native Forests Action Council 195, 200, 203, 204, 210, 212
New South Wales, xvi, 4, 59, 81, 82, 83, 87–90, 101, 102, 126, 137, 139, 148, 192, 198, 199, 200, 201, 205, 215, 299, 306, 308
Nightcap Action Group/Nomadic Action Group (NAG) 205, 207, 210
Nowa Nowa 280, 282, 287

O
Old Port 14, 82, 86, 92, 93, 94
Omeo x, xvi, xxv, 20, 35, 82, 84, 87, 100, 101, 102, 103, 104, 105, 106, 107, 108, 109, 110, 111, 112, 113, 114, 115, 116, 117, 118, 119, 120, 121, 260, 280, 285, 287
Orbost 20, 35, 41, 196, 197, 200, 211, 214, 280, 281, 283, 287, 300, 305, 308

Orbost District Environment Group (ODEG) 200
Overend, Robert Stanton 26, 30, 31–32, 33, 34–35

P

Parishes, *see* government
Parks and reserves 284
    Campaigns for creation of xxix, xxx, 5–6, 188–189, 193, 219, 225, 228, 230, 232, 234, 236, 299, 301, 302, 308–309, 316
    National Parks 215, 284, 287, 301, 302, 308–309
        Alfred National Park 309
        Bulga Park, 181, 185, 225, 226, 227, 228
        Croajingalong National Park 311
        Lind National Park 309
        Mallacoota National Park 299, 301–302, 309, 310
        Morwell National Park 194, 225, 227, 228, 236
        Tarra-Bulga National Park 194, 225, 227, 228, 236, 237
        Tarra Valley National Park 2, 181, 183, 184, 185, 225, 226, 227, 228
        Wilsons Promontory, 1, 2, 4–8, 14, 83, 90, 94, 95, 96, 301
        Wingan Inlet National Park 309
    State Parks
        Holey Plains State Park 194
        Mount Worth State Park 225
    Reserves xxix, 188–189, 193–194
        Crinigan Road Bushland Reserve 194
        Gunyah Rainforest Reserve / Gunyah Timber Reserve 225, 226, 231, 232–233, 235
        Turtons Creek Reserve 225
Paynesville 127, 128, 129, 131
Plantations, xxix, 187, 188, 190, 194, 201, 221, 222, 223, 224, 225, 227, 228, 230, 232, 233, 235, 238, 239, 257, 271, 272, 274, 294, 316
Port Albert (Shipping Point) xxiv, 6, 13, 14, 15, 20, 23, 33, 81–82, 84, 85–89, 91, 92–97, 109, 110, 126, 127, 140
Port Phillip District xxiv, 85, 87, 89–91, 95, 97, 104
Power stations, *see* energy production

R

Rail and tram ways xxiv, xxvi, xxviii, 1, 8, 9, 12, 21, 29, 34, 35, 55, 59, 62, 64, 73, 81, 125, 128, 130, 132, 133, 134, 139, 141, 142, 155, 169, 171, 173, 187, 189, 196, 247, 267, 280, 281, 283, 300, 301, 302, 305, 306, 308

Rainbow, William xxviii, 242–243, 244, 248–249, 250, 253–254
    Impact of coal mining in the region 246–248
    Impact of soldier settlement scheme in the region 251–253
    Modifications to The Pines property 243–244, 246, 252, 253
    Requirements under Victorian Land Acts 245–246, 248, 252
Regional Forest Agreement (RFA) 219, 224, 232, 234, 235, 236, 288
Reserves, *see* parks and reserves
Ridley property xxviii-xxix
    And modifications to property xxix, 260, 261, 262–266, 267–268, 269, 270–272, 273–274
    And income from property 260, 261, 262, 263, 265, 266, 267, 270, 271, 272, 273
    And freehold tenure 261–264, 265
    Requirements under Victorian Land Acts 260, 265–266
    Shift to plantation and nature conservation 270–273, 274
    Shift to scientific agriculture 266–269
Road construction xxv, 6, 9–10, 64, 67, 87, 91, 96, 114–115, 121, 181, 196, 244, 247, 299, 300, 302–303, 305, 308, 310, 316
Rodger River forests 210, 204, 205
Rosedale 33, 188, 194
Ryan, James 26, 30, 31–34

S

Saint, Charles Abraham 29, 31–32, 33, 34, 35
Sale xxvi, xxviii, xxx, 20, 23, 26, 28, 29, 31, 32, 33, 34, 35, 36, 81, 97, 112, 125, 133, 267
Selection xxvii, 21, 26, 39, 52, 53, 146, 147, 150, 152, 154, 157, 158, 164, 165–172, 175, 225, 242, 244, 245–247, 248, 251, 252, 253, 253, 255, 257, 258–261, 262, 265, 274, 299
Shire Councils *see* government
Soldier settlement 2, 21, 251–253
South Gippsland xv, xvi, xxiv, xxvi, xxviii, xxx, 43, 47, 175, 219, 225, 226, 230, 233, 239, 243, 244, 253, 255, 315
South Gippsland Conservation Society 226, 227, 229, 230, 234
Squatting xxiv, 3, 84, 94, 102, 104–105, 106–109, 121, 148, 150, 158, 165–166, 258
State Electricity Commission of Victoria (SECV) *see* energy production
State Parks, *see* parks and reserves
Stockdale 234, 257, 258, 260, 267, 268, 272
Strzelecki, Paul Edmund de 83–84, 104

Strzelecki Forest, xxix, 219, 220, 222, 223, 224, 232, 233, 234, 235, 236, 237, 238, 239, 316
Sydney 58, 81, 82, 83, 86, 87, 89, 90, 92, 95, 104, 298, 299, 302, 304, 310, 316, 317

T

Tambo Forest Management Area 285, 287, 288
Tasmania / Van Diemen's Land 4, 81, 82, 88, 91, 92, 94, 95, 135, 138
Timber and paper/pulp mills xxiii, xxv-xxvi, xxix, xxx, 5, 6, 9, 22, 35, 37, 48, 187, 188, 198, 200–201, 214, 221–222, 225, 227, 238, 239, 246, 267, 268, 278, 280, 283, 284, 285, 286, 287–289, 290, 291, 292–293, 294, 295
Toonallook 125, 127, 128
Traralgon xxiii, xxix, xxx, 20, 32, 35, 94, 186, 187, 189, 190
Tubbut 199, 214
Twofold Bay 4, 81, 83, 87, 88, 90, 92, 198
Tyers 180, 182, 186, 187, 192, 193

V

Victoria, state of xv, xvi, xxii, xxviii, xxix, xxxi, 1, 2, 3, 4, 5, 7, 8, 10, 13, 14, 20, 21, 22, 23, 29, 31, 33, 39, 47, 52, 57, 75, 82, 88, 95, 97, 102, 112, 120, 125, 136, 138, 151, 158, 163, 165, 166, 181, 185, 186, 187, 191, 192, 195, 196, 204, 206, 209, 210, 211, 212, 221, 228, 230, 244, 257, 258, 269, 271, 274, 278, 279, 280, 285, 291, 294, 298, 300, 301, 306, 309, 310, 311, 317
Victorian Association of Forest Industries 289, 292
Victorian Country Roads Board (CRB) 188, 302, 308
Victorian Environmental Assessment Council 237, 238
Victorian Forests Bill 23, 29–30
Victorian Land Acts xxvii, 94, 97, 148, 165, 259–260, 273
    And yeoman ideal xxvii-xxviii, 53, 148–151, 156, 159–160, 164, 259
    Requirements under xxvii, 146, 147–148, 150, 151, 156, 158–159, 164, 166, 171–172, 245–246, 259–260
Victorian Lands Department 146, 171, 301, 308
Victorian Minister/Ministry for Conservation 16, 203
Victorian Minister for Forests 200, 202–203, 204
Victorian National Parks Association (VNPA) 195, 211, 212, 214
Victorian Sawmillers' Association 203, 212, 230

Victorian Timber Industry Inquiry 212–213, 215
Victorian Plantations Corporation (VPC) 222, 232, 233, 234, 235, 239
    see also Hancock Victorian Plantations (HVP)

W

Walhalla xxv, 1, 8–13, 20, 33, 35
Warragul xvi, 20, 35, 41, 44, 225
Weekly Times rural newspaper 37–53
see also environment – modification of, perceptions of
West Gippsland, xxiv
Wonthaggi xxxi, 43

Y

Yallourn 56, 57, 77
Yanakie 4, 6
Yarram 81, 220, 228, 231, 236
Yarram and District Conservation Group 226–228, 229, 230, 231
Yarram Forest District 228